Seismic Design of Buried Pipeline Systems

This book discusses the application of seismic design methods for lifeline earthquake engineering to buried pipelines so that readers can analyze their behavior and structural and functional characteristics, verify the results, and apply appropriate methods to design earthquake-resistant lifelines.

The authors focus on four key areas to provide readers with this understanding. First, they comprehensively explain seismic design methods for lifelines by systematically discussing the seismic design methods of expansion and contraction flexible joints. Second, they describe in detail the theoretical background of design formulas for seismic design of pipelines (the "slip concept") and verify the applicability of the design formulas to the elastoplastic state assumed by the current design method by comparing them with the finite element method (FEM) analysis results. Third, they explain how to verify the validity of FEM numerical analysis in line with the theoretical background of the design formulas. Fourth, they discuss the applicability of analytical models and theoretical formulas, the accuracy of analysis for evaluating load characteristic values, and ensuring the prediction accuracy of resistance strength evaluation formulas. This approach enables readers to deal with the nonlinear behavior of buried pipelines by deriving the formulas for straight and deformed pipe strains. They will then be able to verify the validity of FEM numerical analysis results and ensure the prediction accuracy of their resistance strength evaluations. Overall, the readers' developed spatial, comprehensive, and chronological understanding of earthquake damage to urban lifelines will allow them to contribute to stronger disaster mitigation and faster disaster response in the event of damage to lifelines.

This book is a vital resource for practitioners and students to deepen their understanding of seismic design methods for lifelines. Researchers will also be able to use these findings as a reference for future technological development.

Seismic Design of Buried Pipeline Systems

Takeshi Koike and Lessandro E. Garciano

CRC Press
Taylor & Francis Group
Boca Raton London New York

CRC Press is an imprint of the
Taylor & Francis Group, an **informa** business

Designed cover image: Adobe Stock

First edition published 2026
by CRC Press
2385 NW Executive Center Drive, Suite 320, Boca Raton FL 33431

and by CRC Press
4 Park Square, Milton Park, Abingdon, Oxon, OX14 4RN

CRC Press is an imprint of Taylor & Francis Group, LLC

ISBN: 978-1-041-10528-2 (hbk)
ISBN: 978-1-041-10529-9 (pbk)
ISBN: 978-1-003-65552-7 (ebk)

DOI: 10.1201/9781003655527

Typeset in Times
by KnowledgeWorks Global Ltd.

Contents

Preface

Urban lifelines, such as water, sewer, gas, subway, and highway networks, are indispensable infrastructures that support comfortable urban living. When these lifelines are damaged by earthquakes, city dwellers are immediately forced to live in inconvenient conditions until such time lifelines are restored. Lifeline earthquake engineering was born from an academic system that has been developed to realize earthquake-resistant lifelines. In the 50 years since this discipline was born, it has witnessed a variety of earthquakes, identified new damaging events each time, and developed earthquake disaster prevention technology, seismic design, and analysis technology to overcome these events. Water, sewer, and city gas pipelines are essential facilities of urban lifelines and have the common characteristic of being buried underground, thus paving the way for seismic design methods for buried pipelines to be part of the seismic design technology system.

This book traces the development of seismic design methods for lifelines over the past 50 years to provide a deeper understanding of this design theory and to help identify the direction of future development. When lifeline earthquake engineering was first established, emphasis was placed on the development of methods for analyzing the seismic behavior of pipelines, which are the core facilities of urban lifelines, but later the scope of research gradually expanded to include the analysis of structural and functional characteristics of pipelines during earthquakes. Recent trends in research include: the spatial, comprehensive, and chronological understanding of earthquake damage to urban lifelines to contribute to the speeding up of disaster response; the discussion of the seismic safety of major lifeline subsystems, such as hospitals and manufacturing plants in urban areas; and the understanding of damage to lifelines during disasters from the perspective of a bird's-eye view of the national land, and the prediction of the restoration process.

This book focuses on the following points:

1. Comprehensive explanations of seismic design methods for lifelines. In particular, this book systematically discusses the seismic design methods of expansion and contraction flexible joints, which are the basic elements attached to pipelines and have not been discussed in depth in the past.
2. Theoretical background of design formulas for seismic design of pipelines is described in detail. Especially, the "seismic design method based on the slip concept," which is useful when dealing with the nonlinear behavior of buried pipelines, is explained in detail, including the process of deriving the formulas for straight and deformed pipe strains. This book endeavors to explain how the current design equations were formulated, their background, and their analytical accuracy. Furthermore, the applicability of the design equations to the elastoplastic state assumed by the current design method is verified by comparing them with the results of finite element method (FEM) analysis to clarify the limits of applicability of the design equations and the role of FEM analysis.

3. Although FEM methods are used extensively in practical design these days, the means to verify the validity of FEM numerical analysis results are not yet sufficient. Theoretical background of the design formulas is explained in order to cultivate these abilities.
4. In order to promote the performance design of seismic design methods, the applicability of analytical models and theoretical formulas and the accuracy of analysis are discussed in detail regarding the evaluation method of load characteristic values and the essential prediction accuracy of resistance strength evaluation formulas.

In line with the development of structural design methods from allowable stress design methods to limit state design methods, seismic design methods for lifelines have also been developed. Recently, design systematization based on the performance design method has been progressing, but it is not always clear whether the original purpose of the performance design concept is being realized, how much safety performance has been achieved, and whether the practicing designer can realize it. This book summarizes the current results to pave the way for a clearer and advanced performance design.

To properly set the partial coefficients introduced in the performance design method, it is necessary to accurately understand the background of the design equation, the modeling conditions of the equation, and the adaptive limits, which are specific factors of the imprecision of the design method. This book also focuses on the explanation of performance design methods from this aspect. In addition, as a case study, the book describes in detail the seismic safety verification method for pipelines restarted after the earthquake and the risk mitigation measures to ensure seismic safety during normal operation.

Theoretical approaches will continue to be needed to elucidate new phenomena. The basic design equations for seismic design of lifelines are described in detail, from the theoretical background to the formulation method, in order to serve as a guide for such an approach. These findings should be used as a reference for future technological development.

The authors hope that this book will be of some help in deepening the understanding of seismic design methods for lifelines.

April 2026

Takeshi Koike and Lessandro E. Garciano

About the Authors

Takeshi Koike is Professor (retired) at Kyoto University, Japan. He obtained his Bachelor's, Master's, and Doctoral degrees from Kyoto University in 1971, 1973, and 1976, respectively. He is a Senior Member of the Japan Society of Civil Engineers, a Member of the American Society of Civil Engineers, and an International Science Member of the International Society of Lifeline Earthquake Engineering.

Lessandro E. Garciano is Professor at De La Salle University, Manila, the Philippines. He received his Bachelor's, Master of Science, Master of Engineering, and Doctoral degrees from the University of San Carlos in 1990, De La Salle University in 1997, and Tokyo City University in 2004 and 2007, respectively. Among other positions, he is at present the President of the Association of Structural Engineers of the Philippines, a Life Member of the Philippine Institute of Civil Engineers and a Member of the American Society of Civil Engineers.

1 Lifeline Systems

This chapter describes the historical development of seismic design methods with a history of earthquake damage. In other words, in modern cities hit by major earthquakes, various types of seismic damage occur throughout the infrastructure. By investigating the causes of such damage and the mechanisms of destruction, new seismic design methods have been developed. This book reviews the historical background of the world's first seismic design method, the seismic intensity method, which was born from the experience of the Great Kanto Earthquake, and lifeline earthquake engineering, which was established through a bird's-eye view of the seismic damage patterns of the San Fernando Earthquake, leading to today's lifeline seismic design methods.

1.1 HISTORY OF LIFELINE EARTHQUAKE ENGINEERING

When a modern city is hit by a major earthquake, various above- and below-ground facilities are severely damaged. These damages are summarized as earthquake damage to critical urban infrastructure system, which is called the *urban lifeline*. This led to the creation of lifeline earthquake engineering as a branch of earthquake engineering.

Tables 1.1.1 and 1.1.2 show the history of major earthquakes over the past 100 years and the research and technological development in the field of lifeline earthquake engineering. At the same time, the history of the development of the Japanese Guidelines for Seismic Engineering Methods for Water Supply Facilities[1] is also described as an example. The table shows that new research and technological developments have been conducted based on the lessons learned from earthquake damage, and that these developments have been reflected in the seismic design guidelines from time to time, and have been repeated for each earthquake disaster up to the present day.

From the viewpoint of lifeline earthquake engineering, compared to the practical application of seismic design methods based on acceleration response spectra for above-ground structures, which are governed by inertial forces, the development of seismic design methods for underground structures began only in the late 1960s. Since underground structures subjected to ground shaking are constrained not by inertial forces but by the surrounding ground displacement, the response-displacement method was systematized as a seismic design method for such structures.

In 1969, a design method based on the displacement method was introduced for the first time in the seismic design[2] of BART (subway) constructed in the San Francisco Bay area. In Japan, the response displacement method was developed based on the concept of this design method, and the construction of the Narita Airport fuel pipeline led to the establishment of technical standards for oil pipeline installation[3] based on the response displacement method. Through this process, the response displacement method became established in Japan as a seismic design method for underground structures.

DOI: 10.1201/9781003655527-1

TABLE 1.1.1
Lifeline Earthquake Engineering Narrative

	Historical Events in Lifeline Earthquake Engineering
Year	**Event**
1923	Great Kanto Earthquake, seismic intensity method
1964	Niigata Earthquake, liquefaction damage
1965–1970	Matsushiro cluster earthquakes, accumulation of observation records of seismic wave propagation in the ground
1966	Reliability theory proposed by Freudenthal, Garrets, and Shinozuka
1969	Response displacement method adopted for seismic design of BART subway submerged tunnels
1971	San Fernando Earthquake, Prof. M. Duke proposes lifeline earthquake engineering
1972	Construction of Narita Airport Aviation Fuel Transportation Pipeline begins (practice of Oil Pipeline Standards)
1973	Oil Pipeline Engineering Standards (Ministerial Ordinance) abbreviated as Oil Pipeline Standards)
1975–1977	Construction of Alaska crude oil pipeline (highest level of seismic technology)
1977	ASCE seismic design criteria for pipelines Trans-Alaska pipeline system
	New seismic design method (draft), introduction of response displacement method (propagating waves: SH wave model)
1978	Miyagi Offshore Earthquake, damage to lifelines
1982	Seismic design guidelines for gas pipelines (propagating wave: surface wave model), introduction of slip design method (for Level 1 earthquake ground motion, straight pipes, and deformed pipes)
1983	Narita Airport aviation fuel pipeline complete
1986	WTO agreement ratified, domestic design method established as a global standard
1993	Kushiro Offshore Earthquake, ground liquefaction, lifeline damage
1995	Hyogo-ken Nanbu Earthquake, Level 2 earthquake ground motion proposed
1998	ISO 2394 General Principles for Reliability of Structures, Reliability Design Method
1999	Taiwan Chi-Chi Earthquake, fault displacement exposure, HAZUS99 by FEMA
2005	American Lifeline Alliance (ALA) by FEMA (US version of seismic design for water pipelines)
2011	Great East Japan Earthquake, expansion flexible pipe damage
2016	Kumamoto Earthquake, damage to small diameter pipe crossing fault lines
2018	Northern Osaka Earthquake, case iron pipe strength deterioration damage, material deterioration
	Hokkaido Iburi Eastern Earthquake, large-scale hillside collapse, landslide, ground liquefaction

TABLE 1.1.2
Development of Guidelines for Pipeline Seismic Engineering Methods

	Historical Development of Seismic Design for Various Pipelines
Year	**Event**
1953	Seismic Design of Water Supply Facilities (First Edition)
1966	Seismic Design Methods for Water Supply Facilities (Revised Edition) (stress-based elastic design method)
1973	Technical Standards for Oil Pipelines (Ministerial Ordinance) (abbreviated as Oil Pipeline Standards)
1979	Guidelines and Commentary for Seismic Engineering Methods for Water Supply Facilities, 3rd edition Introduction of response displacement method. and SH wave model as horizontally propagating wave (displacement and strain-based elastic design method, based on Oil Pipeline Criteria) are described
1982	Seismic Design Guidelines for Gas Pipes (propagating waves: surface wave model) Introduction of slip design method (for Level 1 earthquake ground motion, straight pipes, and irregularly shaped pipes)
1997	Guidelines and Commentary on Earthquake-Resistant Construction Methods for Water Supply Facilities (4th edition) In which introduction of Level 2 earthquake ground motion, introduction of slip design method limited to straight pipes are described
2000	Seismic design guideline for high-pressure gas pipelines, corresponding to Level 2 earthquake ground motion, seismic design system for straight and deformed pipes considering slip
2001	Guidelines for seismic design of liquefied high-pressure gas pipelines, introduction of performance design method
2009	Guidelines and Commentary on Seismic Design Method for Water Supply Facilities (5th edition) In which liquefaction investigation, detailed seismic design of facilities, and performance verification methods are adopted
2022	Guideline and Explanation of Seismic Design Method for Waterworks Facilities (6th revision) in which introduction of seismic design of PE pipes, and FEM analysis recommended for seismic safety verification of facilities are described

In the 1970s and 1980s, many major earthquakes occurred, including the 1978 Miyagi-Ken-Oki earthquake, and knowledge for seismic countermeasures has been accumulated from the observation of earthquake damage. Some of them reported that the maximum measured earthquake acceleration was much higher (e.g., more than 1 g) than the design acceleration (0.2 g, where g = gravity acceleration) at that time. However, the results were not reflected in the seismic design guidelines at the time of the 1995 Hyogoken-Nanbu Earthquake. Based on the lessons learned from this earthquake, Level 1 and Level 2 earthquake ground motions were introduced as design seismic loads,[4] and design for elastoplastic behavior was required for underground structures. At that time, the existing response displacement method was systematized based on the assumption of elastic behavior of both the ground and the buried pipe, but a seismic design method that can incorporate nonlinear behavior between the buried pipe and the surrounding ground during a large earthquake was sought. The concept of slip between the buried pipe and the surrounding ground is a solution to this problem, and its introduction has led to the systematization of seismic design methods for underground structures subjected to Level 2 earthquake ground motion.

Subsequently, seismic design methods for underground structures subjected to ground deformation such as liquefaction, landslides, and fault displacement have been developed. First, the seismic design method for gas pipelines crossing liquefied ground was embodied in the Seismic Design Guidelines for Liquefaction Resistance of High-Pressure Gas Pipes.[5] On the other hand, the seismic design method for fault-crossing pipelines did not appear to be in the process of being realized until more than a decade later, but it has gradually been developed since around 2020. The development of effective new pipe materials[6] for fault-crossing countermeasures may have contributed to this development.

Safety verification methods for structural design have evolved from the conventional allowable stress design method to the limit load design method, and recently, the development of the performance design method[7] has been promoted in Japan and abroad. However, although the basic theory of safety verification methods was established in the 1960s as the reliability theory,[8] it took more than half a century for the theory to be put into practical use.

In particular, since 2000, the limit state design method and the seismic design method based on the performance design method have been developed for the purpose of more rational verification of seismic safety, and have been used to this day. After the ratification of the International Trade Organization (WTO) standards, domestic design guidelines have been gradually converted to the performance design method, and major design guidelines and standards, such as the specifications for road bridges,[9] guidelines for seismic design of high-pressure gas, and guidelines for seismic construction methods for water supply facilities, are now written in the style of the performance design method. However, further research and technological development will be required to reach a user-friendly guideline.

Since its first edition in 1953, the Guidelines for Seismic Engineering Methods for Waterworks Facilities have been published approximately every 10 years until the sixth edition in 2022. In each edition, the guidelines have been updated to reflect the lessons learned from the earthquake disasters that occurred during the 10 years before the next edition and to incorporate the results of seismic design techniques that have been developed to improve the level of seismic design techniques. In particular, the use of numerical analysis methods based on the FEM has recently been actively promoted for the seismic design of not only pipelines but also various structures constituting water supply systems.

1.2 LIFELINE CLASSIFICATION AND CHARACTERISTICS

1.2.1 Morphological Classification of Lifelines

The term "urban lifeline" is well established in the field of earthquake engineering, but from a more general point of view, urban lifelines are considered to be "critical urban infrastructures" that have a network shape.

Major lifelines include water supply and sewerage systems, high-pressure gas pipes, energy systems (transmission lines, conduits, and pipes), communication systems (communication lines, conduits, and pipes), community ditches, and road, rail, and subway networks. The first type of lifeline has a hierarchical structure, with the upstream section connected to the source having a simple trunk line shape, while the downstream section shows a network-like form. The third type is a linear type that simply connects the starting and ending points.

When a lifeline is hit by an earthquake, a large amount of damages may occur in a wide area of the lifeline, and a certain section of the network may be out of service, but some lifelines can maintain the transportation of goods through detour routes due to network redundancy. Therefore, when discussing earthquake damage to lifelines, it is necessary to evaluate not only the physical damage in terms of the number of damaged structures but also from the functional perspective of whether or not the transportation function of the network from the starting point to the destination was degraded.

1.2.2 Structural Classification of Lifelines

Major lifelines can be classified into two groups. The pipelines, which include water and gas pipelines, and power and communication conduits are installed at a shallow level of the surface ground. The road and subway tunnels, on the other hand, are installed at deeper level of the surface ground. There are some differences between the pipelines and tunnel structures in their structural components. These pipelines are connected with many pipe elements, while the tunnel structures are connected with axially bolted concrete rings that are jointed with several concrete segments in the circumferential direction. Main discussion in this book is devoted to the pipelines and tunnel structures.

The structural characteristics of pipelines appear in the joints. Taking water pipelines as an example, the mechanical type joints used in many water pipelines in the past, such as cast-iron insert joints and ductile iron pipes, can be classified into steel pipes that are welded together by arc welding at the pipe ends (continuous joints) and polyethylene pipes whose pipe bodies are integrated by a fusion welding method (continuous joints).

Continuous joints are highly safe in that the internal fluid does not leak unless there is a crack in the pipe body, and welded steel pipes are basically used for high-pressure gas pipelines. On the other hand, cast-iron and ductile cast-iron pipes with plug-in and mechanical joints have been used since the Meiji era (1868–1912) and have already been in use for more than 100 years. However, the plug-in type and mechanical joints have basically low tensile resistance, and there is a concern that leakage accidents may occur due to withdrawal during earthquakes. Recently, ductile cast-iron pipes (earthquake-resistant joint pipes) with high pullout resistance have been commercialized.

The long-term use of any of these pipe materials increases the risk of leakage accidents due to symptoms such as corrosion, accumulated fatigue, and pipe body deterioration. Pipelines that transport high-pressure fluids and those that transport low-pressure fluids have dramatically different safety requirements for joints. In general, for high-pressure lines of 1 MPa to 10 MPa, such as high-pressure gas pipelines, the utmost efforts are made to select high-strength pipe materials that can withstand the internal pressure and joints that connect the pipelines to each other in order to prevent leakage accidents that can cause catastrophes. In particular, efforts are required to develop elastoplastic design methods for safe operation under severe operating conditions such as ground shaking caused by earthquakes, ground deformation, and temperature expansion and contraction caused by changes in outside temperature. High-pressure pipelines that must be laid in areas where ground deformation such as ground subsidence or fault displacement is expected are designed to absorb ground deformation through the deformation performance of the pipeline itself, such as meandering or zigzag piping.[10]

On the other hand, low-pressure lines, such as water and sewage pipes and gas medium- and low-pressure pipes in urban areas, employ a variety of fittings and pipe materials that are compatible with the internal pressure level. Again, unequal settlement, liquefaction displacement, or fault displacement may occur in localized areas. In such cases, special expansion joints that absorb axial and perpendicular displacements, such as expansion flexible joints, are usually installed to cope with the ground deformation.

When the safety of a structure is verified according to the limit state design method,[11] it is desirable to develop probability statistical data of structural resistance (strength) and to set the characteristic values of resistance (strength) to be used for safety verification based on probability statistical methods.[12] In the past, even if the nominal value of resistance (strength) was presented to the designer, the probability statistical data that would serve as the basis for the nominal value were often not clearly indicated, making it difficult to implement a rigorous limit state design. As a result, the accuracy level of performance design methods has remained at the level of Level 1 reliability design methods. In order to raise the accuracy level of performance design methods to that of Level 2 or 3 reliability design methods, it is necessary to develop an environment in which designers can easily access probability statistics of resistance (strength) property values.

1.2.3 Structural Characteristics of Pipelines

Pipelines are basically made up of straight pipes, curved pipes, branch pipes, and expansion and contraction flexible joints. For example, in the case of water supply, pipelines can be categorized as conduits, transmission pipes, distribution mains, distribution pipes, water supply pipes, and plant piping in facilities such as water treatment plants. However, seismic damage can occur at the most vulnerable points of the piping system, causing pipeline failure and internal fluid leakage. In particular, seismic damage often occurs at joints between different types of structures, such as joints between conduits and structures, and these joints are considered to be potential hazardous locations.

In the past, many cases of seismic damage have been observed, and design and construction methods have been developed to mitigate or eliminate such damage. However, the next time an earthquake occurs, new cases of seismic damage are encountered, and the search for new solutions continues. The solution to this problem depends on the ability of designers to use their imagination to devise seismic countermeasures against unknown modes of seismic damage, and further improvement in design technology is expected.

1.2.4 Seismic Damage Modes

Seismic damage modes can be classified into three categories: structural damage due to ground shaking, structural damage due to destructive ground deformation, and functional damage to the supply system.

Figure 1.2.1 shows an example of water pipe leakage damage caused by ground shaking during the 2018 Osaka Prefecture Northern Japan Earthquake. Figure 1.2.2(a) shows a case of liquefied ground damage caused by the 2021 Hokkaido earthquake. Figure 1.2.2(b) shows fault displacement (vertical ground

FIGURE 1.2.1 Example of water pipe leakage damage due to ground shaking.[13]

(a) (b) (c)

FIGURE 1.2.2 Examples of destructive ground deformation caused by earthquakes: (a) example of the liquefied damage;[14] (b) example of fault deformation; (c) example of the landslide.

deformation) during the 1999 Taiwan Chi-Chi Earthquake. Figure 1.2.2(c) shows an example of landslide damage caused by the 2011 off the Pacific coast of Tohoku earthquake.

When the ground shakes significantly due to a large earthquake, there have been cases where pipelines attached to bridges have ruptured or leaked due to the effects of large displacement response or failure of the bridge. There have also been cases where pipelines connected to buried pipelines from above-ground structures have been affected by ground subsidence, causing the pipelines at the connection to rupture. When destructive ground deformation occurs, the pipes and joints of various pipelines buried in the ground are reported to rupture and leak internal fluids, causing extensive damage to the surrounding area.

1.3 FUNCTIONAL CHARACTERISTICS OF LIFELINE SYSTEMS

If the pipeline is a single pipeline in series, both structural failure (i.e., pipeline failure) and functional failure (i.e., fluid transport stoppage) occur simultaneously. On the other hand, if the pipeline is part of a network system, the safety of the pipeline function is maintained even if a part of the pipeline is damaged during an earthquake due to network redundancy, as long as fluid transport service is available via another route.

As an example, when evaluating the functionality of a water pipeline system, one way to assess its seismic performance is to evaluate the degree to which water supply service can be maintained from the water source, such as a water treatment plant, to the demand end. However, even in this case, it is necessary to conduct a flow (hydraulic) network analysis to satisfy the flow conditions of the fluid flowing in the network, and to evaluate the seismic function performance based on the degree to which the pressure (water pressure) and flow rate (water volume) reaching the end of demand satisfy the design conditions. On the other hand, there is a simplified method for evaluating the functional performance of supply (water supply) that does not consider the arrival conditions of pressure and flow rate (water pressure and water volume) as long as the connection route from the source to the demand end is secured. Which method should be adopted should be determined based on its suitability for the analysis purpose of the target network system.

When discussing functional characteristics, one of the concerns is that the target function is indispensable to the activities (lives) of consumers (residents), and therefore, the service must never be interrupted, and therefore, absolute seismic performance is required to be 100% seismic performance. However, this is a wrong conclusion because the word "absolute" is a misleading word that demands 100% perfection. The reason for this thinking is the lack of awareness of the unreasonableness of demanding 100% certainty only for the service results of the system when seismic phenomena are inherently uncertain events, and the behavior of consumers and the situation in which facilities and pipeline systems are located are all uncertain at the time of occurrence. This irrational logic is the reason why the system is not working. If unnecessary and excessive investments have been made because of this irrational logic, business operators need to be very careful about this kind of problem.

REFERENCES

1. Japan Water Works Association: Guidelines and Commentary on Earthquake Resistant Construction Methods for Water Supply Facilities, 2022 Edition, 2022.
2. Kuesel, T. R.: Earthquake Design Criteria for Subways, Vol. 95, ASCE, No. ST6, 1969.
3. Ministerial Ordinance: Ministerial Ordinance on Technical Standards for Facilities for Oil Pipeline Projects, Ordinance of the Ministry of International Trade and Industry, Ministry of Construction and Ministry of Home Affairs No. 2, 1972.
4. Japan Society of Civil Engineers: Collection of Proposals Concerning Seismic Standards, etc., Japan Society of Civil Engineers, 1996.
5. Japan Gas Association: Seismic Design Guideline for Liquefaction Resistance of High-Pressure Gas Pipes, JGA Finger-207-01, Japan Gas Association, Committee for Investigation of Technical Standards for Gas Facilities, 2001.

6. Hasegawa, N.: Research on Seismic Design Methodology for Fault-Crossing Pipelines, Thesis, Kyoto University, 2016.
7. Earthquake Engineering Society of Japan: *Performance-Defined Seismic Design, Research Committee on Performance-Defined Seismic Design Methods*, Kajima Publishing Company, 2006.
8. Ang, A. H-S. and W. H. Tang: *Probability Concepts in Engineering Planning and Design*, Volume II: Decision, Risk and Reliability, John Wiley & Sons, 1984.
9. Japan Highway Association: Road Bridge Specifications and Commentary, Japan Highway Association, Bridge Committee, Subcommittee on Seismic Design, 2017.
10. Newmark, N. M.: Seismic Design Criteria for Structures and Facilities, Trans-Alaska Pipeline System, Proc. U.S. Nat. Conf. on Earthquake Engineering, EERI, pp. 94–103, 1975.
11. Melchers, R. E.: *Structural Reliability Analysis and Prediction*, 2nd edition, John Wiley & Sons, 1999.
12. Ang, A. H-S. and W. H. Tang: *Probability Concepts in Engineering Planning and Design*, Volume I: Basic Principles, John Wiley & Sons, 1973.
13. Osaka Regional Waterworks Corporation: Report meeting material related to the earthquake in the northern part of Osaka Prefecture, February 28, 2019.
14. Japan Water Works Association: Survey Report on Damage to Water Supply Facilities, etc., in the Hokkaido Bombs and Earthquakes, 2018.

2 Ground Response Analysis for Lifelines

A large magnitude earthquake often produces two types of ground motions: ground shaking and permanent ground displacement (PGD). Ground shaking is caused by seismic waves propagating horizontally and vertically. Since ground shaking does not produce destructively large ground strains, most lifelines systems have not been broken by a moderate size of ground shaking, which will be encountered in the average service life of lifeline systems. On the other hand, the PGD produced by fault movement, liquefaction, and landslide has brought significant destructive damage to many urban lifelines.

Various types of seismic wave problems are discussed and classified based on the geophysical theories as shown in Table 2.1. In this table, items (1), (2), and (4) are discussed in this chapter, while the other items (3) are not described herein. And items (5)–(9) are discussed in Chapter 3.

Chapter 2 discusses seismic wave propagation, which is the dominant seismic action that acts on lifelines. When an earthquake occurs, the ground near the surface shakes and shows the so-called ground motion. Ground motions that are significant to lifelines are horizontally propagating waves that travel along the lifeline. The characteristics of various types of seismic waves propagating horizontally through the surface ground are best explained using elastic wave theory. Ground motion analysis methods necessary for the seismic design of lifelines are thus discussed in detail in this chapter. Stochastic formulation of seismic ground motion will be developed in the item (4), where spatial variation[1,2] horizontally propagating along a pipeline is discussed.

2.1 SEISMIC WAVE PROPAGATION

2.1.1 Elastic Waves Propagation

The wave equation[3] in a three-dimensional elastic body is expressed as

$$\rho\frac{\partial^2}{\partial t^2}(u,v,w)=(\lambda+\mu)\left(\frac{\partial}{\partial x},\frac{\partial}{\partial y},\frac{\partial}{\partial z}\right)\Delta+\mu\nabla^2(u,v,w), \tag{2.1.1}$$

where ρ is the density, λ,μ are Lame's constant, and $\underline{u}$ is the displacement vector which follows to the coordinate system shown in Figure 2.1.1:

$$\Delta = div\ \underline{u}\quad \Delta=\left(\frac{\partial u}{\partial x}+\frac{\partial v}{\partial y}+\frac{\partial w}{\partial z}\right) \tag{2.1.2}$$

$$\Theta = curl\ \underline{u}\quad \nabla^2=\left(\frac{\partial^2}{\partial x^2}+\frac{\partial^2}{\partial y^2}+\frac{\partial^2}{\partial z^2}\right). \tag{2.1.3}$$

DOI: 10.1201/9781003655527-2

TABLE 2.1
Classification of Seismic Modes

Load	Seismic Mode	Phenomenon	Ground Classification	Motion
Earthquake	Ground shaking	Ground motion in the free field	Infinite elastic ground	(1) Body wave propagation
		Ground motion in the stratified field	Layered ground	(2) Surface wave propagation
			Nonlayered ground	(3) Generalized wave propagation
		Spatial deviation of ground motion	Coherent ground	(4) Coherent displacement between two points
	Permanent ground displacement (PGD)	Fault dislocation	Active fault area	(5) Fault displacement
			Near-fault ground	(6) Near-field ground motion
		Phase transition from solid to liquefied soil	Liquefied ground	(7) Settlement
			Unstable ground	(8) Lateral spreading
		Landslide	Unstable ground	(9) Ground collapse

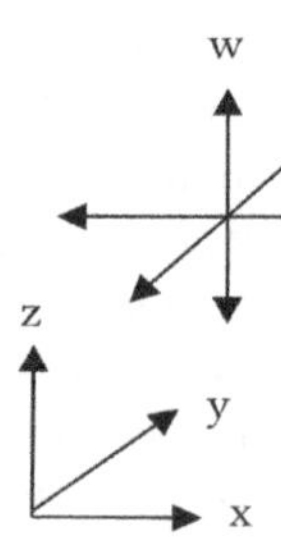

FIGURE 2.1.1 Coordinate system.

Using the vector representation $\underline{u} = (u, v, w)$ shown in Figure 2.1.1, we obtain the following equation:[4]

$$\rho \frac{\partial^2 \underline{u}}{\partial t^2} = (\lambda + \mu)\, grad\Delta + \mu \nabla^2 \underline{u}. \quad (2.1.4)$$

Considering $div\ \underline{u} = \Delta$ as volumetric strain and $div(grad\Delta) = \nabla^2\Delta$,

$$\rho \frac{\partial^2 \underline{u}}{\partial t^2} = (\lambda + 2\mu)\nabla^2\Delta, \quad (2.1.5)$$

and also considering $\Theta \equiv curl\ \underline{u} = (\theta_x, \theta_y, \theta_z)$ as angle change and $curl(grad\Delta) = 0$,

$$\rho \frac{\partial^2 \Theta}{\partial t^2} = \mu \nabla^2 \Theta. \quad (2.1.6)$$

On the other hand, when scalar potential φ and vector potential ψ are used, the displacement component can be expressed as

$$(u, v, w) = grad\varphi + curl\psi. \tag{2.1.7}$$

In this case, the volume change produces coarse and dense waves, or longitudinal waves, and the angle change produces shear waves, or transverse waves, so that P waves (primary waves, longitudinal waves, coarse and dense waves) and S waves (secondary waves, transverse waves, shear waves) can be described by the following equations:

$$\text{P wave (longitudinal wave)} \quad \frac{\partial^2 \varphi}{\partial t^2} = c_p^2 \nabla^2 \varphi, \quad c_p = \sqrt{\frac{\lambda + 2\mu}{\rho}} \tag{2.1.8}$$

$$\text{S waves (shear wave)} \quad \frac{\partial^2 \psi_i}{\partial t^2} = c_S^2 \nabla^2 \psi_i, \text{ for } i = 1\text{: SH waves}, 1 = 2\text{: SH waves}, \quad c_S = \sqrt{\frac{\mu}{\rho}} \tag{2.1.9}$$

The coefficients c_P, c_S in the two wave equations are the propagation velocities of the P and S waves. For S waves, those with only a horizontal displacement component are distinguished as SH waves (ψ_1), and those that also include a vertical in-plane displacement component are distinguished as SV waves (ψ_2).

2.1.2 Reflection and Refraction of Elastic Wave

Ground shaking can be analyzed as an elastic wave propagation problem[5] because the ground can be treated as an elastic body if the strain generated in the ground is less than about 10^{-5}. Let us consider the case where elastic waves (plane waves) propagate in a two-dimensional plane from the left downward as shown in Figure 2.1.2. In this case, the elastic wave is refracted and reflected at the boundary of the medium. The relationship between the two is shown in the figure. Here, the propagation velocity of the wave in the lower medium is c_1 and that in the upper medium is c_2.

When SH waves are incident from the lower medium at an angle of incidence β_1, SH waves that pass through the boundary surface and are refracted at an angle of

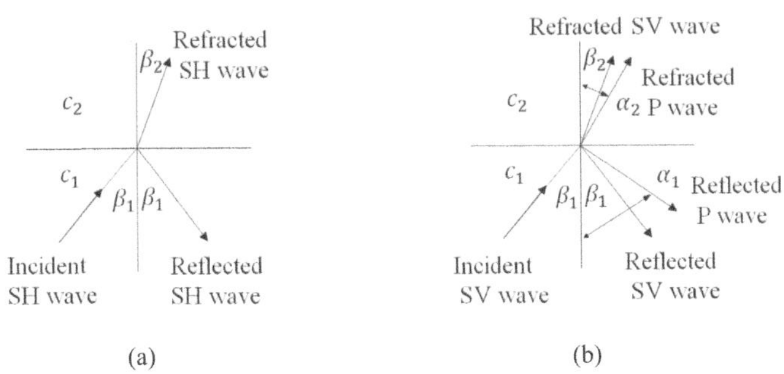

FIGURE 2.1.2 Reflection and refraction of elastic waves: (a) incident SH wave; (b) incident SV wave.

refraction β_2 appear in the upper medium. In addition, SH waves reflected at the boundary surface also appear. On the other hand, in the case of SV waves, refracted P waves and reflected P waves appear simultaneously in addition to the refracted and reflected SV waves. The reflection and refraction angles are α_1 and α_2, which are different from those of β_1 and β_2. The case of incident P waves also shows the same behavior as that of SV waves.

When an incident SH wave enters the boundary at an angle of incidence β_1 and propagates at an angle of refraction β_2 after refraction, the wave heads of two parallel incident SH waves maintain parallelism before and after refraction, which leads to the following relation, called Snell's law:

$$\frac{c_1}{sin\beta_1} = \frac{c_2}{sin\beta_2} = c, \tag{2.1.10}$$

where c is the apparent propagating wave velocity in the horizontal direction.

The angle of incidence at which the refraction angle becomes 90° is called the critical angle $\beta_{cr,1}$. For incident angles larger than this, no refraction angle into the elastic body occurs, and reflected waves of the same amplitude as the incident waves are generated. In other words, total reflection occurs. In this case, propagating waves within the surface layer propagate horizontally between the boundary and the ground surface, repeating total reflection.

$$\beta_{cr,1} = sin^{-1}\left(\frac{c_1}{c_2}\right). \tag{2.1.11}$$

2.1.3 Elastic Surface Waves

2.1.3.1 Surface Wave Model (Rayleigh Wave, Love Wave)

Surface waves are known as waves propagating horizontally near the ground surface. Taking the x-axis in the direction of propagation, the y-axis in the horizontal direction perpendicular to the x-axis, and the z-axis in the direction of depth, the displacement due to Rayleigh waves has only the x- and z-components and that, due to Love waves, has only the y-component.

The propagation velocity of surface waves propagating in the surface soil is expressed as c_R for Rayleigh waves of the dilatational wave type and c_L for Love waves of the shear type. In other words,

Propagation Velocity	Wave Type
C_P	P wave: dilatational wave
C_S	S wave: shear wave
C_R	Rayleigh wave: dilatational-type surface wave
C_L	Love wave: shear-type surface wave

The propagation modes[6] of each elastic wave are shown in Figure 2.1.3.

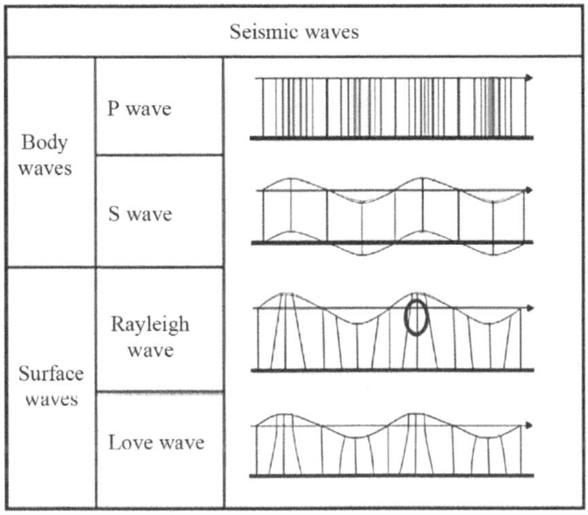

FIGURE 2.1.3 Propagation modes of elastic waves.[4]

2.1.3.2 Characteristics of Surface Waves

Love waves are surface waves formed by SH waves that reflect off the surface of stratified ground any number of times, i.e., they are formed when SH wave propagation is concentrated near the ground surface. Rayleigh waves, on the other hand, are surface waves formed by the interference between P and SV waves.

Surface waves differ from entity waves (P waves and S waves) in that (1) they propagate more slowly than entity waves, (2) their amplitude decreases rapidly from the surface to the depth direction, (3) their amplitude does not decrease with distance, and (4) their propagation speed strongly depends on the period. The characteristics of (4) are called dispersion.

For example, the propagation velocity of a Love wave is between the S wave velocity of the top layer and the S wave velocity of the substrate, and the longer the period, the larger the propagation velocity becomes and the closer to the S wave velocity of the substrate. On the other hand, Rayleigh waves do not exhibit dispersion in a semi-infinite medium, but in stratified ground they exhibit dispersion characteristics similar to those of Love waves.

The seismic design guideline for high-pressure gas pipelines[7] adopts Rayleigh-type surface waves as horizontally propagating seismic waves and presents the dispersion curve shown in Figure 2.1.4 as a dispersion curve representing the dispersion characteristics of their propagation velocity.

2.2 INPUT GROUND MOTION

2.2.1 Strong Ground Motion Records

Seismic motions measured by strong ground motion accelerometers have been gathered as strong ground motion records. The strong ground motion record of the El-Centro Earthquake, which was measured early in the United States, has made an

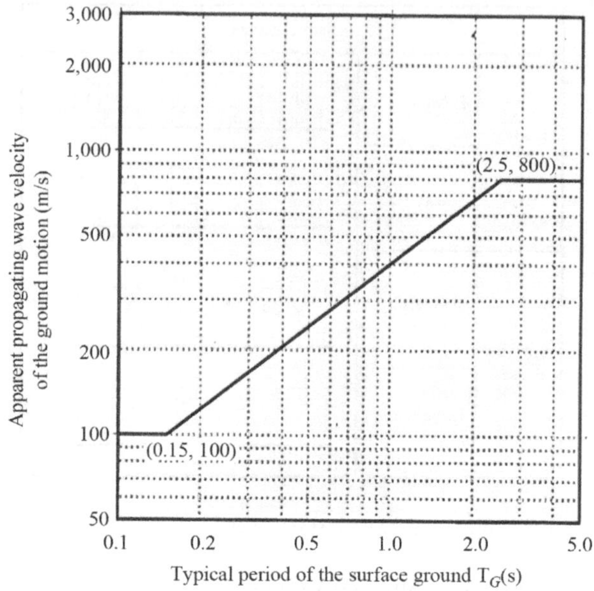

FIGURE 2.1.4 Phase velocity in the seismic design guidelines for high-pressure gas pipelines.[7]

important contribution to the development of subsequent seismic design methods. In Japan, the domestic earthquake observation system has been vigorously developed since the Hyogo-ken Nanbu Earthquake. Figure 2.2.1 shows the acceleration time histories of representative seismic waves.

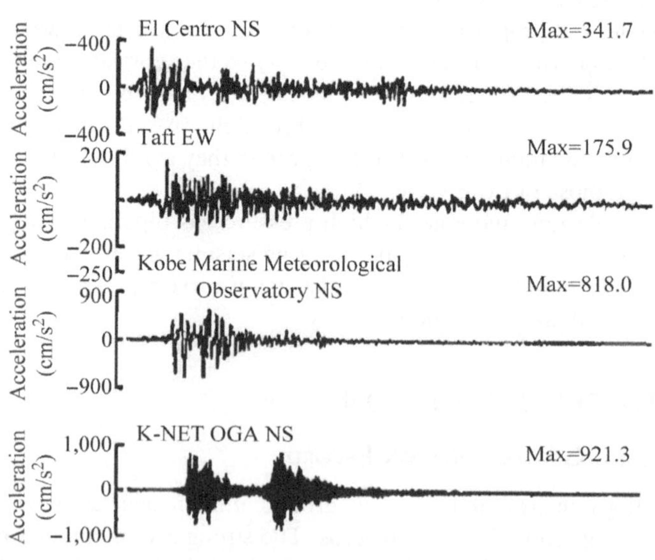

FIGURE 2.2.1 Representative acceleration time history data.

TABLE 2.2.1
Methods of Obtaining Seismic Wave Data Information in Japan and the United States

Item	Information Type	Japan	USA
Seismic Load	Seismic load information for seismic design	Design of earthquake resistance	Seismic hazard information for each site can be obtained using the USGS Seismic Hazard Mapping Utility
		Design acceleration response spectra and velocity response spectra for each ground type are prepared in the design guidelines for Level 1 and Level 2 ground motions	http://earthquake.usgs.gov/provides seismic data for the site
	Seismic data for dynamic design	The strong motion recorded time history waveform data required for dynamic seismic response analysis should be obtained directly from K-net or KIK-net, or the designer should generate seismic time history data compatible with the design acceleration spectrum	Obtain seismic wave data from http://earthquake.usgs.gov/that conforms to the seismic design criteria to be employed

Table 2.2.1 shows how to obtain the seismic wave data from the Data Center in Japan and in USA.

2.2.2 Seismic Hazard Analysis

The probabilistic seismic hazard analysis is useful when information on seismic hazard around the site where the structure is to be constructed is available and the designer needs information on design earthquake motions suited to the ground conditions of the site more precisely than given in the seismic design guidelines, or, conversely, when information on design earthquake motions is limited and the designer needs to find information on earthquake motions necessary to carry out the seismic design. On the other hand, when information on design earthquake ground motions is limited and designers need to search for information on earthquake ground motions necessary for seismic design, the probabilistic method of earthquake ground motion prediction is an effective method. Figure 2.2.2 illustrates the procedure[8] for earthquake motion prediction.

The procedure for predicting the design earthquake ground motions for seismic design follows these steps:

1. Set the probability p of occurrence of the assumed earthquake during the service period (n years) and calculate the seismic recurrence interval (T_R) in years from the following equation:

$$p = 1 - \left(1 - \frac{1}{T_R}\right)^n, \tag{2.2.1}$$

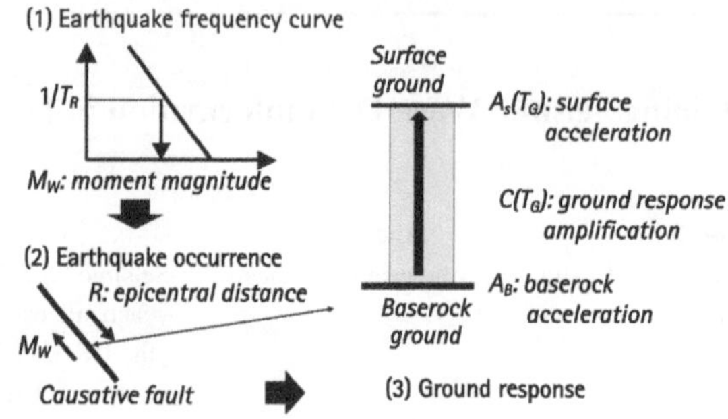

FIGURE 2.2.2 Prediction procedure of earthquake ground motion.

where n is the service period (years) and p is the probability of earthquake occurrence during the service period.

2. Estimate the earthquake magnitude M_W from the earthquake frequency curve obtained near the site.
3. Active faults that match the earthquake magnitude M_W are surveyed near the site, and data on the length of active faults and earthquake occurrence intervals are collected (Figure 2.2.3).
4. Assuming that the earthquake occurs on the target active fault, the epicenter location and the epicentral distance R between the sites are estimated.
5. Estimate the seismic acceleration A_B at the engineering basement of the site from the distance attenuation formula using M_W and R.
6. Estimate the ground acceleration $A_S(T_G)$ by obtaining the ground typical period T_G and ground response amplification $C(T_G)$ from the site ground data.

2.2.3 Response Spectrum

The maximum response of a one-degree-of-freedom vibration system to seismic motion, plotted along the natural period of the system, is called a response spectrum.[10]

Given a strong-motion earthquake time history data, the response spectrum can be created by the following procedure:

1. Prepare several damping constants to analyze the seismic response of a one-degree-of-freedom vibration system. Also, prepare the eigen-periods $T_1, T_2, \cdots, T_n$ of the vibration system to be analyzed.
2. Using the strong-motion time history data as input data, perform a seismic response analysis of the one-degree-of-freedom vibration system for each combination of damping constants and natural period to obtain response results (time history data) for acceleration, velocity, and displacement.
3. Calculate the maximum acceleration, velocity, and displacement from the seismic response analysis results.
4. The response spectrum is obtained by plotting the natural period on the horizontal axis, the maximum response value on the vertical axis, and the damping constant as a parameter.

Figure 2.2.4 illustrates the steps from (1) to (4).

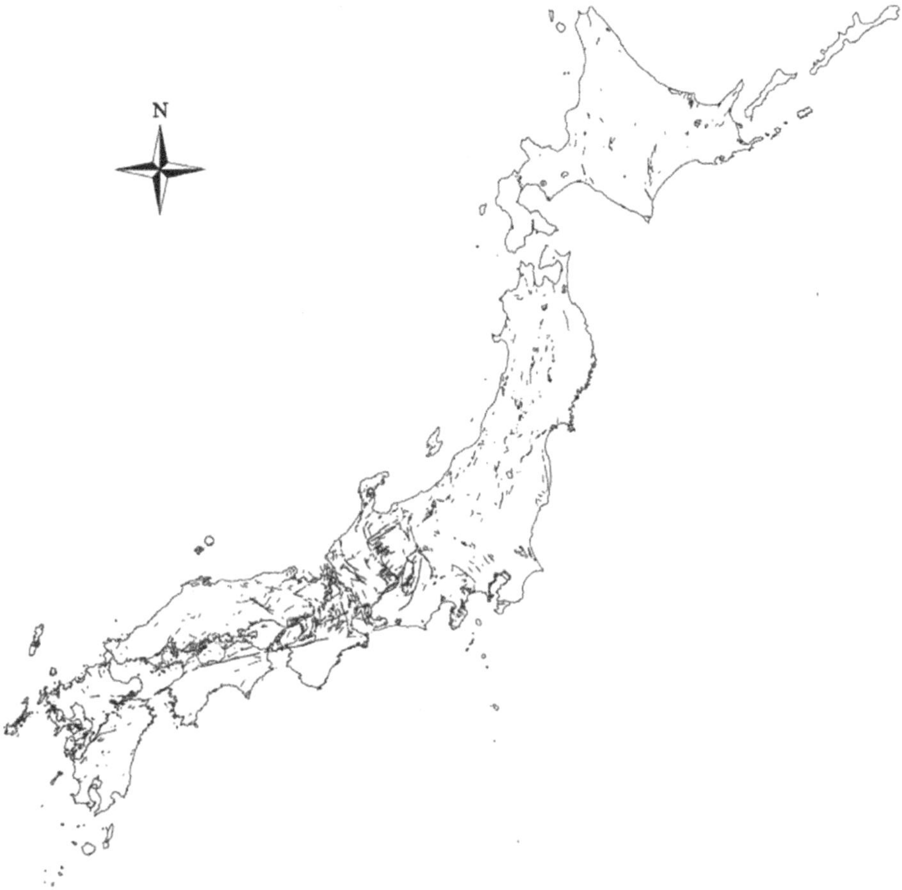

FIGURE 2.2.3 Active fault map of Japan.[9]

2.2.4 Design Earthquake Ground Motions

Based on the experience of the damage caused by the 1995 Hyogo-ken Nanbu Earthquake in Japan, the design earthquake ground motion applied to domestic earthquake-resistant design guidelines[11] has been revised into two types: Level 1 ground motion and Level 2 ground motion. Level 1 ground motion is a load level corresponding to the design earthquake ground motion used in conventional seismic design, while Level 2 ground motion is a newly introduced design earthquake ground motion. Both types of earthquake ground motions are set as follows.

Level 1 ground motions are those caused by earthquakes of medium intensity that have a relatively high likelihood of occurring. Level 1 ground motion is considered to be a variable action because it often occurs during the design service period of target structures.

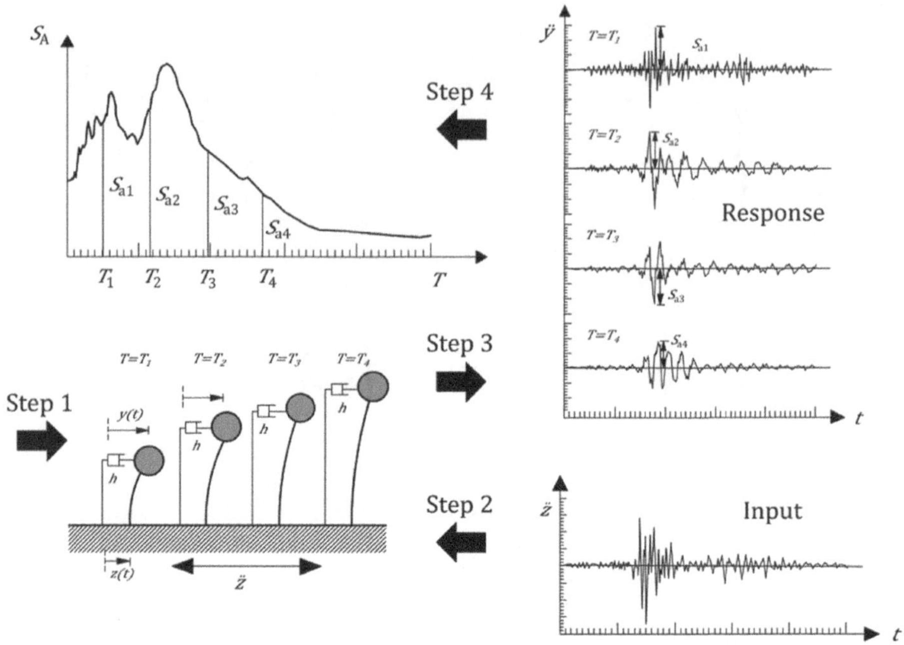

FIGURE 2.2.4 Response spectrum creation procedure.

Level 2 ground motions are classified into two types: Type I, which assumes large-scale oceanic earthquakes that occur infrequently at plate boundaries, and Type II, which assumes earthquake ground motions caused by magnitude-class inland earthquakes of extremely low occurrence frequency, such as the 1995 Hyogo-ken Nanbu earthquake. The standard acceleration response spectra for each ground type are as follows.

2.2.4.1 Design Response Spectrum

Ground structures are subjected to inertia forces during earthquakes. Since the inertia force is the product of the mass of the structure and the seismic acceleration, the value of seismic acceleration is important for seismic design. Since the vibration characteristics of a structure are characterized by its natural period and damping constants, the maximum seismic load acting on the structure can be calculated from the acceleration response spectrum. Therefore, the design acceleration response spectrum is specified as the design load in the seismic design guidelines for above-ground structures.

On the other hand, underground structures[17] are constrained by the surrounding ground displacement during earthquakes and cannot deform freely. In other words, the deformation of an underground structure causes seismic strain within the structure. In general, the strain in an underground structure can be expressed as the ratio of the particle velocity to the phase velocity of the ground, and the design velocity response spectrum is specified as the design load for underground structures in the seismic design guidelines.

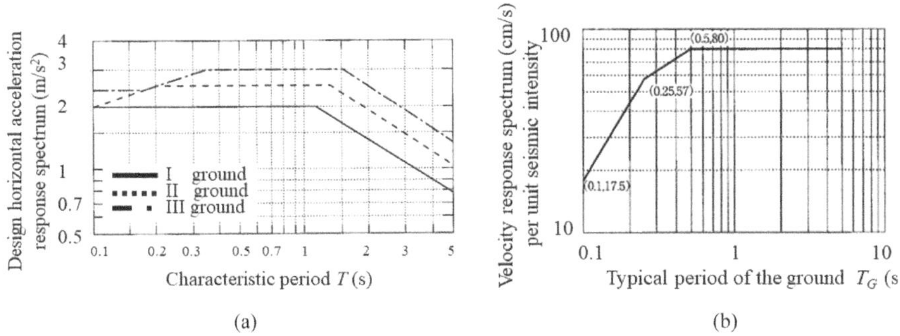

FIGURE 2.2.5 The standard acceleration response spectra for Level 1 ground motion: (a) acceleration response spectrum;[11] (b) velocity response spectrum.[12]

The standard acceleration response spectrum corresponding to Level 1 ground motion is set as shown in Figure 2.2.5(a) for each ground type that is defined as I ground: $T_G < 0.2$s; II ground: $0.2\text{s} \le T_G < 0.6$s; III ground: $T_G \ge 0.6$s with typical period T_G of the ground, respectively. The velocity response spectrum for Level 1 ground motion is the velocity response spectrum due to the input earthquake ground motion in the engineering base and is set as shown in Figure 2.2.5(b).

Level 2 ground motion is classified into two types, and standard acceleration response spectra are set for each ground type as shown in Figure 2.2.6.

The *Guidelines for Seismic Design Methods for Waterworks Facilities*[12] present standard acceleration response spectra and velocity response spectra for Level 2 ground motion corresponding to 90% and 70% non-exceedance probabilities, as shown in Figure 2.2.7.

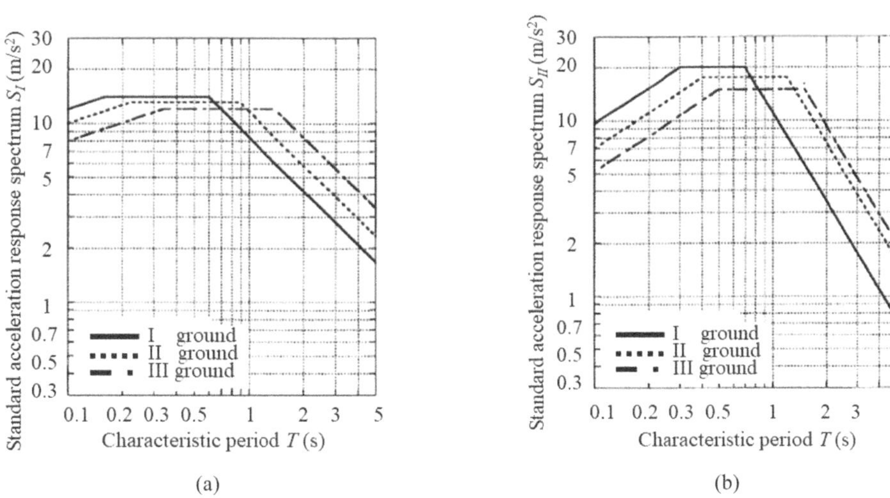

FIGURE 2.2.6 The standard acceleration response spectra for Level 2 ground motion:[11] (a) Type I (offshore-type earthquake); (b) Type II (inland-type earthquake).

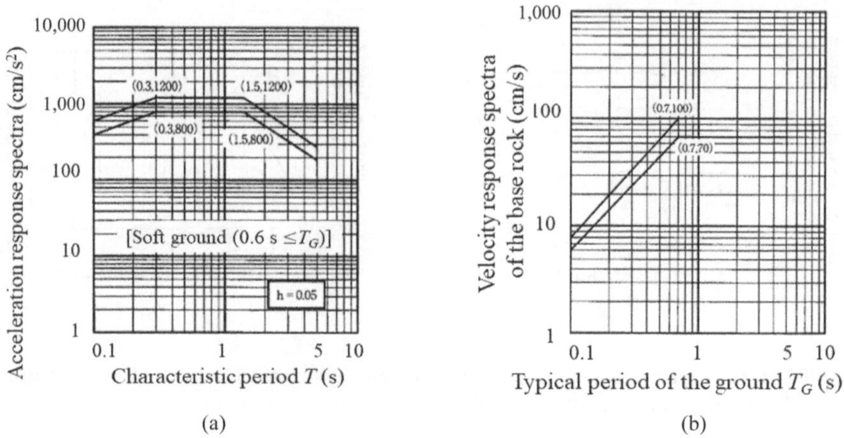

FIGURE 2.2.7 Standard response spectra for 90% and 70% non-exceedance probability for Level 2 ground motion:[12] (a) acceleration response spectrum; (b) velocity response spectrum.

2.2.4.1.1 Prediction Error in Response Spectrum

The prediction error of the acceleration response spectrum $S_A(T)$ is due to the prediction error of the ground response factor and the irregularity of the acceleration time history. The average value and coefficient of variation[7] of the design acceleration response spectrum can be obtained as follows.

Two types of design acceleration response spectra, $S_A(T;0.9)$ and $S_A(T;0.7)$, are expressed by the 90% non-exceedance probability line and the 70% non-exceedance probability line, respectively, for each ground type shown in the Guidelines and Commentary on Seismic Engineering Methods for Waterworks Facilities.[12]

Taking Type III ground as an example, the following equations can be obtained from Figure 2.2.7 using the standard normal distribution function Φ.

$$\begin{aligned} S_A(90\%) &= \mu_{SA}(T) + \Phi^{-1}(0.9)\cdot\sigma_{SA} \\ &= \mu_{SA}(T) + 1.281552\cdot\sigma_{SA}, \end{aligned} \tag{2.2.2}$$

$$\begin{aligned} S_A(70\%) &= \mu_{SA}(T) + \Phi^{-1}(0.7)\cdot\sigma_{SA} \\ &= \mu_{SA}(T) + 0.524401\cdot\sigma_{SA}. \end{aligned} \tag{2.2.3}$$

Therefore, the mean, standard deviation, and coefficient of variation of $S_A(T)$ are the following values:

$$\mu_{SA} = 523\,cm/s^2,\ \sigma_{SA} = 528\,cm/s^2,\ \delta_{SA} = 1.0.$$

Table 2.2.2 lists the mean values and coefficients of variation of the design acceleration response spectra S_A for Type I, II, and III surface soils obtained using these two segment values. The table shows that the coefficients of variation of the design acceleration response spectra for the surface soils are in the range of 0.7 to 1.0.

TABLE 2.2.2
Probability Statistics of the Design Acceleration

Statistics of Response Acceleration		Ground Classification		
Item	**Unit**	**I**	**II**	**III**
S_V (T_G; 0.9)	cm/sec^2	1,000	1,400	1,200
S_V (T_G; 0.7)	cm/sec^2	700	1,000	800
Mean	cm/sec^2	493	723	523
Standard deviation	cm/sec^2	396	528	528
COV		0.80	0.73	1.01

Similarly, the mean value, standard deviation, and coefficient of variation of the design velocity response spectrum can be calculated from the following equations:

$$\begin{aligned} S_v(90\%) &= \mu_{Sv}(T_G) + \Phi^{-1}(0.9)\cdot\sigma_{Sv} \\ &= \mu_{Sv}(T_G) + 1.281552\cdot\sigma_{Sv}, \end{aligned} \tag{2.2.4}$$

$$\begin{aligned} S_v(70\%) &= \mu_{Sv}(T_G) + \Phi^{-1}(0.7)\cdot\sigma_{Sv} \\ &= \mu_{Sv}(T_G) + 0.524401\cdot\sigma_{Sv}. \end{aligned} \tag{2.2.5}$$

Therefore, the mean value, standard deviation, and coefficient of variation of $S_V(T_G)$ for Level 1 and Level 2 ground motion are shown in Table 2.2.3. However, since there is no information on Level 1 ground motion as shown in Figure 2.2.7, the coefficients of variation for Level 2 ground motion are assumed for convenience.

2.2.4.1.2 Procedures for Creating Response Spectrum Adaptive Seismic Waveforms

Seismic acceleration time history data must be prepared for the dynamic numerical analysis of structures. Since the large-scale seismic records assumed in seismic design are not always available in the vicinity of the site, seismic records taken from

TABLE 2.2.3
Probability Statistics of the Design Velocity

Statistics of Response Velocity		Seismic Load	
Item	**Unit**	**Level 1**	**Level 2**
S_V (T_G; 0.9)	cm/sec	12	100
S_V (T_G; 0.7)	cm/sec	8.4	70
Mean	cm/sec	5.9	49
Standard deviation	cm/sec	4.8	40
COV		0.82	0.82

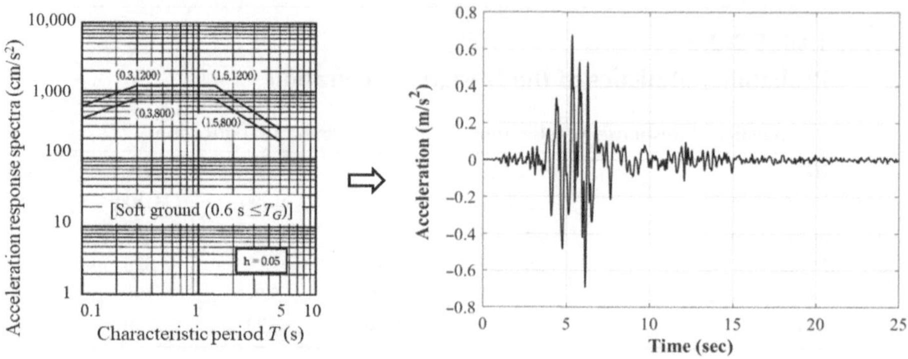

FIGURE 2.2.8 Example of a sample diagram of earthquake ground motion waveforms generated from a design response spectrum.

existing seismic databases under conditions that can approximate the geotechnical and seismic engineering conditions of the site are usually employed as input seismic motion. However, there is no guarantee that the adopted acceleration time history data conform to the acceleration response spectra presented in the seismic design guidelines. In this case, as shown in Figure 2.2.8, there is a method to generate samples of seismic waveforms that conform to the design response spectrum which is obtained based on the flowchart given in Figure 2.2.9.

2.2.4.2 Limit States and Design Earthquake Ground Motion

In the limit state design method[13] (see Appendix A1), the limit states of structures are classified into three limit states: (a) serviceability limit, (b) repairable limit, and (c) ultimate limit. In Japan, design earthquake ground motions, Level 1 and Level 2, are used to verify the seismic safety of each limit state. In the United States, on

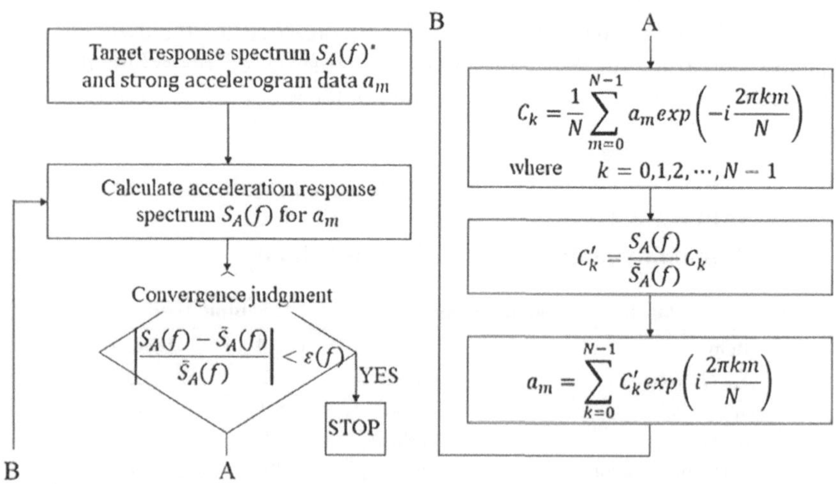

FIGURE 2.2.9 Procedures for creating response spectrum-adaptive seismic waveforms.[11]

TABLE 2.2.4
Relationship Between Limit States and Design Earthquake Ground Motion

Japan	Limit state	Serviceability	Repairable	Ultimate
	Earthquake ground motion	Level 1 ground motion	Level 2L ground motion	Level 2U ground motion
USA	Limit state	IO	LS	CP
	Earthquake	MOE	DBE	MCE

the other hand, the design earthquake maximum operational earthquake (MOE), design-based earthquake (DBE), and life safety (LO) are defined corresponding to the operational limit (IO, instantaneous operational), repair limit (LS, life safety), and critical preparation limit (CP, critical preparation).

In Japan, the design earthquake ground motions are classified into two categories, and it is difficult to simply assign three types of limit states. In this document, therefore, Level 2 ground motions are divided into Level 2L ground motions corresponding to the repair limit and Level 2U ground motions corresponding to the ultimate limit to obtain Table 2.2.4. (See Appendix A2).

2.3 GROUND RESPONSE ANALYSIS

2.3.1 Ground Response Analysis for Surface Layers

2.3.1.1 Ground Strain and Typical Period

Let the ground displacement due to waves propagating horizontally along the ground surface with an apparent velocity of propagation c be expressed by the following equation:[5]

$$u(x,t) = f\left(t - \frac{x}{c}\right) \tag{2.3.1}$$

In this case, the strain $\varepsilon(x,t)$ is given by

$$\varepsilon(x,t) = \frac{\partial u}{\partial x} = -\frac{1}{c} f'\left(t - \frac{x}{c}\right). \tag{2.3.2}$$

The velocity $\dot{u}(x,t)$ of the ground displacement $u(x,t)$ is expressed as

$$\dot{u}(x,t) = \frac{\partial u}{\partial t} = f'\left(t - \frac{x}{c}\right). \tag{2.3.3}$$

Then the strain is expressed as

$$\varepsilon(x,t) = -\frac{1}{c}\dot{u}(x,t). \tag{2.3.4}$$

Now, if we take the maximum absolute value of both sides, the ground strain is generally obtained by the following equation:

$$\varepsilon_G = \frac{\dot{u}_{max}}{c}. \tag{2.3.5}$$

The propagation velocity c is also called the phase velocity, and the velocity $\dot{u}(x,t)$, the time derivative of the ground displacement, is distinguished as the particle velocity.

The typical period T of the surface soil, where the depth of the soil from the ground surface to the base rock is H, is obtained by assuming that *SH* waves propagate with a shear wave velocity V_S from the vertical down to the ground surface, as follows:

$$T = \frac{4H}{V_S}. \tag{2.3.6}$$

In the case of multilayered ground, the thickness of each layer is H_i $(i = 1, \cdots, n)$, and the following formula is proposed:

$$T = \Sigma \frac{4H_i}{V_{S_i}} \text{ or } T = \frac{4\Sigma H_i}{\bar{V}_S}, \text{ where } \bar{V}_S = \frac{\Sigma V_{S_i} H_i}{\Sigma H_i} \tag{2.3.7}$$

2.3.1.2 *SH* Wave Multiple Reflection Analysis[14]

The ground is divided into the base and surface layer, and *SH* waves are assumed to enter and propagate from the base to the surface layer. Figure 2.3.1 shows the surface ground layers and coordinate system used in the analysis.

Consider the case where *SH* waves with an angle of incidence θ_n and a circular frequency ω are incident on a ground base toward the ground surface. The equation of motion of the ground in layer m is given by

$$\frac{\partial^2 u_m}{\partial t^2} = c_m^2 \left(\frac{\partial^2 u_m}{\partial x^2} + \frac{\partial^2 u_m}{\partial z_m^2} \right), \tag{2.3.8}$$

where u_m is the transverse displacement amplitude in the xy-plane at layer m, τ_m is the shear stress at layer m, and c_m is the shear wave velocity at layer m.

The general solution of this equation is expressed as

$$u_m = A_m exp\left[i\left(\nu_m z_m + \omega t - kx\right)\right] + B_m exp\left[-i\left(\nu_m z_m - \omega t + kx\right)\right] \tag{2.3.9}$$

$$\tau_m = i\mu_m \nu_m A_m exp\left[i\left(\nu_m z_m + \omega t - kx\right)\right] - B_m exp\left[-i\left(\nu_m z_m - \omega t + kx\right)\right] \tag{2.3.10}$$

where i is an imaginary unit, A_m and B_m are the amplitudes of ascending and descending waves, μ_m is the shear modulus, and c is the apparent horizontal propagation velocity, and k is wave number, then ν_m, k, k_m are given by the following equations:

$$\nu_m = \sqrt{k_m^2 - k^2}, \quad k = \omega / c, \quad k_m = \omega / c_m. \tag{2.3.11}$$

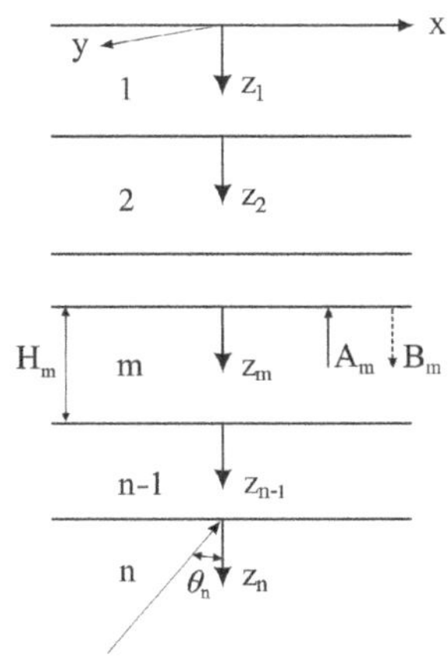

FIGURE 2.3.1 Surface soil layers and coordinate system.

In this case, u_m, τ_m and A_m, B_m have the following relationship, where $i(\omega t - kx)$ in exp is omitted for simplicity:

$$\begin{Bmatrix} u_m(z_m) \\ \tau_m(z_m) \end{Bmatrix} = \begin{bmatrix} exp(i\nu_m z_m) & exp(-i\nu_m z_m) \\ i\mu_m \nu_m exp(i\nu_m z_m) & -i\mu_m \nu_m exp(-i\nu_m z_m) \end{bmatrix} \begin{Bmatrix} A_m \\ B_m \end{Bmatrix}. \quad (2.3.12)$$

The relationship between $z_m = 0$ and $z_m = H_m$ in the *m*oth layer can be expressed as

$$\begin{Bmatrix} u_m(H_m) \\ \tau_m(H_m) \end{Bmatrix} = [T_m] \begin{Bmatrix} u_m(0) \\ \tau_m(0) \end{Bmatrix}, \quad (2.3.13)$$

where

$$[T_m] = \begin{bmatrix} cos(\nu_m z_m) & \dfrac{sin(\nu_m z_m)}{\mu_m \nu_m} \\ -\mu_m \nu_m sin(\nu_m z_m) & cos(\nu_m z_m) \end{bmatrix}. \quad (2.3.14)$$

On the other hand, the continuity condition at the boundary between the *m*th and *m*-1th layers is

$$\begin{Bmatrix} u_m(0) \\ \tau_m(0) \end{Bmatrix} = \begin{Bmatrix} u_{m-1}(H_{m-1}) \\ \tau_{m-1}(H_{m-1}) \end{Bmatrix}. \quad (2.3.15)$$

Therefore, the final equation for the relationship between the base surface layer $(z_m = 0)$ and the ground surface layer $(z_1) = 0$ is obtained by the following equation:

$$\begin{Bmatrix} u_m(0) \\ \tau_m(0) \end{Bmatrix} = [R_{m-1}] \begin{Bmatrix} u_1(0) \\ \tau_1(0) \end{Bmatrix}, \quad (2.3.16)$$

in which

$$\left[R_{m-1}\right]=\left[T_{m-1}\right]\left[T_{m-2}\right]\cdots\left[T_2\right]\left[T_1\right]. \tag{2.3.17}$$

Note that the numerical analysis program SHAKE,[14] a one-dimensional ground vibration analysis method, uses the one-dimensional shear wave theory and fast Fourier transform described above to evaluate the nonlinear behavior of soil. This is achieved by linearizing the nonlinear viscoelastic properties of soil.

2.3.2 Vibration Analysis and Dispersion Characteristic Equations for Surface Waves

2.3.2.1 The Case of Love Waves[15]

As a surface wave, the amplitude of the Love wave is zero in the deep layer sufficiently far from the surface layer. Therefore, the wave solution in the base (layer n) must take the following expression:

$$u_n^L = A_n exp\left[-\nu_n^* z_n + i\left(\omega t - k_L x\right)\right], \tag{2.3.18}$$

where

$$\nu_n^* = \sqrt{k_L^2 - k_n^2},\ k_L = \frac{\omega}{c_L}. \tag{2.3.19}$$

where c_L is the propagating wave velocity (phase velocity) of the Love wave.

On the other hand, since the shear stress is zero at the ground surface, the Love wave must satisfy the following boundary conditions, yielding the following equation. This is the characteristic equation for the phase velocity c_L of the Love wave.

$$R_{n-1}(2,1)+\nu_n^*\mu_n R_{n-1}(1,1)=0. \tag{2.3.20}$$

The dispersion curve of the Love wave is obtained as the solution of the Love wave characteristic equation.

2.3.2.2 Rayleigh Wave Case[16]

For Rayleigh waves propagating horizontally in stratified ground, the characteristic equations for obtaining dispersion curves of Rayleigh waves can be obtained as in the case of Love waves.

Since Rayleigh waves deform in the $x\ z$ plane, the response at the coordinate $\left(x,\ z_m\right)$, i.e., ground strain, direct stress, and shear stress, can be related to the response at the ground surface by the following equation:

$$\begin{aligned}&\left[\frac{\dot{u}_m\left(t,x,z_m\right)}{c}\quad \frac{\dot{w}_m\left(t,x,z_m\right)}{c}\quad \sigma_m\left(t,x,z_m\right)\quad \tau_m\left(t,x,z_m\right)\right]^T\\&=S_m\left(z_m\right)\left[\frac{\dot{u}_m(t,x,0)}{c}\quad \frac{\dot{w}_m(t,x,0)}{c}\quad \sigma_m(t,x,0)\quad \tau_m(t,x,0)\right]^T,\end{aligned} \tag{2.3.21}$$

where c is the propagation velocity (phase velocity) of Rayleigh waves, $\sigma_m(z_m)$, $\tau_m(z_m)$ are the stress in the x-axis direction and the shear stress in the $x\ z$ plane at depth z_m in layer m, and $S_m(z_m)$ is a 4×4 matrix equation, expressed as

$$S_m(z_m) = D_m(z_m) E_m^{-1} \prod_{j=1}^{m-1} \left\{ D_j(H_j) E_j^{-1} \right\}. \tag{2.3.22}$$

The details of the determinants $D_m(z_m)$, E_m in the same equation are given in Appendix A3. From the boundary conditions at the ground surface, i.e., $\sigma_0(t,x,0) = \tau_0(t,x,0) = 0$ and $z = \infty$, where the propagation velocity and stress vanish, we obtain

$$\frac{\dot{u}_0}{\dot{w}_0} = \frac{J_{22} - J_{12}}{J_{11} - J_{21}} = \frac{J_{42} - J_{32}}{J_{31} - J_{41}}, \tag{2.3.23}$$

where $J_{ij}\ (i,j = 1,2,3,4)$ are the i,j components of the 4×4 determinant J. The determinant J is given by

$$J = E_n^{-1} \prod_{j=1}^{n-1} \left\{ D_j(H_j) E_j^{-1} \right\}. \tag{2.3.24}$$

Since the components of the 4×4 determinant J includes the propagation velocity (phase velocity) c and the wave number k, or frequency component, the solution of Equation (2.3.24) gives the relationship between the Rayleigh wave phase velocity c_R and the frequency or natural period T. This is the characteristic equation for dispersion.

2.4 SPATIAL VARIATION AND INCOHERENCE OF SEISMIC GROUND MOTIONS

When the ground properties are uniform and there is no spatial variation, incident wave propagation generates a phase delay of seismic ground motion in a uniform layered ground. On the other hand, if there is spatial variation[1] in the ground properties, the seismic response between two points separated by a distance shows phase differences depending not only on the separation distance[2] but also on incoherent waveform.

Furthermore, if the surface ground has a complex ground structure or if the ground properties have been artificially modified, the seismic response will express more noncorrelated waveforms.

In this section, it is assumed that spatial variation can be given by phase delay of incident wave propagating in the firm rock formation, or by spatial inhomogeneity or incoherency in soil properties in the layered medium.

2.4.1 Stochastic Modeling of Seismic Ground Strains for Various Wave Propagations

In this section, a seismic wave consists of a spectrum of random phased frequency components. The ground strain and motions are, therefore, modeled as stationary Gaussian processes. A known solution for a boundary value problem involving the surface wave,[3,4] and a ground model as previously described has been used to find the

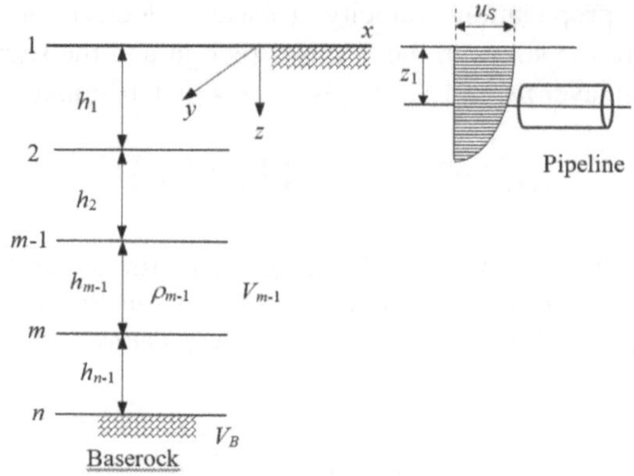

FIGURE 2.4.1 Multilayered model for ground.

ground strain and motion processes. The ground acceleration at the ground surface is also modeled as a stationary Gaussian process with a specified spectral density and is used as one of the boundary conditions. The non-stationarity of the ground motion is accounted for by truncating these processes before time $t = 0$, and after $t = T$, where T = the duration of the ground motion.

2.4.1.1 Ground Model

Consider n-1 parallel, homogeneous, and isotropic soil layers resting on a semi-infinite rock formation (basement layer) as shown in Figure 2.4.1. Shown also in this figure is the seismic wave propagating in the horizontal direction and the underground pipeline buried in the first layer at a depth, z_1. The surface wave produces a horizontal ground surface acceleration, in the direction of wave propagation, which can be expressed in the following form:

$$\ddot{u}_s(t) = \sum_{j=1}^{N} \sqrt{2G_{\ddot{u}_s}(\omega_j)\Delta\omega}\,\exp\left[i(\omega_j t + \phi_j)\right]. \tag{2.4.1}$$

in which $G_X(\omega_j)$ = one-sided power spectral density of a stationary random process denoted by $X{=}X(t);\omega_j{=}j\Delta\omega,\ \Delta\omega{=}\omega_u/N;\ \phi_j$= independent random phase angles distributed over 0 to 2π and N = number of superposed harmonic components.

2.4.1.2 Model of Input Ground Motion

The incident seismic ground motion arrives at a rock formation in the layered medium, in which the ground surface acceleration $\ddot{u}_0(t)$ defined in Equation (2.4.2) is constructed on the basis of fault parameters including the fault size, moment magnitude, corner frequency, and any other site parameters:

$$\ddot{u}_0(t) = \sum_{j=1}^{N} \sqrt{2G_{\ddot{u}_0}(\omega_j)\Delta\omega}\,\exp\left[i(\omega_j t + \phi_j)\right]. \tag{2.4.2}$$

in which the power spectral density $G_{\ddot{u}_0}(\omega)$ developed by Irikura[19] is defined as

$$G_{\ddot{u}_0}(\omega) = \left|CA_s(\omega)A_D(\omega)A_A(\omega)\right|, \tag{2.4.3}$$

where C, A_s, A_D, A_A are the source wave amplification factor, the source spectrum proposed by Brune,[20] the damping characteristics of seismic wave traveling from the source, and the local site amplification factor from the base rock to the ground surface, respectively. All the site parameters in these functions used hereunder are taken from the study of Harada.[21] The duration[6,14] T in which an earthquake accelerogram is greater than 10% of the maximum amplitude together with the one-third fraction as its mean stationary portion is evaluated with a magnitude M as follows:

$$T = 0.33 \cdot 10^{\frac{(M-2.5)}{3.23}}. \tag{2.4.4}$$

2.4.1.3 Ground Response at the Ground Surface

The ground motion at the ground surface shown in Figure 2.4.2 is propagated from an incident seismic wave. The power spectral density $G_{\ddot{u}_s}(\omega_j)$ is also calculated from

$$G_{\ddot{u}_s}(\omega_j) = \left|H_{n1}(\omega_j)\right|^2 G_{\ddot{u}_0}(\omega_j) \quad \text{for} \quad \left|H_{n1}(\omega_j)\right| = \frac{A_1}{A_n}. \tag{2.4.5}$$

2.4.1.4 Ground Strains Based on Various Definitions of Wave Velocities

When an incident SH wave arrives with a delay proportional to the horizontally propagating duration in the base rock ground with the shear wave velocity of V_B, the ground response at the surface layer appears with the same delay in the x direction of Figure 2.4.1, so that the SH wave seems to travel horizontally with a velocity of V_B. If there is a spatial variation in the soil rigidity of the surface layer in the horizontal direction, the seismic response produces not only ground response due to a phase delay of the incident SH wave but also an additional ground response due to spatial variation of soil rigidity.

2.4.1.5 SH Wave Propagating with Base Rock Shear Velocity

In the case of a delay due to an incident SH wave arrival without any spatial variation, the ground response displacement is given in the following form;

$$u_B(t) = \sum_{j=1}^{N} \sqrt{2G_{\ddot{u}_s}(\omega_j)\Delta\omega}\,\left|H^B(\omega_j)\right| / (i\omega_j)^2 \exp\left[i(\omega_j t - k_B x + \phi_j)\right]. \tag{2.4.6}$$

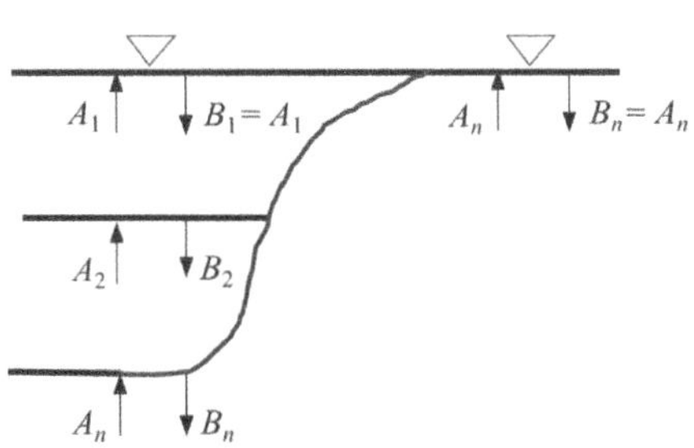

FIGURE 2.4.2 The incident ground motion at the layered medium.

in which k_l and k_B are the wave numbers of the first layer and base rock of the surface ground, respectively:

$$\left|H^B\left(\omega_j\right)\right| = \cos\left\{k_1\left(\omega_j\right)z_1\right\},\ k_B = \frac{\omega_j}{V_B},\ k_1\left(\omega_j\right) = \frac{\omega_j}{\bar{V}_1}. \tag{2.4.7}$$

with the mean value of shear velocity $\bar{V}_1$ at the first layer.

Differentiating Equation (2.4.6) with respect to x, the axial ground strain ε_G in the direction of wave propagation is obtained as

$$\varepsilon_G(t) = \frac{\partial u_B(t)}{\partial x} = \sum_{j=1}^{N}\sqrt{2G_\varepsilon^B(\omega_j)\Delta\omega}\ \exp\left[i\left(\omega_j t - k_B x + \phi_j\right)\right], \tag{2.4.8}$$

with

$$G_\varepsilon^B\left(\omega_j\right) = G_{\ddot{u}_s}\left(\omega_j\right)\cdot\left(k_B/\omega_j^2\right)^2\left|H^B\left(\omega_j\right)\right|^2. \tag{2.4.9}$$

2.4.1.6 SH Wave with Spatial Variation of Soil Rigidity in the Surface Layer

When the effect of spatial variation is taken into consideration, the ground response is evaluated with two components due to a delay arrival and due to spatial variation. A spatial variation[22] of soil rigidity of the first layer along the horizontal direction can be modeled with the mean value of shear velocity $\bar{V}_1$ and its variation v in the following way:

$$V_1(x) = \bar{V}_1(x) + v(x). \tag{2.4.10}$$

with the variation of shear velocity to be modeled as

$$v(x) = \sqrt{2}\sum_{j=1}^{N}\sqrt{2S_v(\kappa_j)\Delta k}\ \cos\left(\kappa_j x + \psi_j\right). \tag{2.4.11}$$

in which the stochastic property of the shear velocity variation is expressed with the stationary two-sided spectral density function $S_v(k)$ with its wave number κ and phase angle ψ. When the autocorrelation of v is assumed with the standard deviation σ_{V_1} and specific distance b,

$$R_v\left(x_1, x_2\right) = \sigma_{V_1}^2\ \exp\left(\frac{\left|x_1 - x_2\right|}{b}\right). \tag{2.4.12}$$

the corresponding spectral density is calculated in the following way:

$$S_v(\kappa) = 2\sigma_{V_1}^2\frac{b}{1 + b^2\kappa^2}. \tag{2.4.13}$$

Then the ground responses have similar expressions with Equations (2.4.6)–(2.4.9) under the spatial variation along the horizontal component denoted by x:

$$u_{SH}(t) = \Sigma_{j=1}^{N}\sqrt{2G_{\ddot{u}_s}\left(\omega_j\right)\Delta\omega}\left|H^{SH}\left(\omega_j, x\right)\right| / \left(i\omega_j\right)^2 \exp\left[i\left(\omega_j t - k_B x + \phi_j\right)\right],$$

$$\varepsilon_{SH}(t) = \sum_{j=1}^{N}\sqrt{2G_\varepsilon^{SH}\left(\omega_j\right)\Delta\omega}\exp\left[i\left(\omega_j t - k_B x + \phi_j\right)\right], \tag{2.4.14}$$

$$G_\varepsilon^{SH}\left(\omega_j\right) = G_{\ddot{u}_s}\left(\omega_j\right)\cdot\left(k_B / \omega_j^2\right)^2 \mid H^{SH}\left(\omega_j, x\right)\mid^2.$$

Since the function $H_{SH}(\omega_j, x)$ has two components of wave response and spatial variation, their combined expression is given as follows:

$$\left|H^{SH}\left(\omega_j, x\right)\right|^2 = \left|H^{SH}\left(\omega_j\right)\right|^2 \left\{ \begin{array}{c} \dfrac{\left|H^{SH}\left(\omega_j | V_1(x)\right)\right|^2}{\left|H^{SH}\left(\omega_j\right)\right|^2} \\ + \left[\dfrac{\partial H^{SH}\left(\omega_j | V_1(x)\right)}{\partial x}\right]^2 \cdot \dfrac{1}{\left(k_B^2\right)\left|H^{SH}\left(\omega_j\right)\right|^2} \end{array} \right\}. \quad (2.4.15)$$

An approximation of Equation (2.4.15) is carried out with the expectation of the maximum value of a stationary random process $v(x)$ in the following way:

$$\left|H^{SH}\left(\omega_j, x\right)\right|^2 \approx \left|H^{SH}\left(\omega_j\right)\right|^2 \left\{ 1 + \frac{\left[\left|E\left[\dfrac{\partial H^{SH}\left(\omega_j | V_1(x)\right)}{\partial V_1}\right]\right| \cdot \left(\sqrt{21\text{n}n} + \dfrac{0.577215}{\sqrt{21\text{n}n}}\right)\sqrt{E\left[\left|\dfrac{\partial V_1}{\partial x}\right|^2\right]}\right]^2}{\left(k_B^2\right)\left|H^{SH}\left(\omega_j\right)\right|^2} \right\}.$$

(2.4.16)

in which n is the sample size to be equal to the duration in second of the earthquake ground motion, and

$$\sqrt{E\left[\left|\frac{\partial V_1}{\partial x}\right|^2\right]} = \sqrt{\sum_{j=1}^{N}\left(\kappa_j\right)^2 \cdot 4S_v(\kappa_j)\Delta\kappa}, \; k_1(\omega_j; x) = \frac{\omega_j}{V_1(x)} \quad (2.4.17)$$

$$\text{and } H^{SH}\left\{\omega_j | V_1(x)\right\} = \cos\left\{k_1(\omega_j; x) z_1\right\}.$$

2.4.1.7 Surface Models with Spatial Variation of Soil Rigidity in the Surface Layer

The surface wave produces a horizontal ground motion in the direction of wave propagation, in which Rayleigh wave type and Love wave type models are adopted as a typical surface wave model for a simplified version of seismic design guidelines for buried pipelines.[15,17] Especially, the Love wave type model, which can produce a transverse component in the direction of wave propagation, is a comparative model with the SH wave model described in the previous section.

2.4.1.8 Rayleigh Wave Type Model

According to Haskell,[18] the ground response and its normal strain for the Rayleigh wave type model are given in Equation (2.4.18) with the phase velocity $V_R(\omega)$ obtained from the dispersion curve of the surface ground of Figure 2.4.1:

$$u_R(t)=\sum_{j=1}^{N}\sqrt{2G_{\ddot{u}_S}(\omega_j)\Delta\omega}\left|H^R(\omega_j)\right|/(i\omega_j)^2\exp\left[i(\omega_j t-k_R x+\phi_j)\right],$$
$$\varepsilon_R(t)=\sum_{j=1}^{N}\sqrt{2G_\varepsilon^R(\omega_j)\Delta\omega}\exp\left[i(\omega_j t-k_R x+\phi_j)\right],\tag{2.4.18}$$
$$G_\varepsilon^R(\omega_j)=G_{\ddot{u}_S}(\omega_j)\cdot(k_R/\omega_j^2)^2\left|H^R(\omega_j)\right|^2.$$

in which k_R is the wave number of the Rayleigh wave to be given by the dispersion curve $V_R(\omega)$, and the detail expression of $H_R(\omega)$ was developed by Shinozuka[16] et al.:

$$\left|H^R(\omega_j)\right|=\left\{S_{m11}(z_m)+\eta S_{m12}(z_m)\right\},\quad k_R=\frac{\omega_j}{V_R(\omega_j)},\quad \eta=\frac{\dot{w}_0}{\dot{u}_0}\tag{2.4.19}$$

where S_{m11}, S_{m11}, η, and k_R are the response functions[16] of multilayered media by Rayleigh wave, ratio of ground motion velocities, and wave number given by the dispersion velocity $V_R(\omega)$.

2.4.1.9 Love Wave Type Model

The similar approach can be taken to obtain the ground response and its strain for the Love wave type model as follows:

$$u_L(t)=\sum_{j=1}^{N}\sqrt{2G_{u_s}(\omega_j)\Delta\omega}\left|H^L(\omega_j)\right|/(i\omega_j)^2\exp\left[i(\omega_j t-k_L x+\phi_j)\right],$$
$$\varepsilon_L(t)=\sum_{j=1}^{N}\sqrt{2G_\varepsilon^L(\omega_j)\Delta\omega}\exp\left[i(\omega_j t-k_L x+\phi_j)\right],\tag{2.4.20}$$
$$G_\varepsilon^L(\omega_j)=G_{\ddot{u}_s}(\omega_j)\cdot(k_L/\omega_j^2)^2\left|H^L(\omega_j)\right|^2.$$

in which k_L is the wave number of the Love wave to be given by the dispersion curve $V_L(\omega)$,

$$k_L=\frac{\omega_j}{V_L(\omega_j)}.\tag{2.4.21}$$

and the amplification factor[6] is derived from those equations hereunder:

$$\left|H^L(\omega_j)\right|=P_k\cos\left\{\tilde{\beta}_k(\omega_j)z_k\right\}+Q_k\sin\left\{\tilde{\beta}_k(\omega_j)z_k\right\},$$
$$V>V_k\quad \tilde{\beta}_k(\omega_j)=\sqrt{\left(\frac{V}{V_k}\right)^2-1}\cdot\frac{\omega_j}{V},\tag{2.4.22}$$
$$V\le V_k\quad \tilde{\beta}_k(\omega_j)=\sqrt{1-\left(\frac{V}{V_k}\right)^2}\cdot\frac{\omega_j}{V},$$

$$\begin{pmatrix} P_k \\ Q_k \end{pmatrix} = \begin{bmatrix} \cos\{\tilde{\beta}_{k-1}(\omega_j)h_{k-1}\} & \sin\{\tilde{\beta}_{k-1}(\omega_j)h_{k-1}\} \\ -\mu_{k-1}\tilde{\beta}_{k-1}(\omega_j)\sin\{\tilde{\beta}_{k-1}(\omega_j)h_{k-1}\} & \mu_{k-1}\tilde{\beta}_{k-1}(\omega_j)\cos\{\tilde{\beta}_{k-1}(\omega_j)h_{k-1}\} \end{bmatrix} \begin{pmatrix} P_{k-1} \\ Q_{k-1} \end{pmatrix},$$

$$\mu_k = \rho_k V_k^2.$$

2.4.1.10 Maxima of Free Field Strain

Since the ground strain is idealized as a stationary Gaussian process, its absolute maximum in the duration T, $\varepsilon_{max}=\{max_{0<t<T}|\varepsilon_G(t)| < a\}$ is a random variable. Thus, the distribution function $F_{\varepsilon_{max}}(\alpha;T)$ of ε_{max} and the distribution function $F_{T0}(T;a)$ of the first passage time T_0 must be considered first. Writing $P(E)$ for the probability of event E, these distribution functions are defined as

$$F_{\varepsilon_{max}}(\alpha;T) = P\left\{ \max_{0\le t\le T} |\varepsilon_G(t,x)| \le \alpha \right\} = 1 - F_{T_0}(T;\alpha) \tag{2.4.23}$$

and

$$F_{T_0}(T;\alpha) = P\left\{ T_0 \le T; |\varepsilon_G(T_0,x)| > \alpha \right\}. \tag{2.4.24}$$

Using Vanmarcke's approximation method,[23] Equation (2.4.23) results in

$$F_{\varepsilon_{max}}(\alpha;T) = erf\left(\frac{\alpha_0}{\sqrt{2}}\right)\exp\left\{ -2\hat{\nu}_0 T \exp\left(-\frac{\alpha_0^2}{2}\right) \frac{1-\exp\left(-\frac{\sqrt{\pi}\hat{q}\alpha_0}{\sqrt{2}}\right)}{1-\exp\left(-\frac{\alpha_0^2}{2}\right)} \right\}, \tag{2.4.25}$$

in which $erf()$ indicates the error function and

$$\alpha_0 = \frac{\alpha}{\hat{\sigma}_\varepsilon}, \hat{\sigma}_\varepsilon = \sqrt{\hat{\lambda}_0}, \hat{q} = \sqrt{1-\frac{\hat{\lambda}_1}{\hat{\lambda}_0\hat{\lambda}_2}}, \ \hat{\nu}_0 = \frac{1}{2\pi}\sqrt{\frac{\hat{\lambda}_2}{\hat{\lambda}_0}} \tag{2.4.26}$$

$$\text{for } \hat{\lambda}_i = \int_0^\infty \omega^i G_{\varepsilon_G}(\omega)\,d\omega, i = 0,1,2.$$

The expected value $E\{\varepsilon_{max}\}$ and variance $Var\{\varepsilon_{max}\}$ of ε_{max} are obtained from

$$\begin{aligned} E\{\varepsilon_{\max}\} &= \int_0^\infty \{1 - F_{\varepsilon_{\max}}(\alpha;T)\}\,d\alpha, \\ Var\{\varepsilon_{\max}\} &= \int_0^\infty \alpha\{1 - F_{\varepsilon_{\max}}(\alpha;T)\}\,d\alpha - [E\{\varepsilon_{\max}\}]^2. \end{aligned} \tag{2.4.27}$$

2.4.2 Numerical Studies of Seismic Ground Strains Having Spatial Variations

2.4.2.1 Ground Models and Soil Conditions

A simple ground model with three layers is adopted in this numerical study.[22] Ground surface models having different thicknesses are shown in Table 2.4.1.

TABLE 2.4.1
Ground Surface Models Having Various Typical Periods

Layer Thickness	Unit	Typical Period (sec) 0.15	0.2	0.3	0.4	0.5	0.6	0.7	0.8	1
h_1	m	5	3.5	5.1	6.75	8.5	16.6	23.1	28.6	30
h_2	m	5	16.5	24.9	33.3	41.5	33.4	26.9	21.4	50

Layer Thickness	Unit	Typical Period (sec) 1.3	1.5	2	3	4	5	Mass kN/m^3	Shear Velocity m/sec
h_1	m	40	52	80	105	127	182	17	80
h_2	m	60	48	20	95	174	118	22	300
Base rock								23	800

Soil properties, mass, and shear velocity have the same values for each layer. The shear velocity of the base rock is assumed to be 800 m/s, which is consistent with the value used in the design guideline of the Japan Gas Association.[7]

2.4.2.2 Phase Velocities

Dispersion curves for Rayleigh and Love waves are calculated for the ground models in Table 2.4.1. Figure 2.4.3 is a schematic illustration of phase velocities of the model ground having the typical period of 1 second. This figure shows that the phase velocity in lower-frequency approaches the shear velocity of base rock, while that in higher frequency reaches the shear velocity of the surface ground. These trends are valid for both wave models, although Rayleigh wave has a larger value of phase velocity in the range of 0.5 to 1.3 Hz.

2.4.2.3 Input Earthquake Motion

A minimum size of ground motion is assumed herein. Attenuation property and spectral density are specified by the stationary power spectral density of Equation (2.4.3). Figure 2.4.4 shows the attenuation characteristics of the input earthquake model adopted herein.

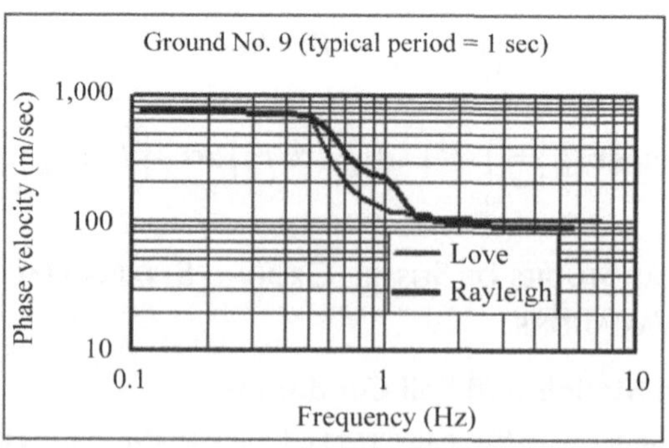

FIGURE 2.4.3 The phase velocities of the model ground.

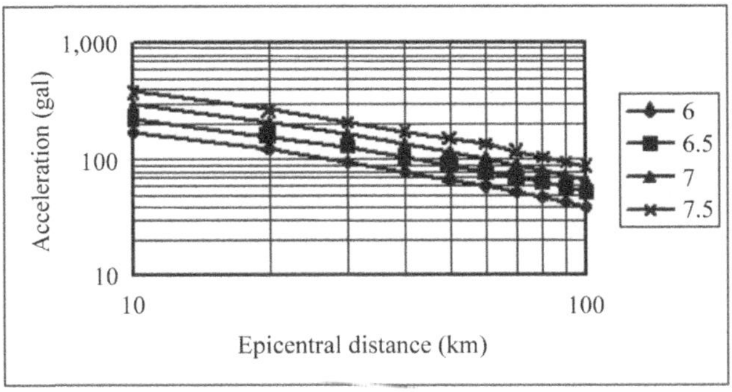

FIGURE 2.4.4 Attenuation relationship of ground acceleration for various magnitudes of medium-size earthquakes.

Figure 2.4.5 shows the power spectral densities for various magnitudes. It should be noted that larger magnitude can provide relatively greater spectral amplitude in a wider frequency range.

In Figure 2.4.6, the effect of spatial variation is investigated, in which a short relative distance ($b = 150$ m) meaning heavier randomness within the short distance can produce larger strain than that in a longer separation distance ($b = 1{,}500$ m).

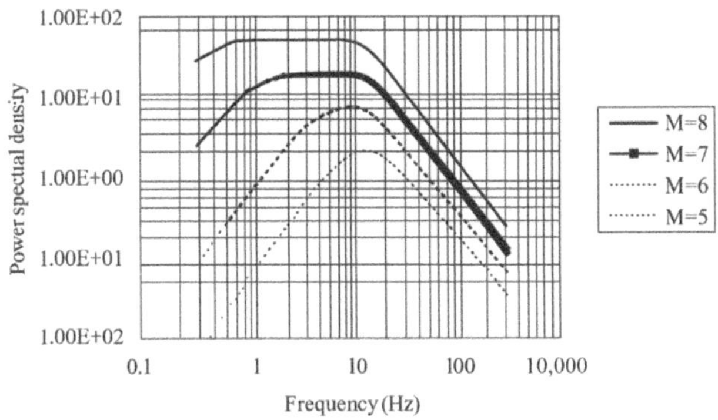

FIGURE 2.4.5 The power spectral densities for various magnitudes.

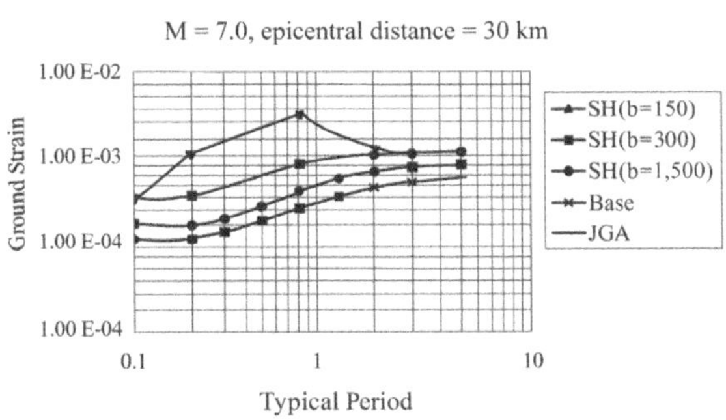

FIGURE 2.4.6 Ground strain for an SH wave with spatial variations.

REFERENCES

1. Zerva, A.: *Spatial Variation of Seismic Ground Motions*, CRC Press, 2009.
2. Kramer, S. L.: *Geotechnical Earthquake Engineering*, Prentice Hall, 1996.
3. Ewing, W. M., W. S. Jardetzky, and F. Press: *Elastic Waves in Layered Media*, McGraw-Hill Book Company, 1957.
4. Sato, Y.: *Theory of Elastic Waves*, IWANAMI Publishing, 1978.
5. Toki, K.: *Analysis of Seismic Resistance of Structures, New System of Civil Engineering 11*, Gekidan Publishing, 1981.
6. Civil Engineering Society: *Methods of Dynamic Analysis, Dynamic Analysis and Seismic Design*, Vol. 2, GIHODO Publishing, 1989.
7. Japan Gas Association (JGA): Seismic Design Guideline of High-pressure Gas Pipelines, JGA-206-03, Japan Gas Association, 2003.
8. Koike, T., No. Hasegawa, M. Hamano, and T. Watanabe: A proposal of seismic design method for submerged pipes for watercourses, *Proceedings of the 42nd Seismic Engineering Research Symposium*, The Civil Engineering Society, 2022.
9. Active Faults Research Society: *A New Edition of Active Faults in Japan – Distribution Chart and Data*, Tokyo University Press, 2002.
10. Chopra, A. K.: *Dynamics of Structures*, 4th edition, Prentice Hall, 2012.
11. Japan Road Association (JRA): Road and Bridge Guidelines and Explanation, V Seismic Design, Japan Road Association, 2017.
12. Japan Water Works Association: *The Guidelines of Seismic Design Methods for Waterworks Facilities*, 2009 edition, Japan Water Works Association, 2009.
13. Melchers, R. E.: *Structural Reliability Analysis and Prediction*, John Wiley & Sons, 1999.
14. Kajima Construction and Design Division: *Seismic Design Method/Performance Design*, New Civil Engineering Design Essentials 10, Kajima Press, 2003.
15. Koike, T.: *Seismic Design Method of Lifelines*, University Press of Kyoto University, 2023.
16. Shinozuka, M., H. Kameda, and T. Koike: Ground strain estimation for seismic risk analysis, *Journal of Engineering Mechanics*, Vol. 109, No. 1, ASCE, 1981.
17. Ogawa Y. and T. Koike: Structural design of buried pipelines for severe earthquakes, *Soil Dynamics and Earthquake Engineering*, Vol. 21, pp. 199–209, 2001.
18. Haskell, N. A.: The dispersion of surface waves on multi-layered media, *Bulletin of Seismological Society of America*, Vol. 43, pp. 17–34, 1953.
19. Irikura, K.: Prediction of strong accelerations motions using empirical Green's function, *Proceedings of 7th Japan Earthquake Engineering Symposium*, pp.151–156, 1986.
20. Brune, J. N.: Tectonic stress and the spectra of seismic shear waves from earthquakes, *Journal of Geophysical Research*, Vol. 75, pp. 4997–5009, 1970.
21. Harada, T., T. Tanaka, and Y. Tamura: Digital simulation of earthquake ground motions using a seismological model, *JSCE*, No. 507/I-30, pp. 209–217, 1995.
22. Koike, T., M. Maruyama, and L. E. Garciano: Ground strain estimation for lifeline earthquake engineering, *Structural Engineering and Mechanics*, Vol. 25, No. 3, pp. 291–310, 2007.
23. Vanmarcke, E. H.: Properties of spectral moments with applications to random vibration, *ASCE*, Vol. 92, No. EM2, pp. 425–446, 1972.

3 Seismic Design Methods for Pipelines

This chapter discusses seismic design methods for pipeline elements that comprise urban lifelines, such as straight pipelines, deformed pipes, and branch pipes used to change the direction of pipe travel vertically and horizontally. The loads acting on lifelines during earthquakes are classified into those due to ground shaking and those due to ground deformation (ground liquefaction, fault displacement, landslides, etc.). Seismic design for ground shaking is based on the response displacement method, while seismic design for ground deformation is based on an extreme limit analysis method using bold analytical modeling to discuss the behavior of both the ground and lifeline structures in a state close to the ultimate limit state. This chapter describes the specifics of this method.

3.1 SEISMIC DESIGN METHODS FOR PIPELINES AGAINST GROUND MOTIONS

3.1.1 Basic Analytical Methods for Seismic Design of Pipelines

3.1.1.1 Horizontal Ground Strain[1]

The ground strain can be expressed in general as

$$\varepsilon_G = \frac{\dot{u}_{max}}{c} \quad \text{or} \quad \varepsilon_G = \frac{2\pi}{L} u_{max}, \tag{3.1.1}$$

which is the same as Equation (2.3.5), and also u_{max} and $\dot{u}_{max}$ are the maximum ground displacement and velocity, respectively, while the latter equation is derived from $\dot{u}_{max} = u_{max} \cdot 2\pi / T$, and $L = cT$.

3.1.1.2 Effect of Angle of Incidence on Pipe Axis

3.1.1.2.1 Calculation of Ground Strain Along the Pipe Axis

Seismic waves are assumed to propagate at an angle θ of incidence to the pipe axis as shown in Figure 3.1.1. The direction of seismic wave oscillation depends on the type of wave (P wave, S wave, or surface wave).

3.1.1.2.2 Apparent Wavelength and Amplitude

Apparent seismic waves propagating along the pipe axis propagate as seismic waves with an apparent wavelength $L_a = L/\cos\theta$ and an apparent wave velocity $c_A = c/\cos\theta$ relative to the angle θ of incidence. The amplitudes can be classified in the

DOI: 10.1201/9781003655527-3

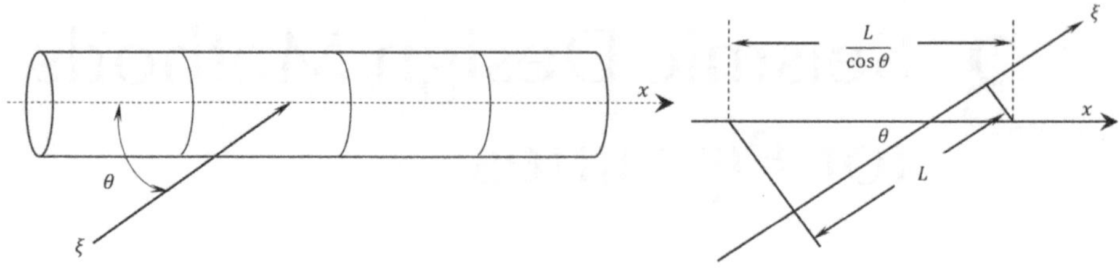

FIGURE 3.1.1 Angle of incidence and apparent wavelength along pipe axis.

direction of oscillation as shown in Figure 3.1.2. The pipe axial strains ε_A^P and ε_A^{SH} in Figure 3.1.2(a) and (b), respectively, are expressed by the following equations:

$$\text{(a) P and R wave-induced axial strain } \varepsilon_A^P = \frac{\dot{u}\cos\theta}{c/\cos\theta} = \varepsilon_G \cos^2\theta \tag{3.1.2}$$

$$\text{(b) SH and L wave-induced axial strain } \varepsilon_A^{SH} = \frac{\dot{u}\sin\theta}{c/\cos\theta} = \varepsilon_G \cos\theta \sin\theta \tag{3.1.3}$$

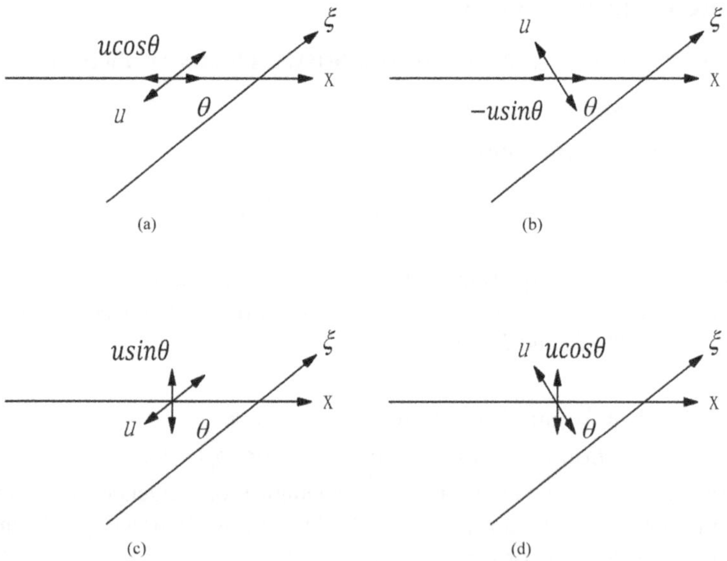

FIGURE 3.1.2 Displacement amplitude along the pipe axis due to seismic waves: (a) pipe axial displacement amplitude (P wave, Rayleigh wave); (b) pipe axial displacement amplitude (SH wave, Love wave); (c) amplitude transverse to the pipe axis (P wave, Rayleigh wave); (d) amplitude transverse to the pipe axis (SH wave, Love wave).

3.1.1.3 Comparison of Ground Strain in the Pipe Axial Direction and Perpendicular to the Pipe Axis Due to Seismic Wave Motion

According to Toki,[1] the ground displacement in the pipe axial direction propagates horizontally with apparent propagation velocity c_A given by $c_A = c/cos\,\theta$, as follows:

$$u_A(x,t) = f(x - c_A t). \tag{3.1.4}$$

The ground strain in the pipe axis direction is calculated by using the particle velocity $v\ (= \dot{u}_A)$, which is the partial derivative of u_A with respect to time, as

$$\varepsilon_A = \frac{v}{c_A}. \tag{3.1.5}$$

On the other hand, the bending strain is given by

$$\varepsilon_B = \frac{M}{EI}\frac{D}{2}. \tag{3.1.6}$$

From the relationship $\left|\frac{\partial^2 u_B}{\partial x^2}\right| = \frac{M}{EI}$ between curvature and bending moment, the bending strain is given by $\varepsilon_B = \frac{D}{2}\left|\frac{\partial^2 u_B}{\partial x^2}\right|$. Then using the wave equation $c_B^2 \frac{\partial^2 u_B}{\partial x^2} = \frac{\partial^2 u_B}{\partial t^2}$ and $a = \frac{\partial^2 u_B}{\partial t^2}$, the bending strain can be finally expressed by using $c_B = c/cos\,\theta$:

$$\varepsilon_B = \frac{Da}{2c_B^2}. \tag{3.1.7}$$

Assuming that the propagation velocity is almost the same $c_A \approx c_B = c$ in the axial and perpendicular directions, the ground strain ratio of the bending strain to the axial strain is given by

$$\frac{\varepsilon_B}{\varepsilon_A} = \frac{a}{v}\frac{D}{2c}. \tag{3.1.8}$$

Example 1: For a water pipeline

Assuming $a = 400\,cm/s^2$, $v = 100\,cm/s$, $D = 100\,cm$, $c = 400\,m/s$,

$$\frac{\varepsilon_B}{\varepsilon_A} = \frac{400}{100}\frac{100}{2*40000} = \frac{4}{800} = 0.005. \tag{3.1.9}$$

For a 1 m diameter pipe, the bending strain is about 0.5% of the axial strain. Furthermore, this axial strain is the dominant strain in the ground around the pipe during an earthquake.

Example 2: Shielded subway tunnel

Assuming $a = 400\,cm/s^2$, $v = 100\,cm/s$, $D = 1000\,cm$, $c = 400\,m/s$,

$$\frac{\varepsilon_B}{\varepsilon_A} = \frac{400}{100}\frac{100}{2*40000} = \frac{4}{80} = 0.05. \tag{3.1.10}$$

In the case of a 10 m diameter tunnel, the bending strain is about 5% of the axial strain. Therefore, bending strain (strain perpendicular to the axis) is also generated in the ground around the tunnel during an earthquake and thus must be taken into account.

3.1.1.4 Displacement Response of Stratified Ground

For a stratified ground, the displacement response is developed using Takata's method[2] as follows:

Assume a ground and pipe model as shown in Figure 3.1.3. The displacement field $u(x, z, t)$ of a three-dimensional wave is defined as

$$u(x,z,t)=U(z)exp\left[-i\left(\omega t-k_x x-k_z z\right)\right], \tag{3.1.11}$$

where

u: ground displacement in the x-axis direction of propagating waves
$U(z)$: ground displacement response along the z-axis
$U_G(t)$: displacement amplitude of input seismic motion at the base rock
ω: circular frequency of a seismic wave
k_x: x-directional component of a wave number
k_z: z-direction component of a wave number
z^*: depth from ground surface to pipe axis
i: imaginary number

The displacement field along the pipe axis is obtained by fixing the z-coordinate to z^*.

$$u(x,t)=u\left(x,z^*,t\right)=U_h\, exp\left[-i\left(\omega t-k_x x\right)\right], \tag{3.1.12}$$

where U_h is an amplitude of the ground displacement response along the pipe axis.

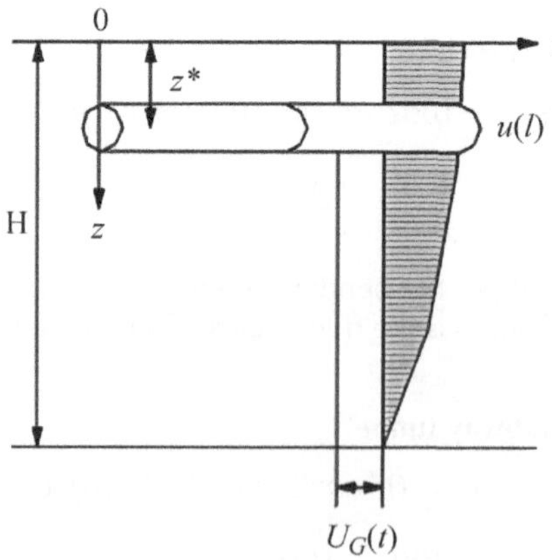

FIGURE 3.1.3 Geotechnical model and coordinate system.

Furthermore, the seismic displacement response of stratified ground is obtained by fixing the x-coordinate to x^*:

$$u(z,t) = u(x^*, z, t) = U_h \, exp\left[-i(\omega t - k_z z)\right]. \quad (3.1.13)$$

In this case, the equation of motion for ground shaking is expressed by the following equation:

$$\frac{\partial^2 (u + U_G)}{\partial t^2} - c^2 \frac{\partial^2 u}{\partial z^2} = 0. \quad (3.1.14)$$

To solve this equation of motion, assume that the ground displacement can be expressed in terms of a series of displacement mode functions and time functions in the z direction:

$$u(z,t) = \sum_i \varphi_i(z) Q_i(t). \quad (3.1.15)$$

The general solution of the mode function is obtained according to the following procedure. First, substitute $u = \varphi_i(z)Q_i(t)$ into the wave equation of Equation (3.1.14) and divide both sides of the equation with $\varphi_i(z)Q_i(t)$, then the following equation is obtained:

$$\frac{1}{\varphi_i}\frac{d^2\varphi_i}{dz^2} = \frac{1}{c^2}\frac{1}{Q_i}\frac{d^2 Q_i}{dt^2}. \quad (3.1.16)$$

Since both sides are functions of z and t, respectively, this equation holds only if both equations are equal to a constant. In this regard, if we let the constant be equal to $-(n_i / c)^2$, we obtain the following two general solutions:

$$\frac{1}{\varphi_i}\frac{d^2\varphi_i}{dz^2} = -\left(\frac{n_i}{c}\right)^2, \quad \frac{1}{Q_i}\frac{d^2 Q_i}{dt^2} = -n_i^2. \quad (3.1.17)$$

The first equation of Equation (3.1.17) is solved by assuming a solution in the form of a cosine function, taking into account the coordinate system of the ground under consideration. Rewriting the first equation as

$$-n_i^2 \varphi_i = c^2 \frac{d^2\varphi_i}{dz^2},$$

the following solution is then obtained as

$$\phi_i = cos\left(\frac{n_i z}{c}\right), \quad (3.1.18)$$

where

$$n_i = c\frac{\pi}{2H}(2i-1). \quad (3.1.19)$$

Furthermore, if the seismic acceleration input is taken into account, the second equation of Equation (3.1.17) can be rewritten as the following equation of motion where μ_i is the stimulation coefficient of the i-th mode:

$$\frac{d^2Q_i}{dt^2}+n_i^2Q_i=-\frac{\int_0^H \varphi_i dz}{\int_0^H \varphi_i^2 dz}\ddot{U}_G(t)=-\mu_i\ddot{U}_G(t) \tag{3.1.20}$$

$$\therefore u(z,t)=\sum_{i=1}^{\infty}\frac{\mu_i\varphi_i}{n_i}S_v, \tag{3.1.21}$$

where S_v is a velocity response spectrum to be excited by the base rock ground motion $\ddot{U}_G$in the following way:

$$S_v=\left\{\int_0^t \ddot{U}_G(\tau)\sin n_i(t-\tau)d\tau\right\}_{max}. \tag{3.1.22}$$

In the case of the first-order mode, $n_1=2\pi/T, \mu_1=4/\pi$ are obtained. Then the maximum value U_h of the ground response displacement at the pipe depth z is given by

$$U_h=\frac{2}{\pi^2}S_v\ T\cos\left(\frac{\pi z}{2H}\right), \tag{3.1.23}$$

where the maximum ground displacement U_h, which is often used in the seismic design guidelines of Japanese Gas Association (JGA)[8], and the Japanese Water Works Association (JWWA)[13] is the same as u_{max} in Equation (3.1.1). The velocity response spectrum S_v is given in Figure 2.2.5(b) for Level 1 ground motion, and in Figure 2.2.7(b) for Level 2 ground motion.

3.1.1.5 Ground Displacement Response of Irregularly Layered Ground

As shown in Figure 3.1.4, irregularly layered ground is a series of ground with varying thicknesses of the surface soil where the engineering base is located near the ground surface with a so-called short ground prevailing period and soft ground with a large surface soil thickness and a long ground prevailing period.

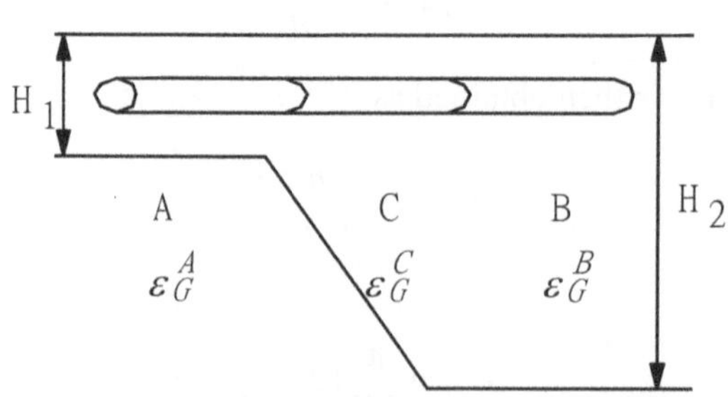

FIGURE 3.1.4 Unevenly layered ground model.

Various analytical methods and simplified formulas have been proposed for earthquake response analysis of irregularly layered ground. In the past, theoretical analysis methods such as those based on wave theory solutions[3] and response analysis in irregularly shaped semi-infinite ground based on elastic wave theory[4] were proposed, but later practical analysis methods such as the combination of boundary element method analysis and wave theory solutions[5] have been proposed. Experimental studies by Koizumi et al.[6,7] have verified the validity of the theoretical approach. Based on the results of these studies, a simplified method that takes practical simplicity into consideration was introduced in the Japan Gas Association's Guidelines for Seismic Design of Gas Pipes.[8] The research in this field has been discussed in numerous research papers and reference books, and interested readers are referred to these references for further information.

3.1.2 Response Displacement Method

3.1.2.1 Formulation

When ground displacement u_G is generated by ground shaking, the method for calculating the displacement u_S of the buried pipe generated by the frictional force from the ground is described below.

As shown in Figure 3.1.5, the axial stress σ acts on the left and right sides of the pipe body, while the shear stress τ_G from the surrounding ground acts on the circumference of the pipe, causing an incremental stress inside the pipe to be balanced. In this case, the following equation is obtained:

$$A\sigma = A(\sigma + d\sigma) + \pi D dx \tau_G, \tag{3.1.24}$$

where D is the pipe diameter and A is the pipe cross-sectional area.

Solving the balance equation, we obtain the following equation:

$$\frac{d\varepsilon_S}{dx} = -\frac{\pi D}{EA}\tau_G. \tag{3.1.25}$$

The difference in stiffness between the pipe and the soil mass surrounding the pipe causes shear stress on the pipe surface, the magnitude of which is proportional to the difference between the ground displacement and the pipe displacement. This relationship is expressed by the following equation using the ground spring coefficient k_1:

$$\tau_G = k_1\left(u_G - u_S\right). \tag{3.1.26}$$

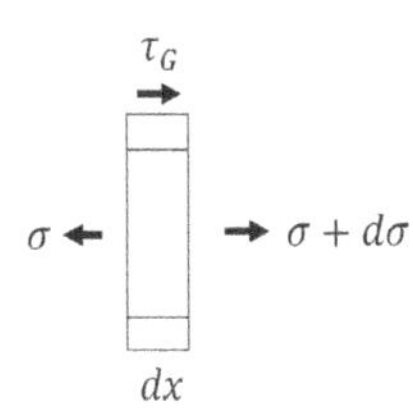

FIGURE 3.1.5 Equilibrium of forces in a pipe element.

Substituting this equation for shear stress and the $\varepsilon_S = \frac{du_S}{dx}$ equation for pipe displacement and strain into Equation (3.1.26), we obtain the following equation:

$$\frac{d^2u_s}{dx^2} = -\frac{\pi D k_1}{EA}\left(u_G - u_S\right) = -\frac{K_1}{EA}\left(u_G - u_S\right), \tag{3.1.27}$$

where k_1 is the ground spring coefficient defined by JGA, and the ground spring is defined as

$$K_1 = \pi D k_1. \tag{3.1.28}$$

Now express the ground displacement as

$$u_G = U_h \sin\left(\frac{2\pi}{L_a}x\right). \tag{3.1.29}$$

Then, by solving this equation, the pipe displacement u_S can be obtained from the ground displacement u_G using the conversion factor α_0 as follows:

$$u_S = \alpha_0 u_G, \tag{3.1.30}$$

in which

$$\alpha_0 = \frac{1}{1+\left(\frac{2\pi}{\lambda_1 L_a}\right)^2}, \qquad \lambda_1 = \sqrt{\frac{K_1}{EA}}. \tag{3.1.31}$$

where

K_1: subgrade spring modulus (N/m^2)
A: cross-section area (m^2)
D: pipe diameter (m)
E: Young's modulus of steel pipe (N/m^2)

Straight pipe strain is also obtained from the pipe displacement by differentiating both sides with x as follows:

$$\varepsilon_S = \alpha_0 \varepsilon_G, \tag{3.1.32}$$

where the ground strain along the pipe axis is rewritten from Equation (3.1.1) and given by

$$\varepsilon_G = \frac{2\pi}{L_a}U_h.$$

3.1.2.2 Formulation of Slippage Factor[9]

Slippage between a buried pipe and the surrounding ground is caused by the inelastic nature of the soil. Assume that the stress–strain relationship of the soil is of a

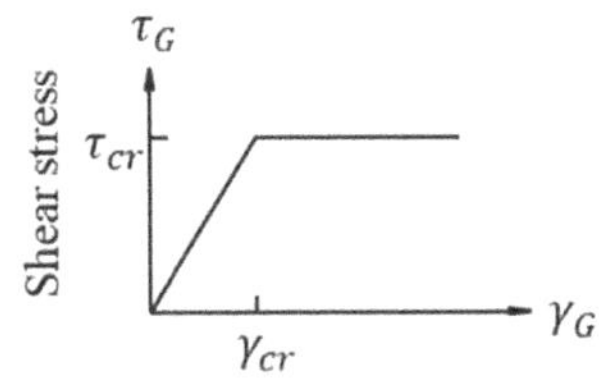

FIGURE 3.1.6 Elastoplastic stress–strain relationship of soil.

perfectly elastoplastic type as shown in Figure 3.1.6. Slippage occurs when the shear stress acting on the pipe surface matches the slip initiation limit shear stress τ_{cr}. Since the stress acting on the pipe surface is a constant value τ_{cr} in the slip initiation zone, the stress generated in the buried pipe does not increase beyond a certain value. On the other hand, the relative deformation between the buried pipe and the surrounding ground increases to the extent that the buried pipe does not extend due to slippage.

In the case of a buried pipe connected to a valve chamber or branch pipe, when the latter is in motion with the ground, large relative deformation is forced on the connection, causing failure.

A full-scale experiment was conducted at a company owning a high-pressure gas pipeline to determine the critical shear stress of the soil around the buried pipe, and the experimental data were generally in the range of 1.0–2.0 N/cm^2. Based on these results, the following equation is used in the *Guidelines for Seismic Design of High-Pressure Gas Pipes*:[8]

$$\tau_{cr} = 1.5\left(\frac{H_{po}}{1.8}\right), \qquad (3.1.33)$$

where

τ_{cr}: critical shear stress (N/cm^2)
H_{po}: burial depth (m)

In the same guidelines, the ground spring constant K_1 and the ground spring coefficient k_1 are given by the following equations:

$$\begin{aligned} K_1 &= \pi D k_1 \quad N/cm^2. \\ k_1 &= 6.0 \quad N/cm^3. \end{aligned} \qquad (3.1.34)$$

Consider the case where a sinusoidal wave of wavelength L_a propagates parallel to the pipe axis. As shown in Figure 3.1.7, the slip section is AB, and l^*, which determines the slip start position, is given by

$$l^* = \frac{L_a}{2\pi} sin^{-1}\left[\frac{\tau_{cr}}{\tau_G}\right]. \qquad (3.1.35)$$

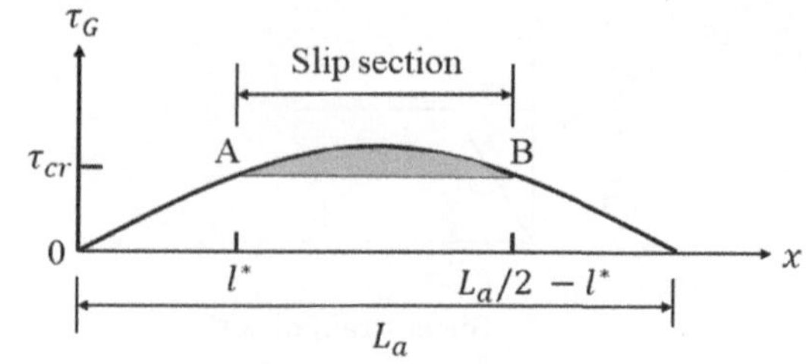

FIGURE 3.1.7 Shear stress distribution in a pipe containing a slip section (AB).

The equation for the balance of the buried pipe in the nonslip sections OA and BC is given by

$$\frac{d^2u_S}{dx^2} + \frac{\tau(t,x)}{Et} = 0. \tag{3.1.36}$$

where u_s is the axial displacement of the buried pipe.

$$\tau(t,x) = \begin{cases} \tau_G \sin\left(\frac{2\pi}{L_a}x\right); & 0 \le x \le l^* \\ \tau_{cr}; & l^* < x < L_a/4 \end{cases} \tag{3.1.37}$$

Furthermore, the continuity condition and symmetry of the pipe strain at $x = l^*$ as a boundary condition leads to the following:

$$\begin{aligned} &\varepsilon_S\left(t,\, l^*_{+0}\right) = \varepsilon_S\left(t,\, l^*_{-0}\right), \\ &\left.\frac{\partial \varepsilon_S}{\partial x}\right|_{x=l^*_{+0}} = \left.\frac{\partial \varepsilon_S}{\partial x}\right|_{x=l^*_{-0}}, \\ &\varepsilon_S\left(t,\, \frac{L_a}{4}\right) = 0. \end{aligned} \tag{3.1.38}$$

Solving the preceding equations yield the following equation for pipe strain:

$$\varepsilon_S(t,x) = \begin{cases} \alpha_0\varepsilon_G\left\{\cos\left(\frac{2\pi}{L_a}x\right) - \cos\left(\frac{2\pi}{L_a}l^*\right) + \frac{2\pi}{L_a}\frac{\tau_{cr}}{\tau_G}\left(\frac{L_a}{4} - l^*\right)\right\}; & 0 < x < l^* \\ \frac{2\pi}{L_a}\varepsilon_{cr}\left(\frac{L_a}{4} - x\right); & l^* \le x \le L_a/4 \end{cases} \tag{3.1.39}$$

in which

$$\varepsilon_{cr} = \tau_{cr} \cdot \frac{L_a}{2\pi} \cdot \frac{1}{Et}. \tag{3.1.40}$$

Since the maximum pipe strain occurs at $x = 0$, the maximum pipe strain is given by

$$\varepsilon_S = \alpha_0 \varepsilon_G \left\{ 1 - cos\left(2\pi \frac{l^*}{L_a}\right) + \frac{\pi}{2} sin\left(2\pi \frac{l^*}{L_a}\right)\left(1 - 4\frac{l^*}{L_a}\right) \right\}. \tag{3.1.41}$$

Now let's introduce the following variable ξ,

$$\xi = 2\pi \frac{l^*}{L_a} = arc\,sin\left(\frac{\tau_{cr}}{\tau_G}\right). \tag{3.1.42}$$

Then the pipe strain is given by

$$\varepsilon_S = \begin{cases} \alpha_0 \varepsilon_G \left\{ 1 - cos\,\xi + \left(\frac{\pi}{2} - \xi\right) sin\,\xi \right\}; \ \tau_G \geq \tau_{cr}. \\ \alpha_0 \varepsilon_G; \ \tau_G < \tau_{cr} \end{cases} \tag{3.1.43}$$

From this we obtain the formulation of the slippage factor q as follows:

$$q = \begin{cases} 1 - cos\,\xi + \left(\frac{\pi}{2} - \xi\right) sin\,\xi; \ \tau_G \geq \tau_{cr}. \\ 1; \ \tau_G < \tau_{cr} \end{cases} \tag{3.1.44}$$

The maximum relative displacement between the buried pipe and the surrounding ground occurs at $x = L_a/4$ and is calculated as

$$\begin{aligned} \Delta_R &= \int_0^{La/4} \left\{ \varepsilon_G\, sin\left(\frac{2\pi}{L_a} x\right) - \varepsilon_s(t,x) \right\} dx \\ &= \frac{L_a}{2\pi} \varepsilon_G \left\{ 1 - \alpha_0 \left[sin\,\xi \left\{ 1 + \frac{\pi^2}{8} - \frac{\xi^2}{2} \right\} - \xi\, cos\,\xi \right] \right\} \\ &= \frac{L_a}{2\pi} \varepsilon_G \cdot \left(1 - \alpha_0 q^*\right) \\ &= \left(1 - \alpha_0 q^*\right) U_h \end{aligned} \tag{3.1.45}$$

in which

$$q^* = sin\,\xi \left\{ 1 + \frac{\pi^2}{8} - \frac{\xi^2}{2} \right\} - \xi\, cos\,\xi. \tag{3.1.46}$$

3.1.3 Seismic Design Methods for Continuous Jointed Pipes

3.1.3.1 Seismic Strain of Straight Pipes

Consider the case where a straight section of pipeline is located in the surface soil as shown in Figure 3.1.8.

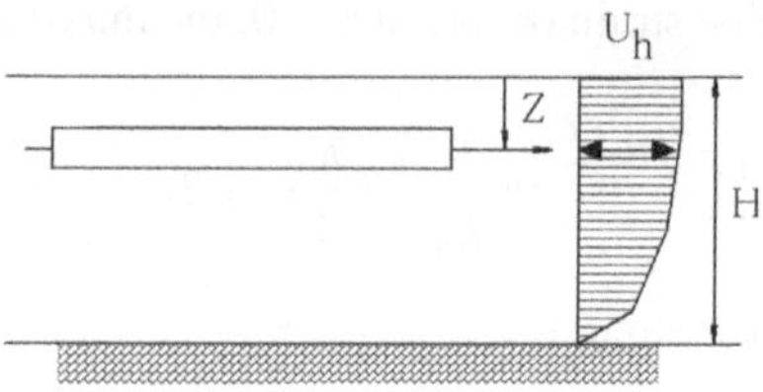

FIGURE 3.1.8 Pipeline in surface ground.

The axial strain of the buried pipe can be calculated by the response displacement method using the following equation:

$$\varepsilon_S = \alpha\varepsilon_G \text{ for } \varepsilon_S \le \varepsilon_y \tag{3.1.47}$$

$$\varepsilon_S = \varepsilon_G \text{ for } \varepsilon_S > \varepsilon_y \tag{3.1.48}$$

and

$$\alpha = q \cdot \alpha_0, \tag{3.1.49}$$

where α_0 is the conversion factor from ground strain to pipe strain, as shown in Equation (3.1.31).

The slippage factor q is given by

$$q = 1 - cos\,\xi + \Omega \cdot \left(\frac{\pi}{2} - \xi\right) sin\,\xi \quad \tau_G \ge \tau_{cr} \tag{3.1.50}$$

$$q = 1 \quad \tau_G < \tau_{cr} \tag{3.1.51}$$

in which

$$\xi = arcsin\left(\frac{\tau_{cr}}{\tau_G}\right), \tag{3.1.52}$$

where

τ_{cr}: shear stress at the limit of slip initiation (N/m^2)
τ_G: shear stress acting on pipe surface (N/m^2) which can be estimated by using Equation 3.1.25 in the following way:

$$\tau_G = \frac{2\pi}{L_a} Et\alpha_0\varepsilon_G \tag{3.1.53}$$

Ω: Modification factor for the slippage factor q, $\Omega = 1.5$

In Equation (3.1.50), a safety adjustment factor Ω was introduced in the FEM analysis. In the *Guidelines for Seismic Design of High-Pressure Gas Pipes*,[8] a correction factor of 1.5 was introduced to ensure a safe side evaluation.

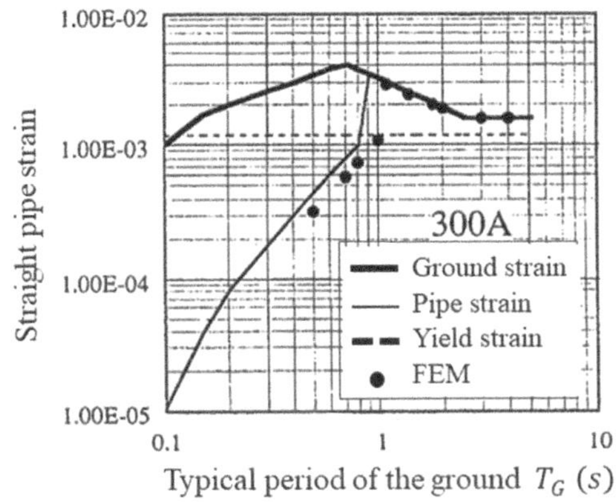

FIGURE 3.1.9 Comparison of accuracy between design calculation formula and FEM analysis results.[8]

3.1.3.1.1 Comparison of Accuracy of Design Equations

Figure 3.1.9 compares the above design equation with the results of FEM analysis[8] using the shell model. The figure compares the ground strain (thick solid line), pipe strain (thin solid line), yield strain (dashed line, $\varepsilon_y = 1.00E-03$), and FEM results (black circles).

As shown in the figure, the pipe strain slightly exceeds the FEM results in the elastic range, while in the plastic range, the pipe strain is slightly higher than the FEM results. In the plastic region, both pipe strains are consistent with the ground strain, indicating the accuracy of the design formulas.

3.1.3.2 Seismic Strain of Geometrical Pipe

3.1.3.2.1 Analysis Method

In the past, many buried geometrical pipes have incurred damage during earthquakes. In particular, buckling and tensile rupture of branch pipes and local buckling of curved pipes have often been observed as failure modes. In this regard, we discuss a method for analyzing the concentrated stresses that accumulate in deformed pipes due to seismic wave propagation.

First, we consider the case where a sinusoidal seismic wave acts on a branch pipe as shown in Figure 3.1.10, i.e., a seismic wave of wavelength L_a enters the pipe along the branch pipe and the maximum relative displacement Δ_R occurs exactly at the branch. This relative displacement must be absorbed by Δ_R, the bending deformation Δ_B on the main pipe side of the branch, the elongation deformation Δ_S on the branch pipe side, and the elastic deformation Δ_F of the branch ($\Delta_F = 0$ if the joint is rigid). In other words, the following relationship must hold:

$$\Delta_R = \Delta_B + \Delta_S + \Delta_F. \tag{3.1.54}$$

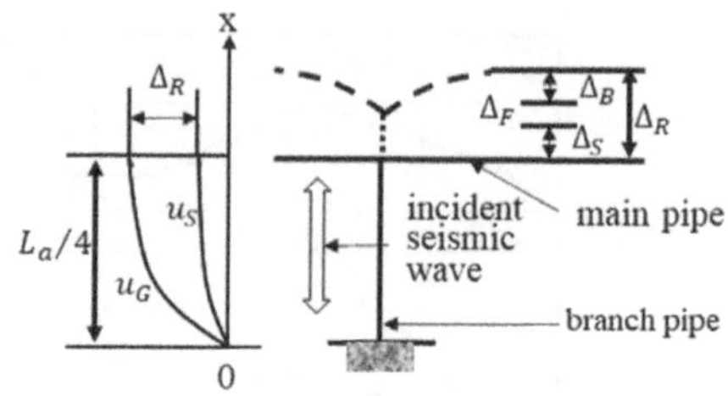

FIGURE 3.1.10 Deformation of T-junction due to relative displacement.

As shown in Figure 3.1.11, the same thing occurs in the curved pipe as in the bifurcated pipe when the seismic wave strikes along one of the pipe axes, and the relationship in Equation (3.1.54) is valid.

In both models, the rigid joints of the buried pipe are located 1/4 seismic wavelength away from the joints. This is because it is assumed that a straight pipe connected to a deformed pipe behaves in the same way as a straight pipe of infinite length without being affected by the stiffness of the joint near the rigid connection point.

The bending deformation of the buried pipe is analyzed using a beam model on an elastic bed. The coefficient of subgrade reaction force K is assumed to be given by $K = \pi D k_0$ using the coefficient of subgrade stiffness k_0.

3.1.3.2.2 Pipe Strain in Curved Pipes

The maximum stress in a curved pipe is expressed as the bending stress locally generated in the curved pipe using the Stress Intensification Factor i as follows:

$$\sigma_B = i\frac{M}{Z}, \tag{3.1.55}$$

where M is the maximum bending moment, Z is the section modulus of the pipe, and the local stress concentration in the curved pipe occurs at its maximum curvature when the center portion of the curved pipe is flattened by the bending moment acting on both ends of the pipe. To evaluate this stress concentration effect, a deflection analysis[10] (flexibility analysis) for the piping system is necessary, and the change in curvature of the curved pipe due to bending deformation must be evaluated. In other

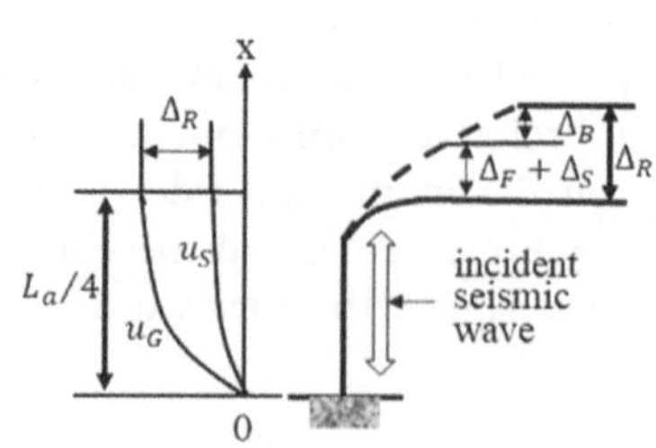

FIGURE 3.1.11 Curved pipe deformation due to relative displacement.

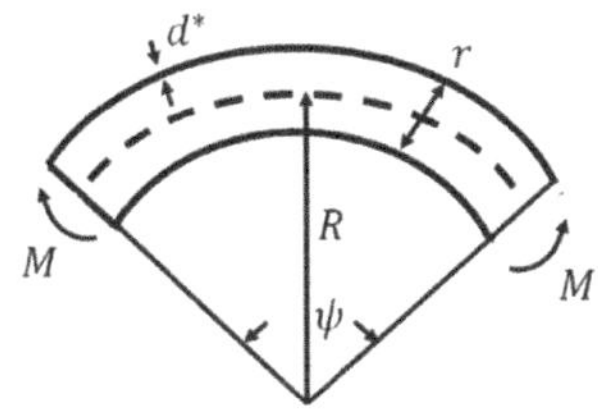

FIGURE 3.1.12 Model of a curved pipe.

words, the change in the bending angle of a curved pipe can be evaluated by using the flexibility factor n as follows:

$$\frac{\Delta\psi}{\psi} = n\frac{MR}{EI^*}, \tag{3.1.56}$$

where $\Delta\psi / \psi$ is the change in bending angle as shown in Figure 3.1.12, where R is the radius of curvature, M is the maximum bending moment, and I^* is the secondary moment of the curved pipe section ($I^* = I$ when the thickness of the curved pipe section equals that of the straight pipe section). The flexibility factor n has the following relationship with the pipe coefficient h:

$$n = \frac{1.65}{h}. \tag{3.1.57}$$

The pipe factor h is given by

$$h = \frac{d^* R}{r^2}, \tag{3.1.58}$$

where d^* is the pipe thickness of the curved pipe and r is the mean radius of the pipe. The stress intensification factor is expressed by the following equation using the pipe coefficient:

$$i = \frac{0.9}{h^{5/6}}, \tag{3.1.59}$$

The formulation process for pipe strain and pipe strain conversion factors for curved and T-junction pipes is described in Appendix A4 and is omitted from the main text. In the formulations that follow, the relative displacement between the pipe and the ground is denoted by Δ instead of Δ_R, in accordance with the notation in the *Guidelines for Seismic Design of High-Pressure Gas Pipes*.[8]

3.1.3.3 Pipe Strain Calculation Formula for Elastoplastic Deformed Pipe

When a Level 2 ground motion occurs, the deformed pipe section is in an elastoplastic state, and the above analysis method that assumes an elastic state cannot be applied as it is.

3.1.3.3.1 In the Case of a Curved Pipe

According to the results of the FEM nonlinear analysis using the shell model for curved pipes, the flattening of the curved pipe cross section progresses further when the pipe is in a totally plastic state due to a Level 2 ground motion, but it was confirmed that the bending rigidity of the curved pipe is not immediately lost. Therefore, a simple evaluation method for the elastic-plastic behavior of a deformed pipe section can be formulated by determining whether the section is in an elastic, partially plastic, or totally plastic state, and then equivalently linearizing the deformed pipe stiffness when it is totally plastic, as follows.

When the curved pipe section is in an elastic or partially plastic state, the curved pipe strain is calculated using the above formula for calculating curved pipe strain (elastic solution).

On the other hand, when the curved pipe section is in a totally plastic state, the bending deformation of the plastically hinged curved pipe and the elastic deflection deformation of the adjacent straight pipe absorbs the relative displacement Δ between the pipe and the ground, as shown in Figure 3.1.13, and the curved pipe strain is calculated by an equivalent linearization method.

In other words, if the decrease in stiffness due to plasticity expressed by the modulus of elasticity of the split line can be evaluated in the form of the curvature strain increase rate C_B, the curved pipe strain in the total plastic state can be obtained by multiplying the above formula for calculating the curved pipe strain (elastic solution) by C_B.

In the *Guidelines for Seismic Design Methods for High-Pressure Gas Pipes*,[8] the value of C_B is set to $C_B = 2$ for curved pipes with nominal diameters of less than 600A and $C_B = 1$ in consideration of compatibility with the results of FEM analysis for each diameter in the range of gas pipe use.

To determine whether the seismic strain of a curved pipe section is in the elastic, partially plastic, or total plastic range, the seismic strain $\beta_B \Delta$ of the curved pipe, assuming that the curved pipe is elastic, is determined by whether it exceeds the

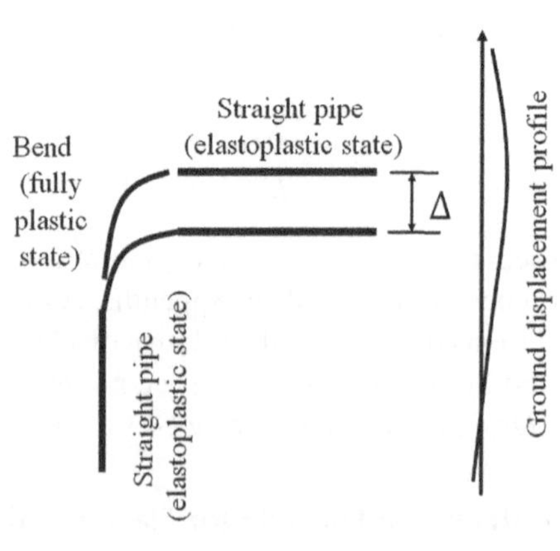

FIGURE 3.1.13 Curved pipe section deformation mode in the all-plastic state.[8]

total plastic strain defined in Equation (3.1.60). That is, the total plastic strain ε_{pl} is defined as a function of M_p (total plastic moment) and M_y (yield moment), as follows:

$$\varepsilon_{pl} = \frac{M_p}{M_y} \cdot \varepsilon_y \cong 1.27\varepsilon_y. \tag{3.1.60}$$

The angles of curvature of curved pipes are 90°, 45°, 22½, and 11¼. Based on the numerical results of *the Guidelines for the Seismic Design Methods for High-Pressure Gas Pipes*,[8] it was confirmed that the maximum seismic strain of curved pipes in the all-plastic region is given by the 90° curved pipe case.[11,12]

3.1.3.2.3 In the Case of T-Junctions

Large-scale deformation tests of T-junctions have reported that failure does not occur at the crotch of the T-junction, but rather that cracking due to tension occurs circumferentially in the branch pipe section of the T-junction.

Therefore, in the seismic design guidelines for high-pressure gas pipelines, the safety of the T-junction section is evaluated based on the bending strain generated in the branch pipe section when seismic waves are incident along the main pipe axis, as shown in Figure 3.1.14.

According to the guideline, the maximum bending strain in the all-plastic state in the branch pipe section of a T-junction can be calculated as $2\beta_T\Delta$, based on the fact that the bending strain in the branch pipe section in the all-plastic strain state is about twice the value of the T-junction seismic strain analysis results (elastic solution) from full-scale experimental results.

3.1.3.4 Design Seismic Strain of Deformed Pipe

3.1.3.4.1 Curved Pipe

When the curved pipe section is in the elastic or partially plastic region, the maximum strain of the curved pipe is

$$\varepsilon_B = \beta_B \cdot \Delta \quad \text{for} \quad \varepsilon_B \le 1.27\varepsilon_y.$$

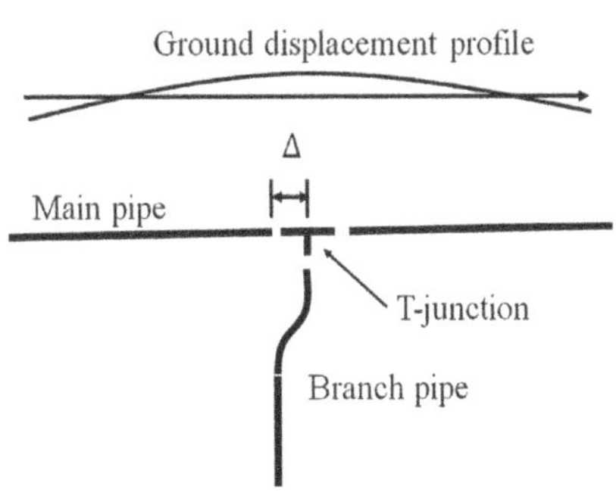

FIGURE 3.1.14 Modeling of T-junction.[8]

When the curved pipe section is in the total plastic range, the maximum strain of the curved pipe is

$$\varepsilon_B = C_B \cdot \beta_B \cdot \Delta \quad \text{for} \quad \varepsilon_B > 1.27\varepsilon_y, \tag{3.1.61}$$

in which

ε_B: seismic strain of the curved pipe section
β_B: conversion factor of the curved pipe section (1/cm)
σ_y: yield strain of the curved pipe
C_B: modification factor for the strain of the curved pipe in the total plastic range

$$C_B = 2 \text{ (Nominal diameter less than 600A)}$$

$$C_B = 1 \text{(Nominal diameter 600A or larger)}$$

Δ: relative displacement of straight pipes adjacent to curved pipes with respect to the surrounding ground

$$\Delta = \left(1 - \alpha^*\right) \cdot U_h, \tag{3.1.62}$$

in which

$$\alpha^* = q^* \cdot \alpha_0, \tag{3.1.63}$$

where

$$\begin{aligned} q^* &= sin\,\xi \cdot \left(1 + \frac{\pi^2}{8} - \frac{\xi^2}{2}\right) - \xi \cdot cos\,\xi \quad && \text{for} \quad \tau_G \geq \tau_{cr} \\ q^* &= 1 && \text{for} \quad \tau_G < \tau_{cr}. \end{aligned} \tag{3.1.64}$$

3.1.3.4.2 T-Junction

When the T-junction is in the elastic or partially plastic range, the maximum strain of the curved pipe is

$$\varepsilon_T = \beta_T \cdot \Delta \quad \text{for} \quad \varepsilon_T \leq 1.27\varepsilon_y.$$

When the T-junction is in the full plastic range, the maximum strain of the curved pipe is

$$\varepsilon_T = 2\beta_T \cdot \Delta \quad \text{for} \quad \varepsilon_T > 1.27\varepsilon_y, \tag{3.1.65}$$

where

ε_T: seismic strain at the T-junction
β_T: conversion factor of the T-junction (1/cm)
Δ: relative displacement of straight pipes adjacent to the T-junction relative to the surrounding ground
σ_y: yield strain of the T-junction

3.1.3.4.3 Strain Conversion Factors for Bend Pipes

The strain conversion factor of a bend pipe is given by

$$\beta_B = \frac{2i_B A\lambda^2 D_o \left|(5+R_r\lambda)b_1\right| + 4\lambda^3 I_r \left|5(1+b_2)-b_1\right|}{10A_r + 5L_A I_r \lambda^3 (1+b_2) + 10A_r b_3}. \quad (3.1.66)$$

in which

$$b_1 = -\frac{1+2R\lambda+(\pi-2)nR^2\lambda^2}{(1+R\lambda)\left|2+\pi nR\lambda+(4-\pi)nR^2\lambda^2\right|}, \quad (3.1.67)$$

$$b_2 = \frac{1-2nR^2\lambda^2-(4-\pi)nR^3\lambda^3}{(1+R\lambda)\left|2+\pi nR\lambda+(4-\pi)nR^2\lambda^2\right|}, \quad (3.1.68)$$

$$b_3 = nR^3\lambda^3\left\{\frac{\pi}{2}+\frac{\pi I^*}{2nA^*R^2}+\left(1-\frac{I^*}{nA^*R^2}\right)b_1+\left(\frac{2}{R\lambda}+\frac{\pi}{2}+\frac{\pi I^*}{2nA^*R^2}\right)b_2\right\}. \quad (3.1.69)$$

where

β_B: strain conversion factor for the curved pipe (1/cm)
i_B: stress concentration factor for the curved pipe

$$i_B = max\left\{\frac{1.95}{h^{2/3}},\ 1.5\right\}. \quad (3.1.70)$$

n: deflection coefficient

$$n = \frac{1.65}{h}. \quad (3.1.71)$$

h: pipe factor

$$h = \frac{tR}{r^2}. \quad (3.1.72)$$

D_o: pipe outer diameter (cm)
t: pipe thickness (cm)
r_m: pipe mean radius (cm)
A^*: pipe cross-sectional area (cm^2)
R: radius of curvature of curved pipe (cm)

I^*: secondary stiffness of cross section (cm^4)
La: apparent wavelength (cm)
λ: parameters

$$\lambda = \sqrt[4]{\frac{K_2}{4EI}}. \tag{3.1.73}$$

K_2: the ground spring constant per unit length perpendicular to the pipe axis (N/cm^2), where

$$K_2 = Dk_2. \tag{3.1.74}$$

k_2: the ground spring coefficient (N/cm^3) which is defined by $k_2 = \sigma_{cr}/\delta_{cr}$ shown in Figure 3.1.15. The ground spring coefficient data are presented in Table 3.1.1 in the *Guidelines of the Seismic Design Methods for High-Pressure Gas Pipes.*[8]
E: Young's modulus of steel pipe, 2.06×10^7 N/cm^2

3.1.3.4.3 Strain Conversion Factors for T-Junctions

The strain conversion factor of a T-junction is obtained by the following equation:

$$\beta_T = \frac{4\bar{\lambda}_1^2 D_1 A_2 (C-1)}{4A_2 + L_a I_1 \bar{\lambda}_1^3 C}, \tag{3.1.75}$$

$$C = \frac{1 + 4\left(\frac{\bar{\lambda}_1}{\bar{\lambda}_2}\right)^3 \left(\frac{D_2}{D_1}\right)}{1 + 2\left(\frac{\bar{\lambda}_1}{\bar{\lambda}_2}\right)^3 \left(\frac{D_2}{D_1}\right)}. \tag{3.1.76}$$

where

β_T: conversion factor of T-junction (1/cm)
D: outer diameter
A: cross-sectional area
I: secondary moment of cross section
L_a: apparent wavelength of seismic motion

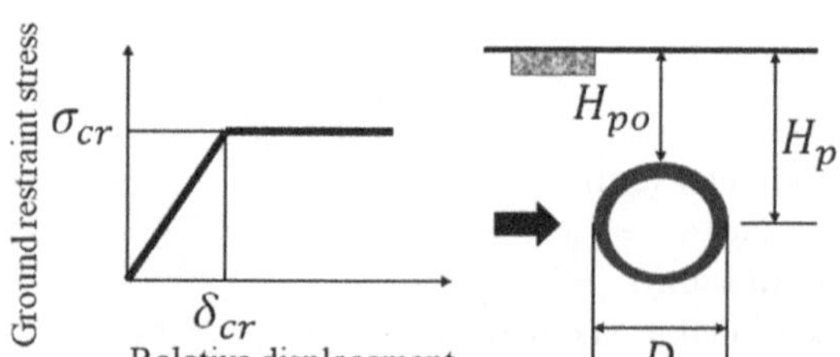

FIGURE 3.1.15 Ground spring coefficient perpendicular to pipe axis.

TABLE 3.1.1
Ground Spring Coefficients

Diameter, D (mm)	Maximum Ground Restraint Stress, σ_{cr} (N/cm²)	Yield Displacement, δ_{cr} (cm)	Ground Spring Coefficient, (N/cm³)
100	53	2.6	20
150	51	2.6	20
200	48	2.6	18
300	42	2.7	16
400	39	2.8	14
500	36	2.8	13
600	34	2.9	12
650	33	2.9	11
750	32	3.0	11
900	30	3.1	10

Note that the subscripts D, A, I, and λ indicate the branch pipe side (subscript 1) and the main pipe side (subscript 2).

$$\bar{\lambda} = \sqrt[4]{\frac{K_2}{4EI}}. \tag{3.1.77}$$

3.1.3.5 Comparison of Accuracy of Pipe Strain Calculation Formula and FEM Analysis Results for Curved Pipes and T-Junctions

In order to confirm the analytical accuracy of the strain calculation formula for irregularly shaped pipes in the *Guidelines of the Seismic Design Methods for High-Pressure Gas Pipes*, a FEM analysis was conducted using a shell model of irregularly shaped pipes to compare the two results.

For curved pipes, the pipe diameter is 150A, and for T-junctions, the pipe diameter is a combination of 150A × 300A for branch and main pipes.

The vertical axis in Figure 3.1.16 shows the maximum pipe strain for the curved pipe and T-junction, and the horizontal axis shows the ground natural period. The solid line shows the calculation results of the design equation, and the black circle shows the FEM results. The dotted line in the figure shows the total plastic strain value of the pipe.

The results of the design equation and the FEM results seem to show good correspondence in general.

3.1.3.6 Critical Pipe Strains Used for Seismic Safety Verification of Straight Pipes

The limit states for pipe strain of pipelines are generally classified into three limit states: elastic limit, buckling limit, and ultimate limit.

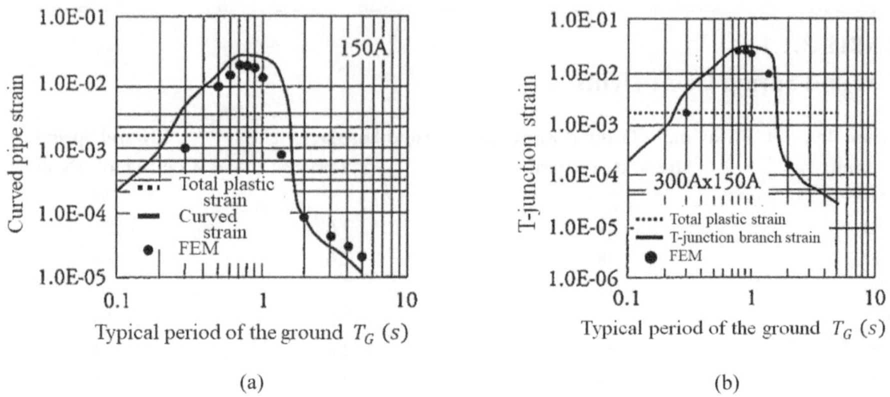

FIGURE 3.1.16 Comparison of accuracy between theoretical solution and FEM numerical analysis results:[8] (a) curved pipe; (b) T-junction branch pipe.

The elastic limit state corresponds to the elastic limit strain ε_y of the pipe material, while the elastoplastic state corresponds to the buckling initiation strain ε_{buckle}, given by

$$\varepsilon_{buckle} = \frac{4}{3}\sqrt{n}\,\frac{t}{D_m}. \tag{3.1.78}$$

where t is the pipe thickness, D_m is the average diameter of the pipe, and the slope n is set to 0.11 in the strain-hardening region of the residual-stressed steel. In the seismic design guidelines for high-pressure gas pipelines, the safety factor is set to 1.25 and the allowable buckling initiation strain is given by the following equation:

$$\varepsilon_{buckle} = 0.35\frac{t}{D_m}. \tag{3.1.79}$$

On the other hand, in the *Guidelines for the Seismic Engineering Methods for Waterworks Facilities*,[13] the following equation was obtained as the allowable buckling initiation strain by assuming a safety factor of 1.25 to the local buckling strain of $0.58t / D_m$ obtained from the full-scale test "Steel pipe buckling test for waterworks" (WSP060-96)[14] conducted by the Japan Water Pipe Association:

$$\varepsilon_{buckle} = 0.46\frac{t}{D_m}. \tag{3.1.80}$$

The ultimate limiting strain of a pipe is given by the breaking strain ε_f. In particular, the seismic design guidelines for high-pressure gas pipelines set the allowable strain for rupture strain based on the results of fatigue failure experiments on straight pipes, respectively, in the following way:

0.01 for Level 1 ground motion
0.03 for Level 2 ground motion

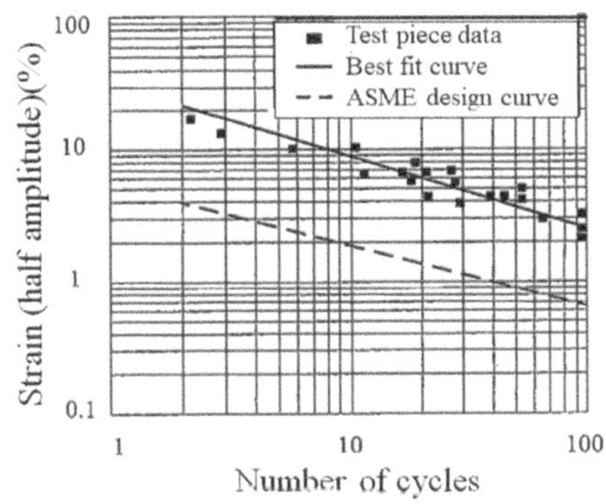

FIGURE 3.1.17 Extreme low-cycle fatigue test data and ASME design fatigue curves for specimens.[8]

It is estimated that the number of repetitions of the maximum strain equivalent to the damage inflicted on a pipe by one Level 2 ground motion is about 3–5. Figure 3.1.17 shows the results of fatigue experiments and their approximate curves. In the ASME design fatigue curve[15] (the dashed line in the Figure 3.1.17), the allowable strain corresponding to this 13–15% (0.13–0.15) strain is about 2.8–3.2% (0.028–0.032). The allowable strain corresponding to Level 2 ground motion is about 2.8~3.2% (0.028~0.032). From this, it can be said that the allowable strain of 0.03 (3%) was set for Level 2 ground motion. Therefore, the ASME design fatigue curve has a factor of safety of about five times against the fatigue strain amplitude.

On the other hand, the *Guidelines for the Seismic Design Methods for Waterworks Facilities* specify the allowable strain based on the buckling initiation strain as follows:

$0.23t / D_m$ for level 1 ground motion
$0.46t / D_m$ for level 2 ground motion

The seismic safety of a pipeline must be verified based on whether the pipe axial strain ε_S generated in a straight pipe by Level 2 ground motion exists within the aforementioned three limit states.

3.1.3.7 Confirmation of Pipe Strain at the Ultimate Limit State by Full-Scale Experiments

In establishing the seismic design guidelines for liquefaction resistance of high-pressure gas pipelines, full-scale cyclic experiments were conducted to investigate the limit values of the ultimate limit state of deformed pipes when subjected to Level 2 ground motion. The cyclic loading and bending deformation of the full-scale curved pipe were measured, and finally the starting point of crack leakage and the maximum curved pipe strain at that point were confirmed. Figure 3.1.18 shows a schematic diagram of the location of crack leakage in the full-scale experiment.

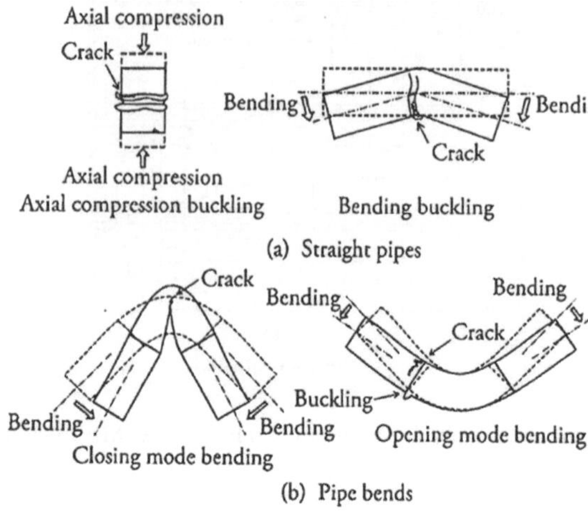

FIGURE 3.1.18 Detailed explanation of deformation test results of full-scale pipeline.[18]

The full-scale curved pipe experimental apparatus shown in Figure 3.1.19 was used to measure its large deformation state, and at the same time, the deformation behavior was traced using a FEM shell model (four-node shell element) with the analysis code ADINA to numerically estimate the local strain of the curved pipe at the point of maximum strain generation. As a result, as shown in Figure 3.1.20, the local strain at the point of maximum bending moment is about 0.03, the curved pipe angle at that time is about 8°, and the bending angle at the final crack leakage is about 50°.

Based on these experimental results, the seismic design guideline for high-pressure gas pipelines sets the limit strain[33] of deformed pipes to 3% (0.03) against Level 2 ground motion.

3.1.4 Seismic Design Method for Mechanically Jointed Pipes

This section discusses the seismic design method[16] for seismic jointed ductile iron pipes as shown in Figure 3.1.21.

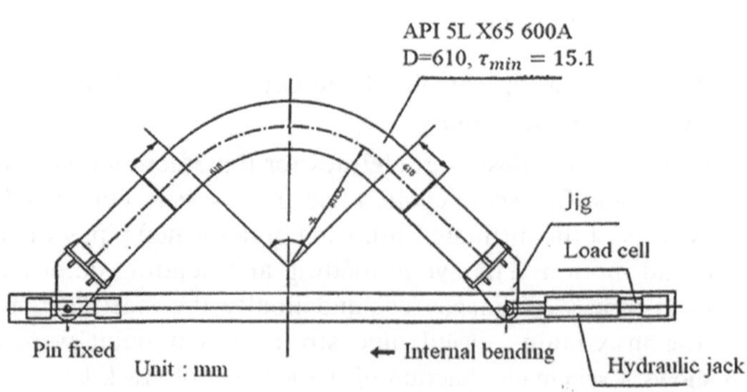

FIGURE 3.1.19 In-plane bending test apparatus for 600A curved pipe.[8]

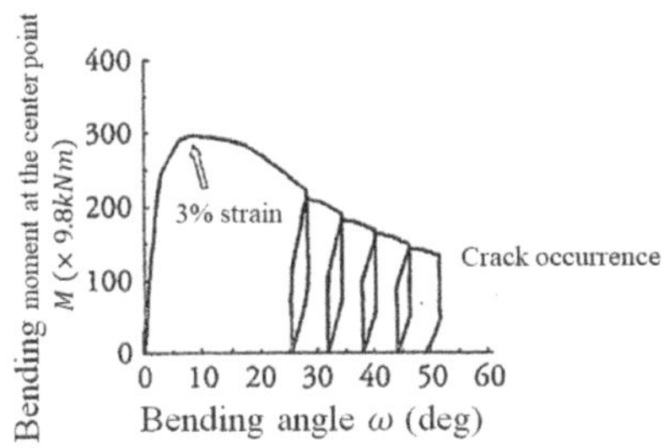

FIGURE 3.1.20 Relationship between moment and bending angle and location of 3% strain in the center section of a 600A curved pipe during an inner bending test.[8]

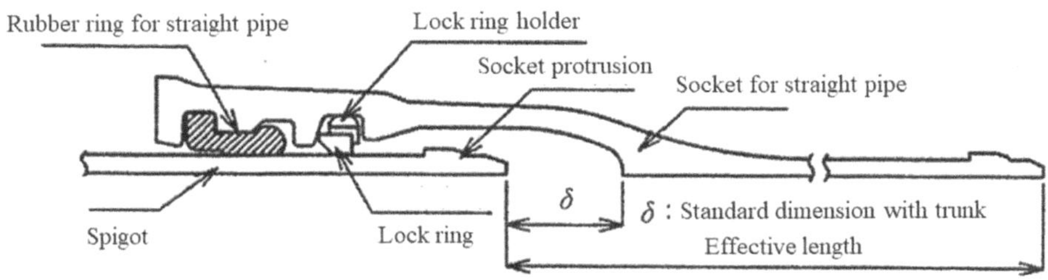

FIGURE 3.1.21 Ductile iron pipe earthquake-resistant pipe joint.[16]

Seismic joint-type ductile iron pipe can avoid joint rupture if the ground displacement acting in the axial direction during an earthquake can be absorbed by multiple joint displacements, and if not, leakage damage will occur due to joint rupture. Therefore, in seismic design, it is necessary to confirm the maximum absorption displacement that a ductile iron pipe with seismic joints can absorb.

3.1.4.1 Number of Single Pipe Connections up to the Rupture Limit

The tensile strength of a seismic pipe joint is defined as the strength at which the locking ring in the joint shears off the insertion butt.

The rupture of a jointed pipe occurs when the joint tensile force caused by the shear force of the soil acting along the circumference of the jointed pipes exceeds the joint tensile strength $3D$ (kN) of the single pipe in which D is the diameter of the pipe and is expressed in mm. The number of connected pipes at this time is the required number of pipes for fracture resistance (ultimate limit), m_{cr}, and is expressed by the following equation:

$$m_{cr} = \frac{3D}{\pi D \ell \tau_{cr}}. \qquad (3.1.81)$$

where

τ_{cr}: critical shear stress at the start of slip in the soil around the pipe
ℓ: length of single pipe

The slip initiation limit shear stress of the soil around the pipe is measured by a shear test using the soil around the buried pipe with a soil cover of 1.8 m. It should be modified according to the depth of the buried pipe.

3.1.4.2 Number of Single Pipes Required to Absorb Ground Displacement

The number of single pipes required is m, which is obtained by dividing the ground displacement U_h by the allowable displacement of the single pipe joint δ. In other words,

$$m = \frac{U_h}{\delta}. \tag{3.1.82}$$

Therefore, for the seismic safety verification of ductile iron pipe with seismic joints, it is sufficient to determine whether the seismic displacement m exceeds the required number of pipes to be ruptured, m_{cr}.

3.1.5 Seismic Design Method for Segmented Shield Tunnel Structures

3.1.5.1 Structural Model

When an earthquake ground motion acts on the tunnel, ground displacement U_X acts on the tunnel in the axial direction, ground displacement U_Z in the direction perpendicular to the axis, and relative displacement ΔU_z at the upper and lower ends in the cross section, as shown in Figure 3.1.22. In the tunnel direction, axial strain ε_A and the resulting axial force F_X is produced. In the direction perpendicular to the tunnel axis, bending strain ε_B and the resulting bending moment M_y are generated, and in the tunnel lining in the cross-sectional direction as shown in Figure 3.1.23, circumferential stress σ_r and shear stress $\tau_{r\phi}$ are generated.

The axial forces and bending moments generated in the tunnel correspond to the case where the tunnel behaves as a beam, whereas the stresses in the tunnel lining are the stresses generated by the deformation of the tunnel section.

Recent large-section tunnels are constructed by assembling the segments shown in Figure 3.1.24 to form a circular cross section with a shield machine. Tunnel deformation during earthquakes is absorbed by the bolting of each bolted segment and by the expansion and contraction of the segments themselves.

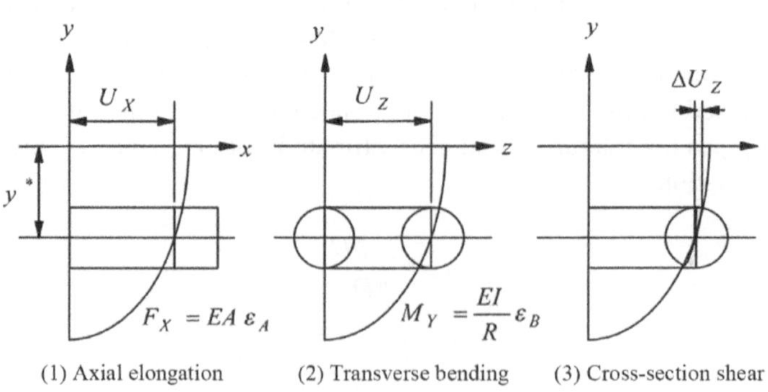

FIGURE 3.1.22 Three modes of tunnel deformation under earthquake excitation.

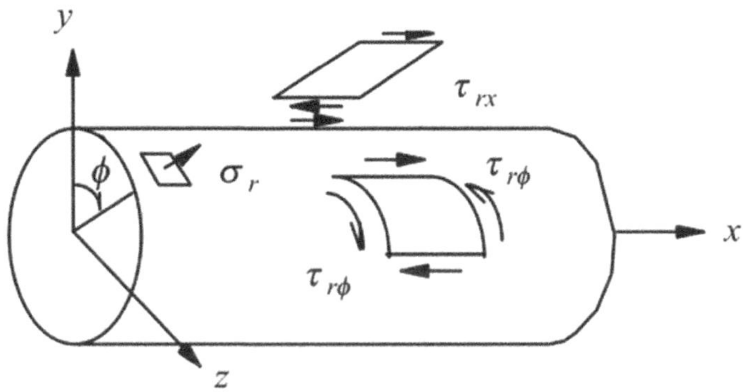

FIGURE 3.1.23 Stresses acting on the tunnel surface.

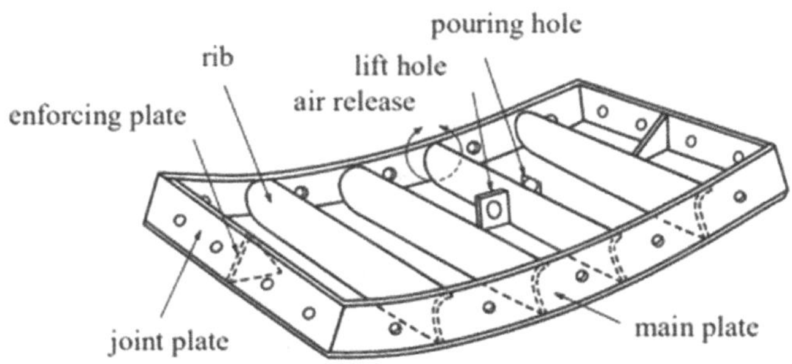

FIGURE 3.1.24 Segment of shield tunnel.[36]

In addition to elastic elongation of the segment members and joint bolts when subjected to tensile forces, tunnels composed of segments and bolts also exhibit elastoplastic elongation characteristics when a higher tensile force is applied. On the other hand, when compressive forces are applied, they distort in a manner that is not necessarily symmetrical to that in tension, depending on the properties of the segmental constituent materials. As a result, as shown in Figure 3.1.25, the tensile and

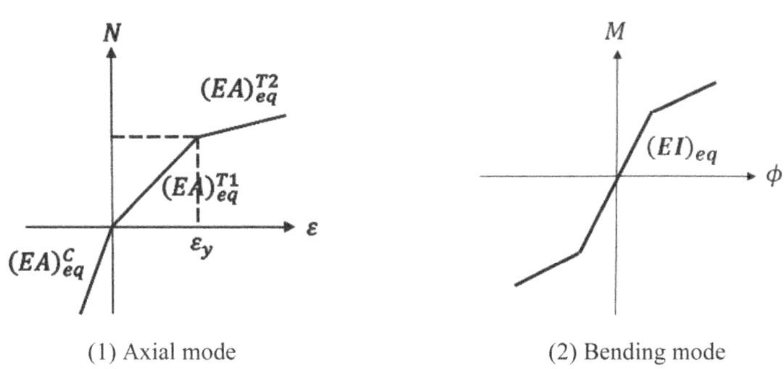

FIGURE 3.1.25 Relationship between axial forces and strains, and bending moment and angle.

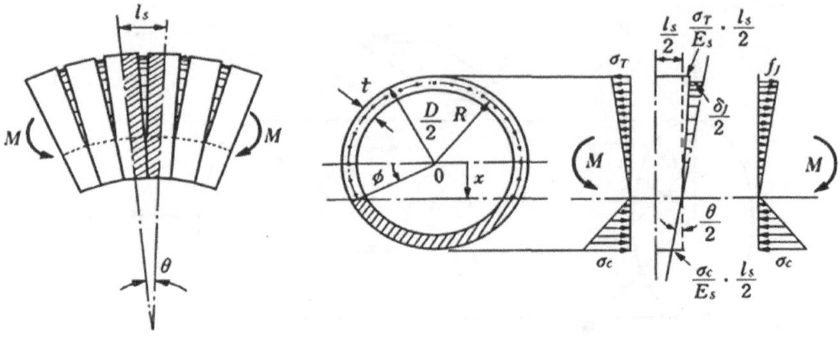

FIGURE 3.1.26 Stress distribution of s segment cross section in an elastic stage.[36]

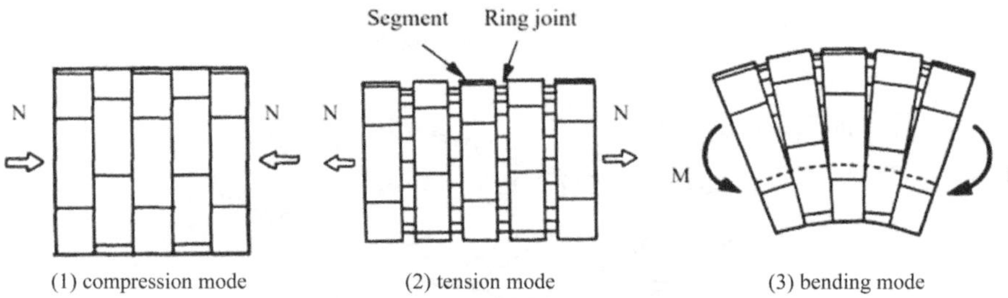

FIGURE 3.1.27 Displacement modes of shield tunnel.[36]

compressive properties are generally asymmetrical with regard to the axial deformation characteristics of tunnels, while it should be noted that the bending mode is symmetrical.

When bending moments act on the tunnel, asymmetries in tensile and compressive properties also appear with the bending centerline eccentric downward from the pipe center position, as shown in Figure 3.1.26, where the following parameters are used as l_S: width of one segment ring; f_J: tensile force of the segment ring joint; and δ_J: elongation of the segment ring joint, respectively.

Kawashima[36] describes the characteristics of structural behavior of shield tunnel, where the tunnel deformations are different in compression mode and tensile mode, as shown in Figure 3.1.27. This difference also makes the neutral axis deviate to the compression side in bending mode.

Based on this consideration, the cross-sectional area and bending rigidity of shield tunnel are estimated in the equivalent form with Young's modulus. Two different cross-sectional areas are defined in the following way[37] for compression and tension modes, respectively:

$$(EA)_{eq}^{C} = E_S A_S, \quad (EA)_{eq}^{T} = \frac{\cos^3\phi}{\cos\phi + \left(\frac{\pi}{2} + \phi\right)\sin\phi} E_S A_S \text{ with } \phi = \sin^{-1}\left(\frac{d}{r}\right), \quad (3.1.83)$$

in which

$(EA)_{eq}^{C}$: equivalent cross-sectional rigidity in compression
$(EA)_{eq}^{T}$: equivalent cross-sectional rigidity in tension
A_S: cross-sectional area of a segment ring
E_S: Young's modulus of a segment
r: radius of a tunnel
d: length from the tunnel center to the cross-sectional center

Consider a seismic wave to be horizontally traveling. The peak ground displacement U_h of the ground response displacement at the pipe axis position is previously given from Equation (3.1.23) by

$$U_h = \frac{2}{\pi^2} S_v \ T \ cos\left(\frac{\pi y^*}{2H}\right),$$

where H is the depth of the ground surface.

When a wave type W is assigned such as $W = R$ (Rayleigh wave type) or $W = L$ (Love wave type), and incident to the tunnel with an angle ϕ, the wave motion is classified into the axial component $U_{h,AW}$ or transverse component $U_{h,BW}$.

Using the ground displacements $U_{h,AW}$, $U_{h,BW}$, the axial ground displacement along the pipe axis is given in the following way:

$$\begin{aligned} &\text{for an axial ground displacement } U_{h,AW}, \quad u_{AW}(\phi) = U_{h,AW}\chi_{AW}(\phi) \\ &\text{for a transverse ground displacement } U_{h,BW}, \quad u_{BW}(\phi) = U_{h,BW}\chi_{BW}(\phi) \end{aligned} \tag{3.1.84}$$

where

$\chi_{\nu W}(\phi)$: directional parameter given for $\nu = A$ for axial displacement and $\nu = B$ for transverse displacement, and also $W = R$ for Rayleigh-type surface waves and $W = L$ for Love-type surface waves.

$$\chi_{AR}(\phi) = cos\,\phi,\ \chi_{BR}(\phi) = sin\,\phi, \chi_{AL}(\phi) = -sin\,\phi, \chi_{BL}(\phi) = cos\,\phi.$$

Then, the ground strain at the free field is rewritten from Equation (3.1.1) and given by

$$\varepsilon_G = \frac{2\pi}{L} U_h,$$

where L is a wavelength and U_h is a peak ground displacement at the free field.

When a surface wave type W and a ground shaking mode ν are assigned, the ground strains in the free field are given by

$$\varepsilon_{GAW} = \frac{2\pi}{L} U_{h,AW}\chi(\phi) \quad and \quad \varepsilon_{GBW} = \frac{2\pi}{L} U_{h,BW}\chi(\phi). \tag{3.1.85}$$

Then the ground strains incident to the tunnel with an angle ϕ along the tunnel axis are expressed as

$$\varepsilon_{GA} = \frac{2\pi}{L_a} u_{AW}(\phi) \quad and \quad \varepsilon_{GB} = \frac{2\pi}{L_a} u_{BW}(\phi), \tag{3.1.86}$$

in which the apparent wavelength L_a is given by

$$L_a = \frac{L}{\cos\phi}. \tag{3.1.87}$$

Since the peak ground motion for $W = R$ occurs at $\phi = 0$ and for $W = L$ occurs at $\phi = \pi/4$, the discussion of tunnel strains hereunder is fixed to these incident angles providing the peak value for its simplicity.

When the ground displacement in the pipe transverse direction propagates horizontally with propagation velocity c_B,

$$u_{BW}(x,t) = f(x - c_B t). \tag{3.1.88}$$

The ground strain in the transverse direction is calculated by using the particle velocity $\dot{u}_{BW}$, which is the time partial derivative of u_{BW}, as

$$\varepsilon_{GB} = \frac{\dot{u}_{BW}}{c_B}. \tag{3.1.89}$$

On the other hand, the bending strain of the ground is given by

$$\varepsilon_B = \frac{M}{EI}\frac{D}{2}. \tag{3.1.90}$$

From the relationship $\left|\frac{\partial^2 u_{BW}}{\partial x^2}\right| = \frac{M}{EI}$ between curvature and bending moment, the bending strain is given by $\varepsilon_{GB} = \frac{D}{2}\left|\frac{\partial^2 u_{BW}}{\partial x^2}\right|$. Then using the wave equation $c_B^2 \frac{\partial^2 u_{BW}}{\partial x^2} = \frac{\partial^2 u_{BW}}{\partial t^2}$, the bending strain of the ground can be finally expressed as

$$\varepsilon_B = \frac{D\ddot{u}_{BW}}{2c_B^2} = \left(\frac{D}{2}\frac{2\pi}{L_a}\right)\varepsilon_{BW}. \tag{3.1.91}$$

Similarly,

$$u_B = \left(\frac{2\pi}{L_a}\cdot\frac{D}{2}\right)u_{BW}. \tag{3.1.92}$$

The equation for equilibrium of forces in the transverse direction to the tunnel is obtained by

$$\rho A \frac{\partial^2 u_{SB}}{\partial t^2} + EI\frac{\partial^4 u_{SB}}{\partial z^4} = K_{Bh}(u_B - u_{SB}), \tag{3.1.93}$$

where K_{Bh} is the spring modulus in the transverse direction to the pipe axis in horizontal plane. Solving this equation, the bending strain of the tunnel along the vertical axis is given by

$$\varepsilon_{SBh} = \alpha_{Bh}\left(\frac{D}{2}\frac{2\pi}{L_a}\right)\varepsilon_{GB}. \tag{3.1.94}$$

And in the same way, the bending strain of the tunnel along the horizontal axis is given by

$$\varepsilon_{SBv} = \alpha_{Bv}\left(\frac{D}{2}\frac{2\pi}{L_a}\right)\varepsilon_{GB}, \tag{3.1.95}$$

in which

D: external diameter of the tunnel
α_A^T: conversion factor to tunnel axial strain in tension
α_A^C: conversion factor to tunnel axial strain in compression
α_{Bh}: conversion factor to tunnel bending strain along the vertical axis
α_{Bv}: conversion factor to tunnel bending strain along the horizontal axis

where

$$\alpha_A^T = \frac{1}{1+\left(\dfrac{2\pi}{\lambda_A^T L_a}\right)^2},\ \alpha_A^C = \frac{1}{1+\left(\dfrac{2\pi}{\lambda_A^C L_a}\right)^2},\ \alpha_{Bh} = \frac{1}{1+\left(\dfrac{2\pi}{\lambda_{Bh} L_a}\right)^4},\ \alpha_{Bv} = \frac{1}{1+\left(\dfrac{2\pi}{\lambda_{Bv} L_a}\right)^4} \tag{3.1.96}$$

and

$$\lambda_A^T = \sqrt{\frac{K_A}{(EA)_{eq}^T}},\ \lambda_A^C = \sqrt{\frac{K_A}{(EA)_{eq}^C}},\ \lambda_{Bh} = \sqrt[4]{\frac{K_{Bh}}{(EI)_{eqh}}},\ \lambda_{Bv} = \sqrt[4]{\frac{K_{Bv}}{(EI)_{eqv}}} \tag{3.1.97}$$

K_A: spring modulus in the axial direction
K_{Bh}: spring modulus in the transverse direction in the horizontal plane
K_{Bv}: spring modulus in the transverse direction in the vertical plane
$(EA)_{eq}^T$: equivalent axial section rigidity in tension
$(EA)_{eq}^C$: equivalent axial section rigidity in compression
$(EI)_{eqh}$: equivalent transverse section rigidity along the vertical axis
$(EI)_{eqv}$: equivalent transverse section rigidity along the horizontal axis
β_T: eccentricity factor in tension, simply $\beta_T = 1$
β_C: eccentricity factor in compression, simply $\beta_C = 1$

If more accurate estimation of eccentric factors are necessary, the following formula[36] is prepared:

$$\beta_T = \frac{(EA)_{eq}^T}{(EA)_{eq}^C}\cdot\frac{\alpha_A^T}{\alpha_A^C}\left\{1-\frac{\cos\left(\dfrac{2\pi\eta}{L_a}\right)}{\cosh\left(\lambda_A^T\eta\right)}\right\},\ \beta_C = 1+\frac{\cos\left(\dfrac{2\pi\eta}{L_a}\right)}{\cosh\left\{\lambda_A^C\left(\dfrac{L_a}{2}-\eta\right)\right\}} \tag{3.1.98}$$

in which the parameter η can be obtained as the solution of the following equation:

$$\frac{2\pi}{\lambda_A^T L_a}\alpha_A^T \cos\left(\frac{2\pi\eta}{L_a}\right)\tan h\left(\lambda_A^T\eta\right)+\frac{2\pi}{\lambda_A^C L_a}\alpha_A^C \cos\left(\frac{2\pi\eta}{L_a}\right)$$
$$\tan h\left\{\lambda_A^C\left(\frac{L_a}{2}-\eta\right)\right\}-\left(\alpha_A^T-\alpha_A^C\right)\sin\left(\frac{2\pi\eta}{L_a}\right)=0 \quad (3.1.99)$$

The axial and bending strains of shield tunnel is summarized in Table 3.1.2.

In small-diameter pipelines, the dominant seismic structural strain is only axial strain. However, in large-diameter tunnels, the dominant structural strains are not only axial strain but also bending strain. Furthermore, the structural strains in both tension and the compression sides are not always identical because of structural characteristics of segmental bolted system that are different in both sides.

Since soil pressure depends on soil depth, the soil spring modulus at the top of the tunnel is different at the bottom or at mid-height. So the apparent structural strain of the tunnel can be classified as discussed in the following subsection.

3.1.5.2 Structural Strain Analysis for Large Ground Shakings[36,37]

Whether a tunnel with the segmental bolted system predicted at the time of design behaves as expected or not during an earthquake may only become clear after investigating damage to tunnels that have experienced major earthquakes.

TABLE 3.1.2
Axial and Bending Strains of Shield Tunnel

Seismic Wave Type		Rayleigh Wave (W = R) or Love Wave (W = L)	
Wave Motion		**Axial Elongation, A**	**Transverse Deflection, B**
Free field motion	Ground displacement	$U_{h,AW}$	$U_{h,BW}$
	Free field strain	Axial strain $\varepsilon_{GAW}=\frac{2\pi}{L}U_{h,AW}$	Shear strain $\varepsilon_{GBW}=\frac{2\pi}{L}U_{h,BW}$
Ground motion along the tunnel axis	Ground displacement	Axial component $u_{AW}(\phi)$	Transverse component $u_{BW}(\phi)$
	Ground strain	Axial mode $\varepsilon_{GA}=\frac{2\pi}{L_a}u_{AW}(\phi)$	Bending mode $\varepsilon_{GB}=\frac{2\pi}{L_a}u_{BW}(\phi)$
Tunnel strains	Tensile axial strain	$\varepsilon_{SA}^T=\alpha_A^T\beta_T\varepsilon_{GA}$	———
	Compressional axial strain	$\varepsilon_{SA}^C=\alpha_A^C\beta_C\varepsilon_{GA}$	———
	Bending strain along the vertical axis	———	$\varepsilon_{SBh}=\alpha_{Bh}\varepsilon_{GB}\left(\frac{2\pi}{L_a}\cdot\frac{D}{2}\right)$
	Bending strain along the horizontal axis	———	$\varepsilon_{SBv}=\alpha_{Bv}\varepsilon_{GB}\left(\frac{2\pi}{L_a}\cdot\frac{D}{2}\right)$

The 1995 Hyogo-ken Nanbu Earthquake also caused extensive seismic damage in underground tunnels. Based on the results of this investigation, a method for analyzing the seismic behavior of tunnels is presented here.

During the 1995 Hyogoken-Nanbu Earthquake, segmented shield tunnels were cracked at the concrete lining wall but were not damaged at the joints in the axial direction. However, based on the current design method that was established on the basis of allowable stress criterion, all the joints of the shield tunnel in the axial direction cannot be designed without expansion joints for severe ground motion, which is given as Level 2 ground motion established in Japan since 1995. This great discrepancy between the non-damage observations of actual joints and the design requirement of expansion joints for stress release results from the stress-based design, although the underground structure should be designed and thus based on the structural strains against externally forced ground displacements. In order to realistically assess the seismic risk of a shield tunnel, the accurate estimate of the structural strains, which depends upon the structural details, segment materials, properties of the surrounding soil, the nature of the propagating wave, and so on, is critical. Emphasis in this study, therefore, has been placed on the analysis of structural strains for segmented shield tunnel under severe earthquake ground motions.

When the earthquake intensity is severe and the free field strain reaches the order of magnitude of 10^{-3} to 10^{-2}, the chance of a slip taking place between the tunnel and the surrounding soil significantly increases, causing large strain concentrations at various joints and connections in the shield tunnel system. The simplified mechanism of slip assumed in this study is that the slip initiates when the shear stress τ exceeds the critical value of where c is the cohesion stress, σ is the normal stress at the interface between the tunnel and the soil, and φ = angle of friction between the tunnel and the soil. Therefore, the critical shear stress increases as the depth of soil cover increases. The possible upper limit of normal stress is controlled by the shadow zone in Figure 3.1.28, the normal soil stress of which is effectively acting to the tunnel surface as shown in Figure 3.1.29. The preceding discussion suggests that the critical shear stress behaves not linearly but at least bilinearly, as expressed in Figure 3.1.6. If the ground condition installing the shield tunnel can expect a

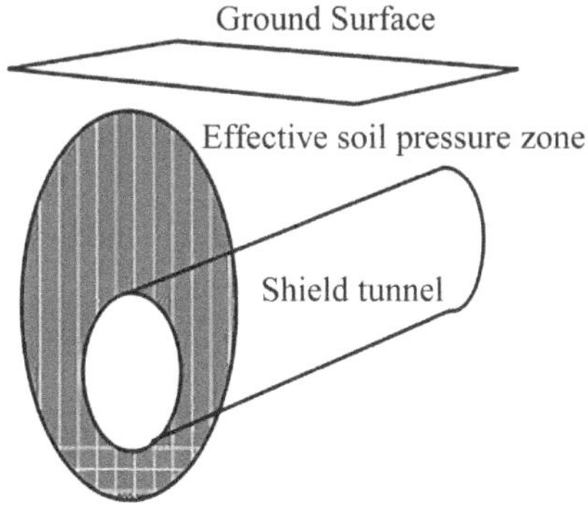

FIGURE 3.1.28 Vertical soil pressure acting on the tunnel.

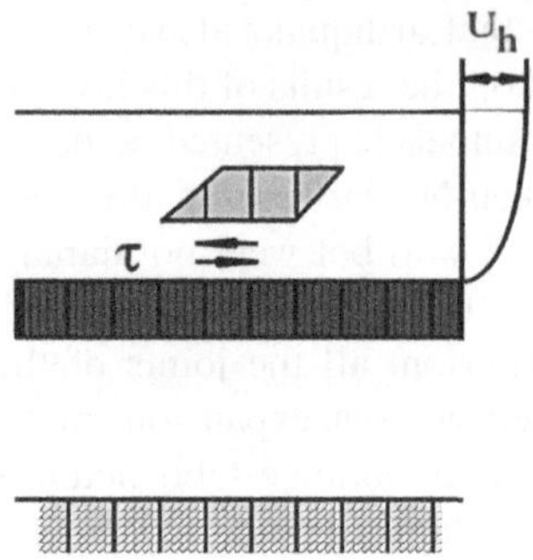

FIGURE 3.1.29 Shear stress acting on the tunnel surface caused by free field motion.

TABLE 3.1.3
Axial and Bending Strains of Shield Tunnel Considering the Slippage Effect

Description	Mode	Formulation
Axial strain	Tension	$\varepsilon^{T}_{SAQ} = q^{T}\varepsilon^{T}_{SA}$
	Compression	$\varepsilon^{C}_{SAQ} = q^{C}\varepsilon^{C}_{SA}$
Relative displacement between the free field and a pipe	Tension	$\Delta^{T} = \left(1 - q^{*T}\alpha^{T}_{A}\beta_{T}\right)u_{A}$
	Compression	$\Delta^{C} = \left(1 - q^{*C}\alpha^{C}_{A}\beta_{C}\right)u_{A}$

sufficient lateral force, the height of the effective soil pressure zone in Figure 3.1.28 is less than several times of the tunnel diameter, so that the possible height of soil cover to the tunnel is approximately 5–10 meters. According to the JGA, the critical shear stress is recommended to use the value of 0.01N/mm^2 for the gas pipeline, which is often installed under the soil cover of 1.2–1.5 m. Based on these data, the critical shear stress of the shield tunnel installed in several 10 m depth should increase up to 0.05–0.1 N/mm^2 in proportional to the effective soil depth of 5–10 m. The value of 0.05–0.1 N/mm^2 is assumed for numerical purposes as the critical shear stress in this study.

The formulations in Table 3.1.3 include the following components:

ε^{Z}_{SAQ}: axial strain in the mode Z in slippage, $Z = T$ in tension, and $Z = C$ in compression

Δ^{Z}: relative displacement between the free field and a pipe in the mode Z

q^{Z}: slippage factor for axial strain in the mode Z

q^{*Z}: slippage factor for relative displacement in the mode Z.

$$
\begin{aligned}
q^{Z} &= 1 - \cos\xi^{Z} + \frac{\pi}{2}\left(1 - \frac{2\xi^{Z}}{\pi}\right)\sin\xi^{Z},\ \ \tau^{Z}_{G} > \tau_{cr};\ \ q^{Z} = 1,\ \tau^{Z}_{G} \le \tau_{cr} \\
q^{*Z} &= \sin\xi^{Z}\left(1 + \frac{\pi^{2}}{8} - \frac{\xi^{2}}{2}\right) - \xi^{Z}\cos\xi^{Z},\ \ \tau^{Z}_{G} > \tau_{cr};\ \ q^{Z} = 1,\ \tau^{Z}_{G} \le \tau_{cr},
\end{aligned}
\tag{3.1.100}
$$

TABLE 3.1.4
Formula to Obtain Forces from Tunnel Strains

Forces	Formulation
Tension force	$N^T = (EA)_{eq}^T \varepsilon_{SA}^T$
Compression force	$N^C = (EA)_{eq}^C \varepsilon_{SA}^C$
Bending moment	$M = \left(\frac{2}{D}\right)(EI)_{eq} \varepsilon_{SB}$
Shear force	$Q = \frac{2\pi}{L_a}\frac{2}{D}(EI)_{eq} \varepsilon_{SB}$

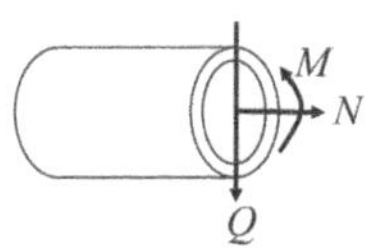

FIGURE 3.1.30 Various forces acting on a tunnel as a beam.

where

$$\xi^Z = arc\,sin\left(\frac{\tau_{cr}}{\tau_G^Z}\right) \tag{3.1.101}$$

in which

τ_G^Z: a shear stress in the mode Z
τ_{cr}: the critical shear stress

Once the tunnel strains are obtained, various forces acting on a tunnel segment are given in Table 3.1.4. These forces are formulated on the basis of the assumption that a tunnel behaves as a beam as shown in Figure 3.1.30.

3.1.5.3 Axial, Bending, and Shear Stresses at the Cross Section of the Tunnel Lining

The actual tunnel, on the other hand, is constructed with several segments for a single cross section, in which several forces such as axial, bending, and shear components are generated at the cross section of the tunnel lining as shown in Figure 3.1.31. According to Koike,[37,38] these forces can be formulated in the following way: when a lining displacement is given by w, v as shown in Figure 3.1.31 (1), various forces generated in the lining are formulated as

$$\begin{aligned} N(\varphi) &= -\frac{E\overline{A}_s}{R}\left[\kappa\frac{d^2w}{d\varphi^2} + (\kappa - 1)\frac{dv}{d\varphi} + w\right],\ Q(\varphi) = \frac{1}{R}\frac{dM}{d\varphi}, \\ M(\varphi) &= -E\,\overline{A}_s\,\kappa\left(\frac{d^2w}{d\varphi^2} + w\right), \end{aligned} \tag{3.1.102}$$

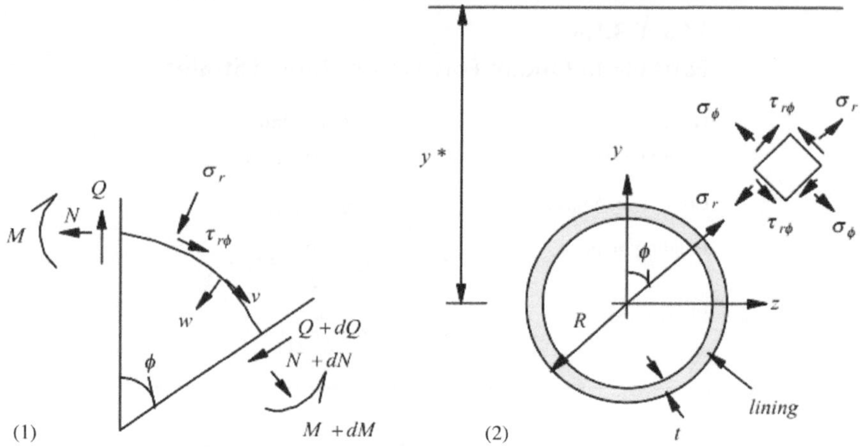

FIGURE 3.1.31 Various forces acting on the tunnel lining; (1) Forces acting on the tunnel lining, and (2) Tunnel notation.

in which

$$\bar{A}_S = \frac{A}{1-\nu^2}, \quad \bar{I}_S = \frac{I_S}{1-\nu_S^2}, \quad \kappa = \frac{\bar{I}_S}{R^2 \bar{A}_S}, \tag{3.1.103}$$

where A_s, I_s, *and* κ are the cross-sectional area, cross-sectional bending rigidity, and poison ratio for one tunnel segment.

The equilibrium equation between cross-sectional forces, N *and* Q, generated in tunnel lining and external forces induced by ground stresses, σ_r, σ_ϕ, and $\tau_{r\phi}$, are related in the following way:

$$\begin{aligned} N &= -\frac{dQ}{d\phi} - R\sigma_r, \\ &at\ r = R \\ Q &= \frac{dN}{d\phi} + R\tau_{r\phi}. \end{aligned} \tag{3.1.104}$$

It should be noted that the displacement, $w(\phi)$, in the tunnel lining is given as the difference between the free field ground displacement, $w_g(\phi)$, and the ground response near the tunnel surface, $w_i(\phi)$, generated by self-excitation as shown in the following way, and also other variables, v, σ_r, and $\tau_{r\phi}$, are given in the same way:

$$\begin{aligned} w(\phi) &= w_g(\phi) - w_i(\phi) \\ v(\phi) &= v_g(\phi) - v_i(\phi) \\ \sigma_r(\phi) &= \sigma_g(\phi) - \sigma_i(\phi) \\ \tau_{r\phi}(\phi) &= \tau_g(\phi) - \tau_i(\phi). \end{aligned} \tag{3.1.105}$$

The ground displacement $w_i(\phi)$ and $v_i(\phi)$ generated by self-excitation is defined to have the following characteristics:

1. The displacement generated by self-excitation vanishes at an infinite distance.
2. The displacement generated by self-excitation is equal to the displacement of the tunnel lining at the tunnel surface.

The analytical solution of the ground displacement by self-excitation is described in detail in reference 38.

And these variables, w, v, σ_r, $\tau_{r\phi}$, can be formulated in the following series system:

$$\begin{aligned} w(\phi) &= \sum_{n=0}^{\infty} w_n sin(n\phi), v(\phi), v(\phi) = \sum_{n=0}^{\infty} v_n cos(n\phi), \\ \sigma_r(\phi) &= \sum_{n=0}^{\infty} \sigma_n sin(n\phi), \tau_{r\phi}(\phi) = \sum_{n=0}^{\infty} \tau_{r\phi n} cos(n\phi). \end{aligned} \tag{3.1.106}$$

The sinusoidal components of each variable are given in reference 38. Finally, tunnel strains, hoop, bending, and shear modes are expressed in the following way:

For hoop strain,

$$\begin{aligned} \varepsilon_{hoop}(\phi) &= \frac{N(\phi)}{E\bar{A}_s} = -\frac{1}{R}\left[\kappa \frac{d^2 w}{d\phi^2} + (\kappa - 1)\frac{dv}{d\phi} + w\right] \\ &= -\frac{1}{R}\sum_{n=1}^{\infty}\left[\left(1 - \kappa n^2\right) w_n + n(1-\kappa) v_n\right] sin(n\phi). \end{aligned} \tag{3.1.107}$$

For bending strain,

$$\begin{aligned} \varepsilon_{bend}(\phi) &= \frac{MR}{E\bar{I}_s} = -\frac{1}{R}\left(\frac{d^2 w}{d\phi^2} + w\right) \\ &= -\frac{1}{R}\sum_{n=1}^{\infty}\left(1 - n^2\right) w_n sin(n\phi). \end{aligned} \tag{3.1.108}$$

For shear strain,

$$\begin{aligned} \varepsilon_{shear}(\phi) &= \frac{Q}{\mu \bar{A}_s} = \frac{E\kappa}{\mu}\frac{d\varepsilon_{bend}}{d\phi} \\ &= -\frac{E\kappa}{\mu R}\sum_{n=1}^{\infty} n\left(1 - n^2\right) w_n cos(n\phi). \end{aligned} \tag{3.1.109}$$

3.2 OVERVIEW OF SEISMIC DESIGN METHODS FOR PIPELINES AGAINST DESTRUCTIVE GROUND DEFORMATION

Destructive ground deformation is a condition in which the ground collapses due to an earthquake. Ground failure may take the form of ground liquefaction, or a portion of the sloping ground may slide along a slip line. In other cases, fault displacement that generates the earthquake itself may appear above-ground and cause extensive damage to structures and pipelines that cross the fault. The succeeding sections describe in detail the methods for analyzing pipeline behavior and design calculations for ground liquefaction, fault displacement, and landslides, which are typical destructive ground deformations.

Table 3.2.1 lists a summary of existing design methods for destructive ground deformation in Japan and overseas. The table lists in parallel the seismic design guidelines for pipelines of the JWWA and the JGA and the seismic design guidelines for pipelines of the American Lifeline Alliance (ALA, USA). The guidelines provide policies for safety verification against liquefaction displacement, fault displacement, landslide, and unequal settlement.

The following sections describe the specifics of these design guidelines.

For pipelines affected by the three types of destructive ground deformation, the design calculation method for straight pipes is shown below. For deformed pipes, design calculation methods for soil liquefaction are described.

In particular, for pipelines crossing liquefied ground, design calculation methods are discussed for calculating pipe strains in the axial and perpendicular directions, depending on the direction of crossing liquefied ground, and in regions where

TABLE 3.2.1
Summary of Existing Design Methods for Destructive Ground Deformation

Hazard Mode		Japan Water Works Association JWWA	Japan Gas Association JGA	American Lifeline Alliance ALA
Liquefaction damages	Risk assessment	Based on the code	Based on the code	Based on HAZUS
	Estimation of liquefied displacement	Proportional to the thickness of liquefied layers	5% of liquefied layers	Predict using regression formula
	Assessment of straight pipes	Check for the axial displacement	Assessment based on formula check	None
	Assessment of bent pipes	None		
Fault crossing	Straight pipes	None	None	Assessment based on formula check
Other ground deformations (land slide, settlement)	Straight pipes	None	None	FEM analysis or frame analysis

Note: HAZUS[17]: A book published by the Federal Emergency Management Agency (FEMA).

destructive ground deformation is large, design calculation methods are limited to the bending angle at the plastic hinge point generated in the pipe. In addition, for deformed pipes, especially for curved pipes, calculation methods for bending deformation angles at the corners will be discussed.

3.2.1 Seismic Design Method for Pipelines Crossing Liquefied Areas[18]

The two photographs shown in Figure 3.2.1 illustrate the changes in the ground surface before and after ground liquefaction during the 2018 Hokkaido Iburi East Japan Earthquake.[35]

Figure 3.2.2 shows a model of the deformation of a curved pipe buried in liquefied ground before and after ground liquefaction.

(Before) (After)

FIGURE 3.2.1 Example of destructive ground deformation due to liquefaction during the 2018 Hokkaido IBURI East Japan Earthquake.[36]

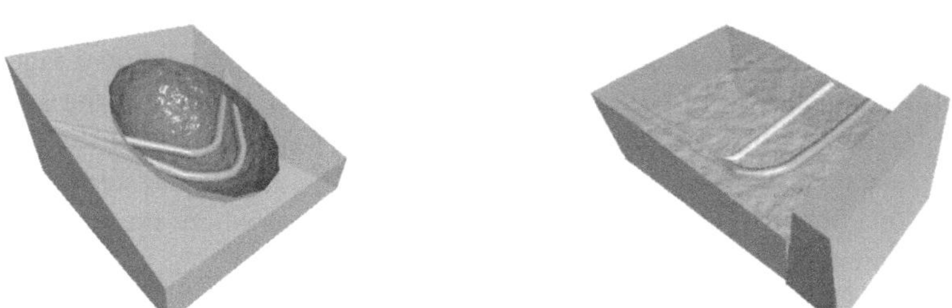

(a) (b)

FIGURE 3.2.2 Typical liquefied ground deformation model: (a) slip deformation in sloping ground; (b) destructive ground deformation due to seawall movement near the seawall.

The ground susceptible to liquefaction can be summarized as follows:

1. Ground that satisfies the liquefaction conditions (i.e., strata with a negative *FL* value and a *PL* value of 5 or more)
2. Sloping ground
3. Near seawalls (within 100 m separation distance)
4. Amount of ground subsidence due to liquefaction: approximately 5% of liquefied layer thickness

3.2.1.1 Estimation of Liquefaction Displacement

Several methods have been proposed for predicting liquefaction displacement that will occur in these target areas.

a. Hamada, et al. (1986) method[19]
   This is a prediction method based on the analysis of measured ground displacement data in liquefaction-prone areas. The horizontal displacement *D* of the sloping ground is estimated by the following equation:

$$D = 0.75\sqrt{H} \cdot \sqrt[3]{\theta}, \tag{3.2.1}$$

   where *H* is the liquefied layer thickness, and θ is the maximum gradient (%) of the subsurface gradients of the liquefied layer.
b. Bartlett and Youd (1992) Method[20]
   This method predicts liquefied ground surface displacement and proposes the following equation:

$$\begin{aligned}\log(D+0.01) = b_0 + b_{off} + b_1 M + b_2 \log(R) + b_3 R + b_4 log(W)\\ + b_5 \log(S) + b_6 \log(T_{15}) + b_7 \log(100 - F_{15}) + b_8 D50_{15},\end{aligned} \tag{3.2.2}$$

   where *D* is the horizontal displacement of the ground surface (m), *M* is the moment magnitude of the earthquake, *R* is the epicentral distance (km), *S* is the ground surface slope (%), T_{15} is the total thickness of the saturated sand layer with a modified N value $(N_1)_{60}$ less than 15 (m), F_{15} is the average fine grain content (%) of the sand layer contained in T_{15}, $D50_{15}$ is the average diameter of soil in the sand layer in T_{15}, and *W* is the slope of the ground surface (%), which is estimated by the ratio of the horizontal distance from the revetment to the revetment height at the point where the ground displacement is estimated:

$$\begin{aligned}b_0 = -15.787, b_{off} = 0,\ b_1 = 1.178,\ b_2 = -0.927,\ b_3 = -0.013,\\ b_4 = 0,\ b_5 = 0.429,\ b_6 = 0.348,\ b_7 = 4.527,\ b_8 = -0.922.\end{aligned} \tag{3.2.3}$$

c. Method[21] of Jean-Pierre Bardet et al. (2002)
Based on the work of Bartlett and Youd, the following equation is proposed for ground surface displacement on sloping ground with revetments in ground liquefaction:

$$log(D+0.01) = b_0 + b_{off} + b_1 M + b_2\, log(R) + b_3 R + b_4\, log(W) + b_5\, log(S) + b_6\, log(T_{15}), \tag{3.2.4}$$

where

$$b_0 = -15.787, b_{off} = -0.579,\ b_1 = 1.178,\ b_2 = -0.927,\ b_3 = -0.013,\ b_4 = 0.657,\ b_5 = 0.429,\ b_6 = 0.348,\ b_7 = 0,\ b_8 = 0. \tag{3.2.5}$$

3.2.1.2 Formulation of a Simplified Analytical Method

This section describes a simplified analytical method for estimating the deformation behavior of buried pipes when the ground around the pipes becomes liquefied. The content of this section follows the method[22] of Suzuki et al.

3.2.1.1.1 Plastic Hinge Analysis Method as a Simplified Analysis Method

a. Plastic Hinge Formation
Steel pipes used for pipelines are made of ductile materials and can withstand large plastic deformation. Under the extreme condition of sufficiently advanced yielding, the influence of elastic deformation is relatively small, and plastic deformation is considered to be dominant. In the limit analysis of steel structures, the loading conditions leading to collapse are investigated by modeling the structure as a rigid-plastic body, ignoring elastic deformation. In other words, by solving the problems of the location of the plastic hinge and the collapse mechanism, it is said that structural safety can be examined. Here, a rigid-plastic model is applied to the lateral flow analysis of buried steel pipes to induce deformation equations. In plastic hinge analysis, an important issue is how to set the moment at the hinge. In particular, the curved pipe section is a section where the stiffness changes rapidly, so an appropriate estimation equation is required.
The results of the plastic hinge analysis were compared with the FEM analysis (shell model), which can well explain the experimental results of actual damage and large deformation, and a good agreement was confirmed. The plastic hinge analysis method presented in this section captures the basic deformation mechanism of buried steel pipes subjected to lateral flow due to liquefaction, and is considered a useful method for design practice of buried steel pipes in liquefied soils.
b. Analytical Model of a Pipeline in Liquefied Ground
In the seismic design of underground pipes, the surrounding ground is generally modeled with springs, and the spring coefficients are determined

based on the results of burial experiments. However, when the ground deforms greatly, as in the case of liquefaction, the reaction force from the ground is not proportional to the relative displacement between the pipe and the ground, and Shimamura et al.[34] have confirmed that the coefficient of ground reaction force is constant when the relative displacement between the pipe and the ground exceeds a certain value. Based on these experimental results, buried pipes subjected to large deformation of the ground are often represented by a beam model supported by a nonlinear spring, and this section uses an analytical method assuming a fully elastoplastic type of ground spring.

Figure 3.2.3 shows an analytical image of a buried steel pipe subjected to liquefaction lateral flow.

As shown in Figure 3.2.3(a), when the pipe passes through the liquefied ground in a straight line, the relative displacement between the pipe and the ground becomes large at the boundary with the non-liquefied ground, and the pipe is considered bent due to the equally distributed load and fulcrum displacement. The relative displacement between the pipe and the ground becomes small at some distance from the boundary, and the buried pipe follows the ground and moves. On the other hand, when there is a curved pipe section as shown in Figure 3.2.3(b), the pipe movement is restrained by the axial force and can be regarded as an apparent fixed point. The forces acting on the right side of the figure are shown in the equilibrium relationship between the fulcrum displacement, the equally distributed load, and the axial force of the bent pipe.

Assuming that the buried steel pipe is a circular ring with an outer diameter D and pipe thickness t, and that the yield stress σ_y is a perfectly elastoplastic

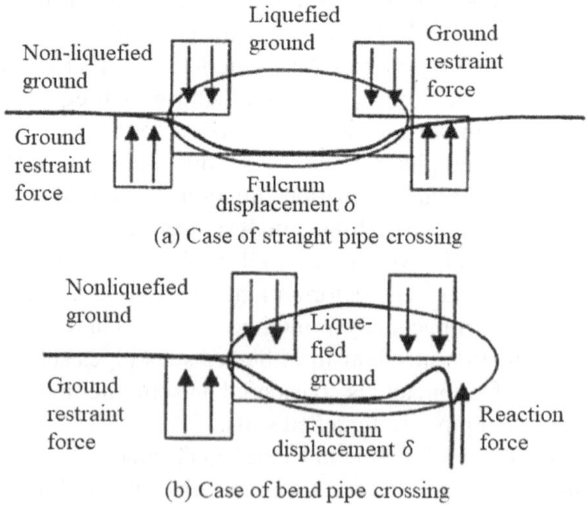

FIGURE 3.2.3 Analysis image of liquefied ground.[22]

material that becomes a constant stress after yielding, the yield moment is expressed as

$$M_y = \frac{\pi\left\{D^4 - (D-2t)^4\right\}}{32D}\sigma_y. \tag{3.2.6}$$

On the other hand, the plastic moment at which the entire cross section is plasticized is given by the plastic moment at which the entire cross section becomes plastic is given by

$$M_p = \frac{4}{\pi}M_y. \tag{3.2.7}$$

3.2.1.1.2 Analysis Method for Deformation Behavior of Straight Pipe in Liquefied Ground

To apply the simplified analysis method, the model to be analyzed is shown in Figure 3.2.4.

The case in which bending strain occurs in the pipe due to liquefaction displacement is considered.

As shown in Figure 3.2.4(a), if the pipe axis is aligned parallel to the flow direction of the liquefied ground, the pipe strain can be calculated by the following equation:

$$\varepsilon_S = \frac{\tau_{cr}}{Et}W_h, \tag{3.2.8}$$

where

ε_S: axial strain in liquefied ground
τ_{cr}: slip initiation limit shear strain
W_h: liquefaction section parallel to the pipe axis
E: modulus of elasticity of pipe material
t: pipe thickness

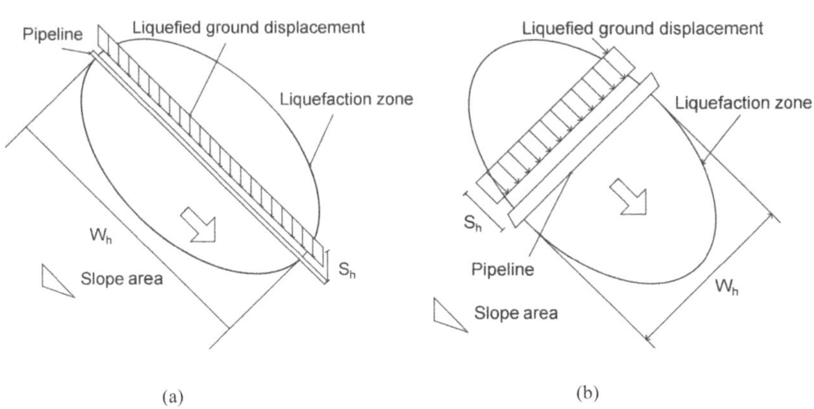

FIGURE 3.2.4 Analytical model of a straight pipeline crossing a liquefied slope:[18] (a) pipe model in the pipe axial direction; (b) pipe model due to loads perpendicular to the pipe axis.

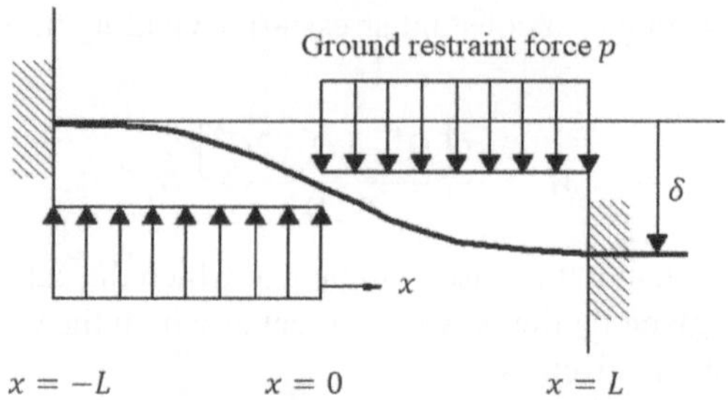

FIGURE 3.2.5 Analytical model of straight pipe subjected to bending.[22]

On the other hand, as shown in Figure 3.2.4(b), in a configuration where the pipe axis is perpendicular to the flow direction of the liquefied ground, the bending stress generated in the pipe near the boundary between the liquefied and the non-liquefied ground becomes large, and if the liquefaction displacement is large, the pipe near the boundary will become totally plastic and form a plastic hinge due to the plasticization progress. In the case of large liquefaction displacement, plasticization of the pipe near the boundary will develop, resulting in a total plastic state and the formation of a plastic hinge. Assuming such a situation, a simplified analysis method[22,23] is developed as follows:

Assume that a ground constraint force p is applied and a ground liquefaction (flow) displacement δ is generated. Figure 3.2.5 shows an analytical model of a straight pipe subjected to bending, and Figure 3.2.6 shows an image of the progressive deformation of the straight pipe section.

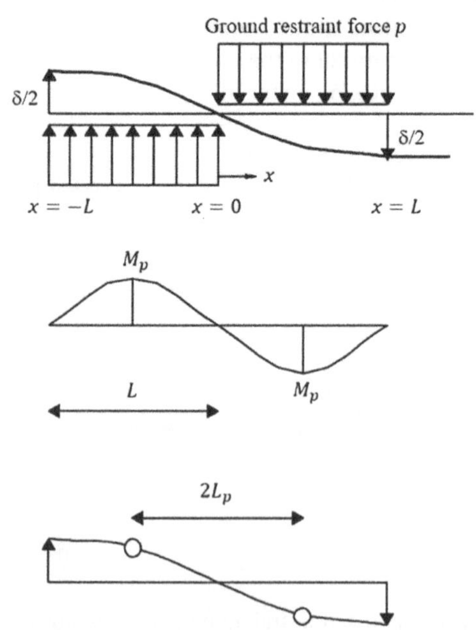

FIGURE 3.2.6 Image of progressive deformation of straight pipe section.[22]

As the ground displacement δ increases, the maximum moment occurs at the midpoint of the interval L (not yet determined), and a plastic hinge is generated when the maximum moment reaches the total plastic moment M_p. A plastic hinge analysis that follows this process is described in this section.

Using rotational symmetry at $x = 0$, the following differential equation is solved in the $(0,\ L)$ section:

$$EI\frac{d^4w}{dx^4} = p \quad 0 \le x \le L, \tag{3.2.9}$$

and the boundary conditions are

$$\begin{aligned} x &= 0, \quad w(0) - \frac{\delta}{2}, \quad \frac{d^2w}{dx^2}(0) = 0, \\ x &= L, \quad w(L) = \delta, \quad \frac{d^2w}{dx^2}(L) = 0. \end{aligned} \tag{3.2.10}$$

Solving this yields the following results for displacements and bending moments:

$$w = \frac{pL^4}{48EI}\left\{2\left(\frac{x}{L}\right)^4 - 3\left(\frac{x}{L}\right)^3 + \left(\frac{x}{L}\right)\right\} + \frac{\delta}{4}\left\{-\left(\frac{x}{L}\right)^3 + 3\left(\frac{x}{L}\right) + 2\right\}, \tag{3.2.11}$$

$$M = \frac{pL^2}{48}\left\{-4\left(\frac{x}{L}\right)^2 + 3\left(\frac{x}{L}\right)\right\} + \frac{3EI\delta}{2L^2}\left(\frac{x}{L}\right). \tag{3.2.12}$$

The length of section L, on which the distributed force due to the yielding spring acts, is determined:

$$L = \sqrt[4]{\frac{12EI\delta}{p}}. \tag{3.2.13}$$

The maximum bending moment occurs at $Q(x) = M'(x) = 0$ with respect to shear force Q and bending moment M, yielding the following relationship:

$$\frac{x}{L} = \frac{3}{8} + \frac{3EI\delta}{2L^4} = \frac{1}{2}. \tag{3.2.14}$$

The maximum bending moment M_{max} is given by

$$M_{max} = \frac{\sqrt{3}}{4}\sqrt{EIp\delta}. \tag{3.2.15}$$

If the displacement of the fulcrum when the bending moment reaches the plastic hinge is δ_p and the length of the beam is, L_p, the following formulas for the various quantities are obtained.

The deformation δ_p at the onset of the total plastic moment is

$$\delta_p = \frac{16}{3}\left(\frac{M_p^2}{EIp}\right). \tag{3.2.16}$$

The length of the beam up to the plastic hinge point, L_p, is given by

$$L_p = 2\sqrt{2}\sqrt{\frac{M_p}{p}}. \tag{3.2.17}$$

Note that the deflection angle of a straight pipe can be defined as

$$\theta = \frac{M_{max}}{EI}\ell, \tag{3.2.18}$$

where ℓ is the distance for calculating the deflection angle, and $\ell = 20D$ is assumed here.

Therefore, the maximum bending angle at the plastic hinge point can be estimated by the following equation:

$$\theta = \frac{M_{max}}{EI}\ell = \frac{\sqrt{3}}{4}\frac{\sqrt{EIp\delta}}{EI}\cdot 20D = 8.66D\sqrt{\frac{p\delta}{EI}}.$$

The *Seismic Design Guidelines for Liquefaction Resistance of High-Pressure Gas Pipes*[24] conducted full-scale experiments on the limit displacement and bending angle of straight pipes and obtained the following load-displacement curve for straight pipes (Figure 3.2.7).

Based on the experimental results of axial compression buckling and bending buckling of straight pipes, the limit displacement is defined as the sum of the buckling crest formation and the deformation at the final collapse stage during the deformation process that leads to straight pipe failure (crack penetration), and the displacement limit equation is derived as follows:

Critical axial compressive displacement

$$\lambda_{sc} = \frac{44t_s}{100D}\cdot L_s + \frac{3.44}{\sqrt{2}}\cdot t_s\cdot\sqrt{\frac{D}{t_s}}\cdot\left(1+\frac{\varepsilon_f}{2}\right). \tag{3.2.19}$$

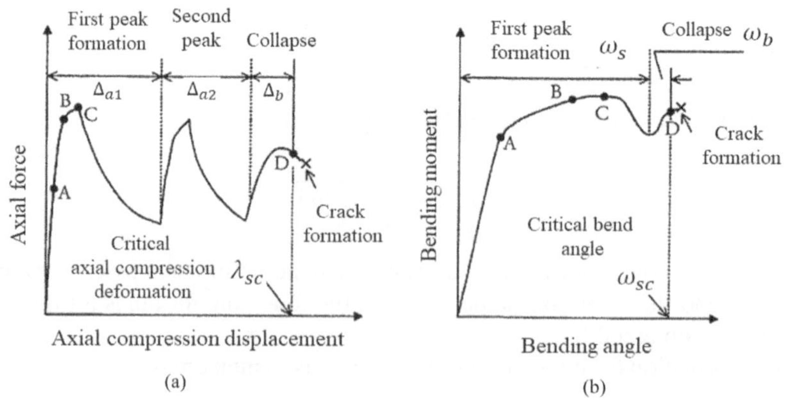

FIGURE 3.2.7 Load-displacement curve of straight pipe:[18] (a) axial compression mode; (b) bending mode.

Critical bending angle

$$\omega_{sc} = \left\{ \frac{44t_s}{100D}\left(8k - \frac{2k^2}{3}\right) + \frac{3.44}{\sqrt{2}\sqrt{D/t_s}}\left(1+\frac{\varepsilon_f}{2}\right)\right\} \cdot \frac{180}{\pi} \quad (3.2.20)$$

where

λ_{sc}: limiting axial compressive displacement of straight pipe (cm)
ω_{sc}: limiting bending angle of straight pipe (degree)
D: outer diameter of straight pipe (cm)
t_s: nominal pipe thickness of straight pipe (cm)
L_s: reference length for determining the axial compressive displacement and bending angle of the straight pipe (cm)

$$L_s = 6.4D$$

ε_f: critical strain at break, 0.35
k: ratio of $L_s/2$ to outside diameter D (=3.2)

3.2.1.1.3 *Analysis Method for Deformation Behavior of Deformed Pipes in Liquefied Soil*

a. Analysis Model

The case shown in Figure 3.2.3(b), in which there is a curved pipe in liquefied ground, is analyzed as follows. As shown on the right side of the figure, deformation is constrained by the effect of piping in different directions, and the curved pipe section is considered to behave like an apparent fixed point. The analytical model shown in Figure 3.2.8 assumes that the pipe follows the ground flow displacement at some distance from the curved pipe but does not move due to resistance in the vicinity of the curved pipe. Point B, the fixed end, represents the curved pipe section. Point A is also a fixed condition, but this is an approximate result of the condition supported by the ground spring. As in Section 3.2.1.1.2, the model is subject to fulcrum displacement and distributed load, but the beam length is an unknown number determined by the balance of forces as in the previous section.

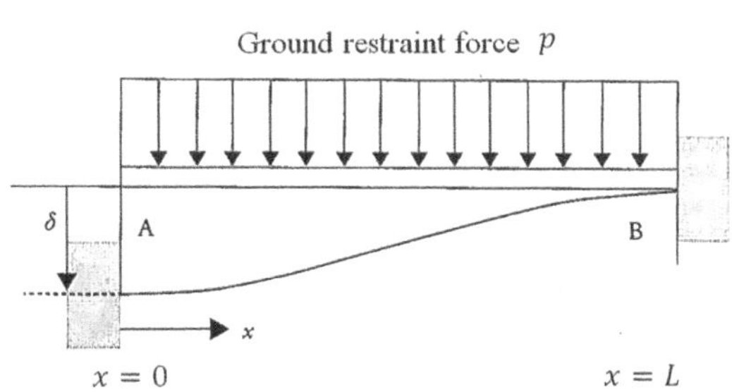

FIGURE 3.2.8 Analytical model of piping including a curved pipe section.[22]

The differential equation is given by

$$EI\frac{d^4w}{dx^4} = p \ \ (0 \le x \le L). \tag{3.2.21}$$

b. Formation of the First Hinge

The boundary conditions up to the formation of the first hinge are fixed at both ends, and the results of solving the differential equations are shown in the upper part of Table 3.2.2 for displacement and bending moment. As in the case of a straight shear, $x = 0$ (point A) is the boundary leading to the spring-supported section, and since the bending moment is zero, the beam length can be determined. The beam length L shown in Figure 3.2.9 can be calculated as

$$L = \sqrt[4]{\frac{72EI\delta}{p}}. \tag{3.2.22}$$

and the relationship increases with an increase in δ. In this case, the extreme values of the bending moment occur at $x = L/3$ and $x = L$, and the values are

$$M\left(\frac{L}{3}\right) = \frac{\sqrt{2}}{3}\sqrt{EIp\delta}, \quad M(L) = -\sqrt{2}\sqrt{EIp\delta}. \tag{3.2.23}$$

TABLE 3.2.2
Deflection and Bending Moment of Piping Including Curved Pipe[22]

Ground Displacement	Boundary Conditions
	$w(0)=\delta, \ \frac{dw}{dx}(0)=0 \qquad w(L)=0, \ \frac{dw}{dx}(L)=0$
	Analytical Results
$0 \le \delta \le \delta_1$	$w = \frac{pL^4}{24EI}\left\{\left(\frac{x}{L}\right)^4 - 2\left(\frac{x}{L}\right)^3 + \left(\frac{x}{L}\right)^2\right\} + \delta\left\{2\left(\frac{x}{L}\right)^3 - 3\left(\frac{x}{L}\right)^2 + 1\right\}$
	$M = \frac{pL^2}{12}\left\{-6\left(\frac{x}{L}\right)^2 + 6\left(\frac{x}{L}\right) - 1\right\} + \frac{6EI\delta}{L^2}\left\{-2\left(\frac{x}{L}\right) + 1\right\}$
	$w(0)=\delta, \ \frac{dw}{dx}(0)=0 \qquad w(L)=0, \ \frac{d^2w}{dx^2}(L)=0$
	Analytical Results
$\delta_1 \le \delta \le \delta_2$	$w = \frac{pL^4}{48EI}\left\{2\left(\frac{x}{L}\right)^4 - 5\left(\frac{x}{L}\right)^3 + 3\left(\frac{x}{L}\right)^2\right\} + \frac{\delta}{2}\left\{-\left(\frac{x}{L}\right)^3 + 3\left(\frac{x}{L}\right)^2\right\}$
	$M = \frac{pL^2}{16}\left\{-8\left(\frac{x}{L}\right)^2 + 5\left(\frac{x}{L}\right) - 2\right\} + \frac{3EI\delta}{L^2}\left\{-\left(\frac{x}{L}\right) + 1\right\} + M_{pb}\left\{-\frac{3}{2}\left(\frac{x}{L}\right) + \frac{1}{2}\right\}$

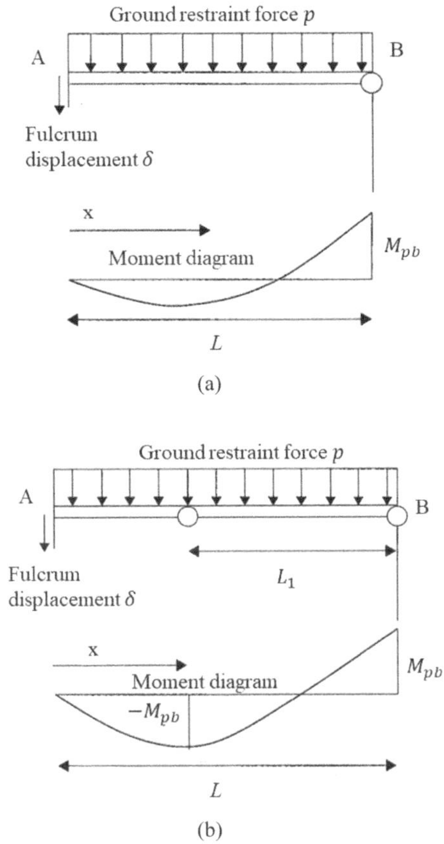

FIGURE 3.2.9 Image of progressive deformation of curved pipe section:[22] (a) analysis model with curved pipe section as hinge and bending moment; (b) curved tube section hinged.

and the curved pipe section ($x = L$) is always larger. Therefore, if the plastic moment of the curved pipe section is M_{pb} and the ground displacement δ_1 in Figure 3.2.10 is given by

$$\delta_1 = \frac{1}{2}\left(\frac{M_{pb}^{\ 2}}{EIp}\right), \tag{3.2.24}$$

a plastic hinge is formed in the curved pipe section, and the length of the beam is

$$L = \sqrt{6}\sqrt{\frac{M_{pb}}{p}}, \tag{3.2.25}$$

So, the length is determined from the plastic moment and the distributed ground. For more than a ground displacement of δ_1, the curved pipe section becomes a hinge and its deformation proceeds.

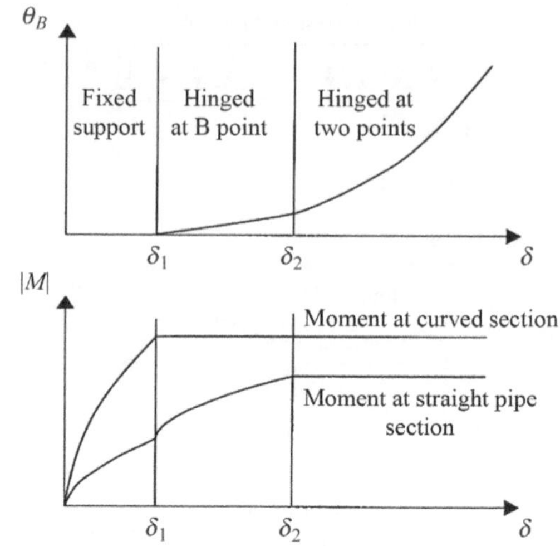

FIGURE 3.2.10 Relationship between ground displacement, bending angle, and bending moment.[22]

c. Formation of the Second Hinge

If the fulcrum displacement continues to increase after the hinge is formed in the curved pipe section, the analytical model becomes a model in which external forces due to distributed forces and fulcrum displacement are applied. The boundary conditions up to the formation of the second hinge and the solutions of the differential equations are shown in the lower part of Table 3.2.2. In the bending moment calculation results, the plastic moment $-M_{pb}$ is applied to the plastic hinged $x = L$. $x = 0$ is the boundary leading to the spring-supported section, and the equation of equilibrium for the bending moment is as follows:

$$\frac{1}{8}pL^2 = \frac{3EI\delta}{L^2} + \frac{1}{2}M_{pb}. \tag{3.2.26}$$

The following equation is obtained by solving the equation for the length of the beam, L:

$$L = \sqrt{2}\sqrt{\frac{M_{pb}}{p} + \sqrt{\left(\frac{M_{pb}}{p}\right)^2 + \frac{6EI\delta}{p}}}. \tag{3.2.27}$$

Given the ground displacement, the beam length L can be calculated, and the distribution of bending moments can be computed.

The ground displacement is increased, and the condition that the maximum moment of the straight pipe section reaches M_{ps} is calculated. Then the ground displacement is obtained as

$$\delta_2 = \frac{-M_{pb}^2 + 4M_{ps}\left(\sqrt{M_{pb} + M_{ps}} + \sqrt{M_{ps}}\right)^2}{6EIp}. \tag{3.2.28}$$

The length of the beam for δ_2 is given by the following equation:

$$L = \sqrt{2}\,\frac{\sqrt{M_{pb} + M_{ps}} + \sqrt{M_{ps}}}{\sqrt{p}}. \tag{3.2.29}$$

The distance between the plastic hinge of the curved pipe section and the hinge of the straight pipe section is calculated as

$$L_1 = \sqrt{2}\,\frac{\sqrt{M_{pb} + M_{ps}}}{\sqrt{p}}. \tag{3.2.30}$$

This means that the plastic moment acting on the hinge point is balanced by the rotational force due to the distributed load, and the rotation of the two hinges follows the ground displacement for a forced displacement of δ_2 or greater. The preceding analytical model and the progress of deformation are shown in Figure 3.2.9.

d. Ground Displacement and Bending Angle of Curved Pipe Section
Based on the preceding calculations, the maximum bending angle of a beam subjected to forced displacement and flow force due to ground flow can be calculated. Figure 3.2.10 shows the progress of deformation and the change in bending moment. The rotation angle θ_B of the curved pipe section begins to increase gradually from the amount of ground displacement δ_1, where the hinge is formed in the curved pipe section, and the deformation increases rapidly at ground displacements above δ_2, where the straight pipe section also forms a hinge. The bending moment increases first in the curved pipe section, and after the curved pipe section becomes a plastic hinge, the moment in the straight pipe section increases rapidly. The calculated δ_1 and δ_2 for a typical high-pressure gas pipe (pipe diameter 100A–750A, D/t=15–60, 45°or 90° bend, radius of curvature *1.5D* or *3D*) are less than several tens of centimeters, which is much smaller than the ground displacement caused by liquefaction. In other words, when the ground displacement is somewhat large and $\delta > \delta_2$, the maximum bending angle occurs at the hinge and can be calculated by the following equation:

$$\theta = sin^{-1}\left(\frac{\delta}{L_1}\right). \tag{3.2.31}$$

Therefore, the deformation analysis of buried steel pipes subjected to liquefied ground displacement can be organized and studied with a model that bends at two hinge points.

e. Calculation of the Bend Angle at the Top of the Curved Pipe Using the Three-Point Hinge Model
Figure 3.2.11 shows a deformed pipe placed in liquefied ground. Assuming that a plastic hinge is formed at the top of the curved pipe, the bending angle is analyzed as follows.

The two-point hinge model is applied to an actual piping system bent at the curved pipe section. When the ground displacement increases to

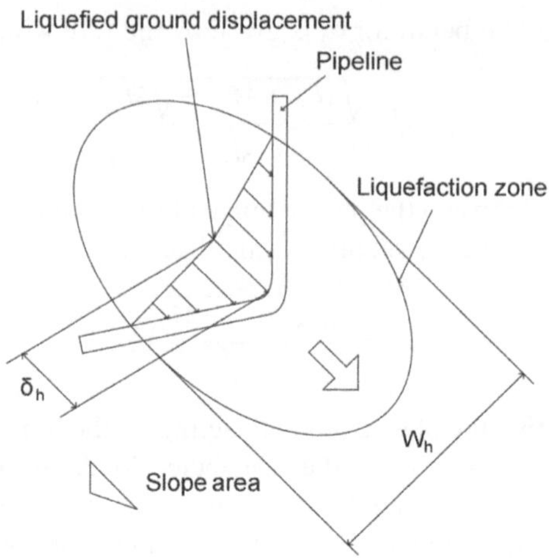

FIGURE 3.2.11 Analytical model of a curved pipe in a liquefied slope area.[18]

some extent, plastic hinges are formed at three locations in the curved pipe section and the straight pipe sections on both sides, and it can be approximated by the deformation of two rigid bars joined by a hinge.

The length of this rigid bar can be calculated from the plastic moments at both ends and the equally distributed load determined from the upper limit of the ground reaction force, σ_{cr}.

Figure 3.2.12 shows an analytical model of a curved pipe. The movement of a rigid rod is calculated under the condition that the ground displacement δ acts on a curved pipe with an opening angle of 2α at the curved pipe section and an inclination angle β from the center line. In this case, the bending angle θ_y at the center point C can be formulated as follows from its geometric relationship.

$$sin\left(\theta_y\right)=\frac{\overline{A'B'}}{2L_1}=\frac{1}{2L_1}\sqrt{\left(\delta_A-\delta_B\right)^2+sin^2\alpha+\left\{2L_1 sin\alpha-\left(\delta_A+\delta_B\right)cos\alpha\right\}^2}. \quad (3.2.32)$$

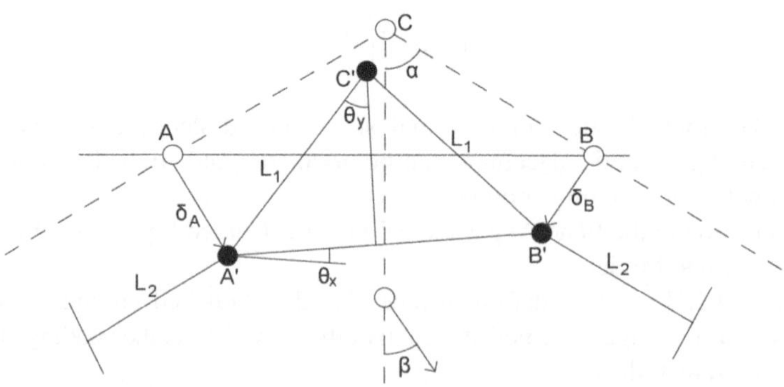

FIGURE 3.2.12 Deformation mechanism by three-point hinge.[18]

The displacements at points A and B are given by the following equations, respectively. S_{cr} is the elastoplastic limit stress of the soil around liquefaction. Furthermore, M_{maxs} and M_{maxb} are the maximum bending moments of straight and curved pipe sections, respectively.

$$\delta_A = \delta sin(\alpha+\beta),\ \ \delta_B = \delta sin(\alpha-\beta).$$
$$L_1 = \sqrt{\frac{2(M_{maxs}+M_{maxb})}{p}},\ \ p = DS_{cr}. \tag{3.2.33}$$

where p: ground reaction force, δ: ground displacement (m), L_1: distance between hinges (m), α: 1/2 of original curved pipe angle (radian), and β: direction of ground deformation (radian).

Inner bending mode: Outer bending mode:

$$\beta = -\alpha,\ \ \beta = \frac{3}{2}\pi - \alpha. \tag{3.2.34}$$

M_{maxs}: the maximum bending moment of pipe (kNm) $M_{maxs} = M_p$.

M_{maxb}: the maximum bending moment of bend (kNm) $M_{maxb} = M^i_{max}$ or M^o_{max}.

(3.2.35)

In this case, the amount of angle change at point C is obtained by the following equation:

$$\theta_C = |2\theta_y - 2\alpha|. \tag{3.2.36}$$

In the seismic design guidelines for liquefaction-resistant high-pressure gas pipelines, full-scale experiments of inner and outer bending of curved pipes were conducted to obtain the load-deformation curves shown in Figure 3.2.13.

Considering the consistency with the results of the full-scale experiments and the FEM numerical analysis, the limit bending angle calculation equation for curved pipes is derived by regression analysis within the range where the equation is valid, as follows:

The bending angle within the limit is

$$\omega_{bsc} = 0.9\frac{\sqrt{D/t_b}\cdot\sqrt{\varphi}}{R_c/D} + \omega_{sc}. \tag{3.2.37}$$

The out-of-limit bending angle is

$$\omega_{bsc} = 2.24\frac{\varphi}{\sqrt{D/t_b}\cdot(R_c/D)^{0.25}\cdot\eta},\ \ (\varphi \geq 22.5).$$
$$\omega_{bsc} = 2.24\frac{22.5}{\sqrt{D/t_b}\cdot(R_c/D)^{0.25}\cdot\eta},\ \ (\varphi < 22.5). \tag{3.2.38}$$

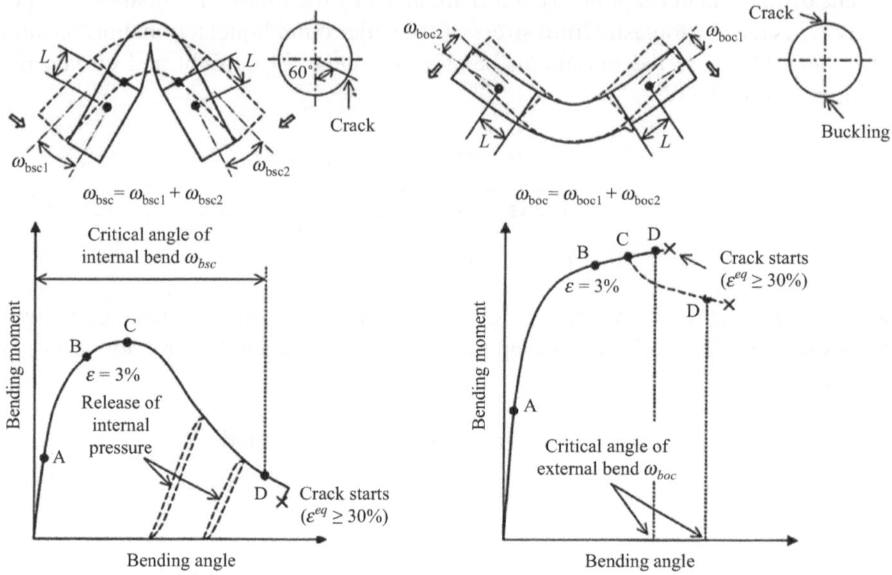

FIGURE 3.2.13 Load-deformation curves of full-scale experiments of bent pipes in inner and outer bending.[18]

where

ω_{bsc}: angle of bending within the limit of the curved pipe in degrees (71° is the maximum value)
ω_{sc}: the value obtained by setting $k = 1$ in (Equation 3.2.20)
ω_{boc}: limiting outer bending angle of the curved pipe (degrees) (71° is the maximum value)
D: outer diameter of the pipe (cm)
t_b: nominal pipe thickness of the curved pipe (cm)
φ: angle of the curved pipe (degrees) (this is the actual angle of the curved pipe)
R_c: radius of curvature
η: according to Table 3.2.3

TABLE 3.2.3
Values of η

Pipe Material (API 5L)	η	Pipe Material (JIS G)	η
X42	0.77	STPG 370	0.64
X46	0.81	STPG 410	0.66
X52	0.87	STPG 370	0.64
X56	0.87	STPG 410	0.66
X60	0.88	STPG 480	0.63
X65	0.93	STPG 400	0.62

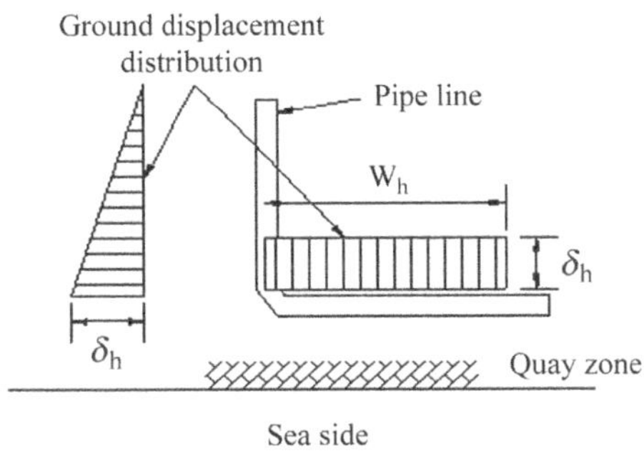

FIGURE 3.2.14 Analytical model of curved pipe deformation in a liquefied seawall.

Liquefaction-induced ground deformation in a seawall area deforms seaward, with the largest deformation along the shoreline and smaller deformations further inland. The ground deformation distribution shown in Figure 3.2.14 is modeled as a triangle. The preceding analytical equation is formulated by modeling the distribution of ground deformation as a triangle. The maximum bending angle due to forced displacement is given by Equation (3.2.39). This equation was developed based on experimental results with a standard pipe diameter of 600 mm. Now, when the revetment area is liquefied, the bending angle (degrees) in the curved pipe is given by the following equation:

$$\omega_{bs} = \frac{180}{\pi}\left|\frac{3}{4}\pi - 2sin^{-1}[b]\right|, \tag{3.2.39}$$

in which

$$b = \frac{1}{2}\sqrt{b_1 + b_2},$$
$$b_1 = 5000\left\{\frac{\gamma_\delta \delta_h}{L_{ps}}\left(\frac{D}{D_{600}} + 0.25\right)\right\}^2, \tag{3.2.40}$$
$$b_2 = -241\frac{\gamma_\delta \delta_h}{L_{ps}}\left(\frac{D}{D_{600}} + 0.25\right) + 3.42.$$

where

ω_{bs}: internal bending angle of curved pipe (degree)
γ_δ: load coefficient for horizontal displacement of the ground due to lateral flow of the ground behind the revetment
δ_h: horizontal displacement of the ground due to lateral flow of the ground behind the revetment (m)
D_{600}: standard outer diameter (= 61.0 cm)
D: outer diameter of the conduit (cm), where $D = D_{600}$ when $D > D_{600}$
L_{ps}: length of the hinge section of the inner bending (cm)

On the other hand, for the displacement of a curved conduit in the ground behind a revetment, when the pipe diameter/thickness ratio D_o/t_b is less than 50, the outside bending angle (degrees) of the curved conduit is given by the following equation:

$$w_{bo} = \frac{180}{\pi}\frac{150\gamma_b\delta_h}{L_{pol}}\left(0.81\frac{D_o}{D_{600}} + 0.43\right), \tag{3.2.41}$$

in which

$$L_{pol} = \sqrt[4]{\frac{1200EI_r\gamma_\delta\delta_h}{P_1}}, \tag{3.2.42}$$

$$P_1 = D\gamma_k\sigma_c.$$

where

ω_{bo}: outer bending angle of the curved pipe (degree)
L_{pol}: hinge section length of the outer bending tension (cm)
E: modulus of elasticity of the conduit ($= 2.06\times10^7\ N/\text{cm}^2$)
I: secondary moment of the cross section (cm^4)
σ_c: maximum ground restraint force per unit projected area of pipe in the non-liquefied layer above the liquefied layer $\left(N/\text{cm}^2\right)$

3.2.2 Seismic Design Method for Pipelines Crossing Faults

3.2.2.1 Fault Displacement Patterns

Figure 3.2.15 shows the modes of fault displacement. There are two types of longitudinal displacement faults: normal fault and reverse fault. In the case of a lateral strike-slip fault, there are two types: Dextral (right lateral) strike-slip fault and Sinistral (left lateral) strike-slip fault.

3.2.2.2 Prediction of Fault Displacement

Earthquakes occur at the epicenter on a fault plane, and their magnitude depends on the size of the fault plane and, therefore, the length of the fault. Fault displacement is the amount of displacement by which the fault plane moves.

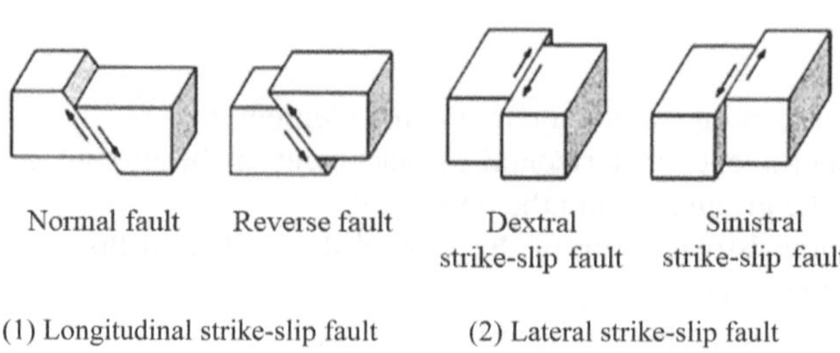

FIGURE 3.2.15 Displacement patterns of fault movement.

Therefore, the amount of fault displacement can be estimated in principle by the following equation:

$$D = \frac{M_o}{\mu S},$$
$$\mu = \rho V_S^2, \tag{3.2.43}$$
$$M_o = 10^{(1.5M_w + 16.1)},$$
$$M_w = \frac{Log_{10} L + 3.55}{0.74}.$$

where

D: fault mean displacement (m)
μ: rigidity (N/m^2)
ρ: density $(= 2.8tf/m^3)$
S: fault area (km^2)
V_S: S wave shear wave velocity (= 3.3 km/s)
M_o: seismic moment[25] $(N \cdot m)$, which is related to the JMA magnitude M_{JMA} and a constant C by the following equation:

$$Log_{10} M_o = 1.5 M_{JMA} + C. \tag{3.2.44}$$

According to Irikura,[26] the fault area and seismic moment are related by the following equation:

$$S = 4.59 \times 10^{-11} M_o^{1/2}. \tag{3.2.45}$$

M_w: moment magnitude
L: surface rupture length (km)

Table 3.2.4 shows the existing prediction equations[24] for fault displacement based on observation records. It shows a method for estimating the magnitude from the fault zone area A and finally predicting the mean and maximum values of fault displacement (m).

3.2.2.3 Simplified Analysis Method

In the United States, seismic design guidelines for the Trans-Alaska Crude Oil Pipeline crossing a fault zone have provided the following design equation[30] as a simplified design formula for the fault crossing model shown in Figure 3.2.16. In other words, when permanent ground displacement (PGD) occurs due to fault crossing, the strain generated in the pipeline is calculated by the following equation:

$$\varepsilon_{pipe} = 2\left[\frac{PGD}{2L_a}\cos\beta + \frac{1}{2}\left(\frac{PGD}{2L_a}\sin\beta\right)^2\right], \tag{3.2.46}$$

TABLE 3.2.4
Prediction Equations for Fault Displacement[24]

Fault Type	Prediction Formula	Prediction Error (Standard Deviation)	Source
Average Surface Fault Displacement *AD* (m)			
Normal fault	$lnAD = -6.32 + 0.90M$	$\sigma_{logAD} = 0.28$	Wells and Coppersmith (1994)[27]
All types	$lnAD = -4.80 + 0.69M$	$\sigma_{logAD} = 0.36$	
Maximum Surface Fault Displacement *MD* (m)			
Normal fault	$lnMD = -3.90 + 0.48M$	$\sigma_{logAD} = 0.26$	Bonilla et al. (1984)[28]
Magnitude M			
Normal fault	$lnM = lnA + 3.98$ for $A \le 537$ km^2	$\sigma_M = 0.03$	Hanks and Bakun (2002)[29]
	$lnM = 1.33lnA + 3.07$ for $A > 537$ km^2	$\sigma_M = 0.04$	

where β and L_a are the effective length of the pipeline for the angle of incidence and fault displacement, and *PGD* is given by a fault displacement of *AD* or *MD*.

For example, when a pipeline crosses a fault plane that undergoes reverse fault displacement, forced displacement acts in the direction of pipe axis compression, as shown in Figure 3.2.17, and crack failure due to buckling may occur on both sides relative to the fault plane.

Hasegawa[31] proposed a fault countermeasure method and its seismic design method for this situation. When axial compressive force acts on a pipeline under a reverse fault, a pair of plastic hinges is generated on both sides from the fault line as shown in Figure 3.2.17. The following method of installing a steel pipe for faults has been proposed at the location where these hinges occur.

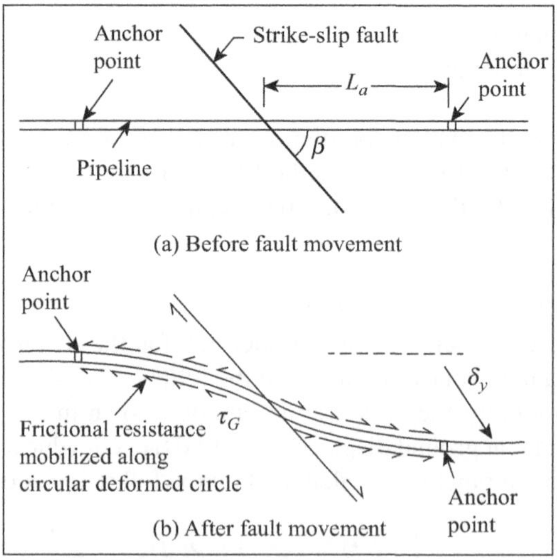

FIGURE 3.2.16 Analytical model of a pipeline crossing a fault.[30]

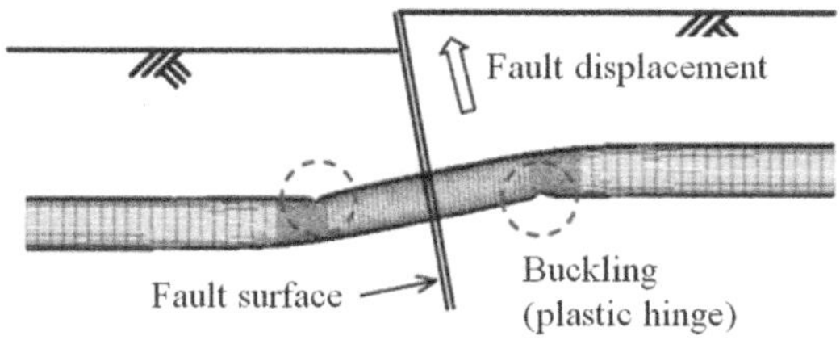

FIGURE 3.2.17 Buckling deformation of a pipeline crossing a reverse fault.[31]

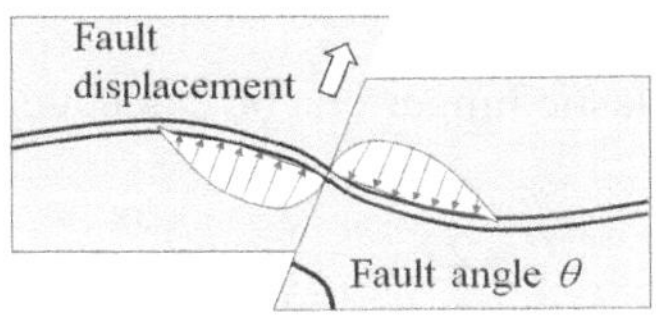

FIGURE 3.2.18 Earth pressure distribution model acting on a pipeline near a fault.[31]

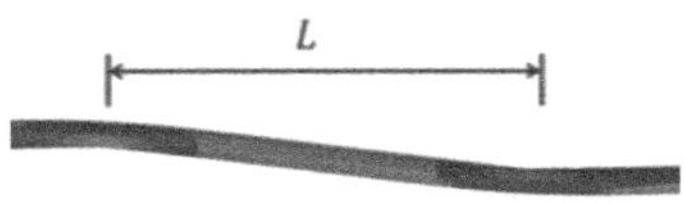

FIGURE 3.2.19 Spacing of plastic hinge occurrence L[31] due to compression buckling of fault crossing pipeline.

To determine the installation location, information on the buckling point spacing L is required. The spacing L can be obtained by the following procedure, assuming the earth pressure distribution shown in Figure 3.2.18 and using the analytical model shown in Figure 3.2.19.

First, the following equation is obtained as the equilibrium equation between the pipe and the external force acting from the oblique direction in the x > 0 section as shown in Figure 3.2.20:

$$EI\frac{d^4y}{dx^4}+N\frac{d^2y}{dx^2}+ky=0, \tag{3.2.47}$$

where N is the axial force and k is the soil elasticity constant.

Solving this equation yields

$$y(x)=\delta_p\cdot e^{-\alpha x}\cdot\left(\cos\gamma x+\frac{\alpha^2-\gamma^2}{2\alpha\gamma}\sin\gamma x\right). \tag{3.2.48}$$

in which

$$\begin{aligned}&\alpha=\sqrt{\lambda-\epsilon},\quad \gamma=\sqrt{\lambda+\varepsilon},\\ &\varepsilon=\frac{N}{4EI},\quad \lambda=\sqrt{\frac{k}{4EI}}.\end{aligned} \tag{3.2.49}$$

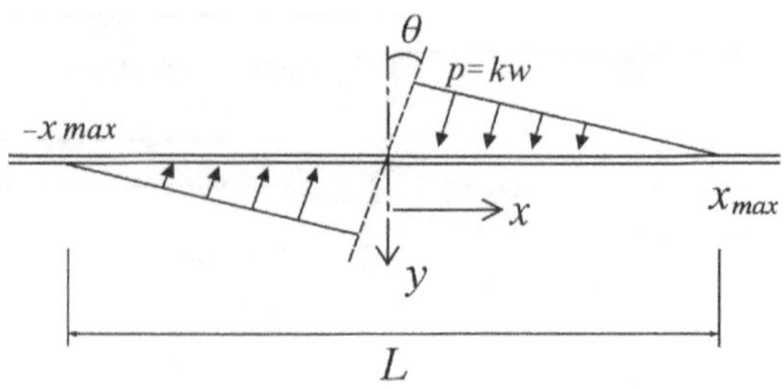

FIGURE 3.2.20 Analytical model of a fault-crossing pipeline subjected to axial forces.[31]

The distance between plastic hinges can be calculated by the following equation:

$$L = 2x_{max} - \frac{2}{\gamma} tan^{-1} \frac{\alpha}{\gamma} . \tag{3.2.50}$$

3.2.3 Seismic Design Method for Pipelines against Landslides

Figure 3.2.21 shows a typical pattern of a pipeline crossing a landslide zone.

Pipelines crossing a landslide zone should be designed to be seismic-resistant to landslides, based on the relationship between the direction of the landslide and the direction of the pipeline, as shown in Figure 3.2.21.

Three types of combinations occur. The three types of pipelines are a case in which the PGD acts perpendicular to the pipeline, a case in which the PGD acts parallel to the axis of the pipeline, and a case in which the PGD acts obliquely to the pipeline.

3.2.3.1 Case of Landslide Displacement Parallel to the Pipe Axis

Figure 3.2.22 shows an example of a case where landslide displacement acts parallel to the pipe axis.

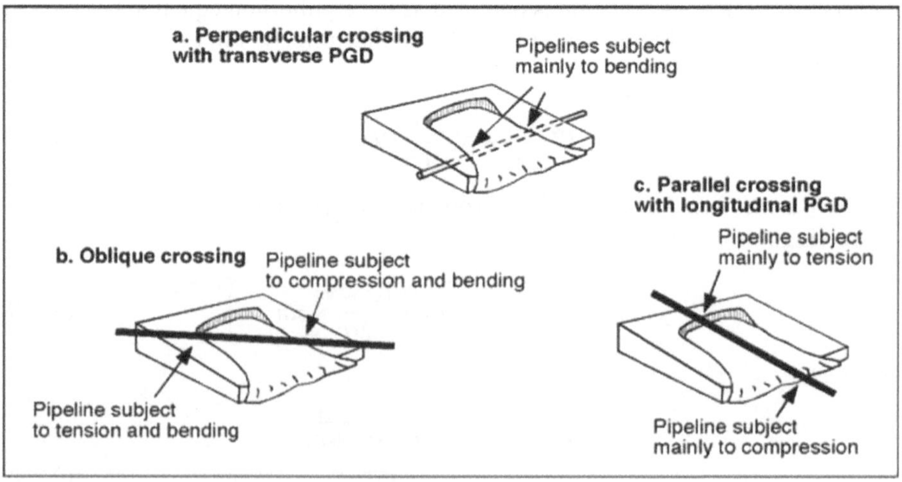

FIGURE 3.2.21 Pipelines crossing a landslide zone.[30]

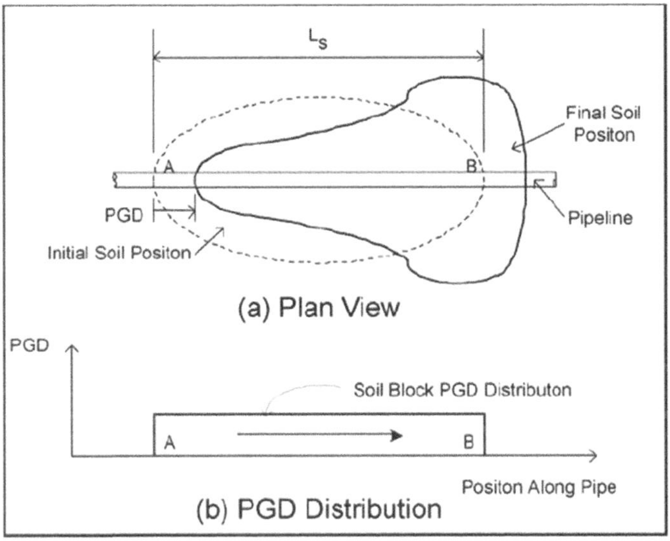

FIGURE 3.2.22 Landslide displacement acting parallel to the pipe axis.[30]

In this case, the force acting on the pipe can be calculated from the following equation:

$$F = \min(F_1, F_2), \tag{3.2.51}$$

where

$$F_1 = \sqrt{A_r E t_u \delta}\text{: Elastic force acting on the pipe} \tag{3.2.52}$$

where δ is the landslide displacement PGD. On the other hand,

$$F_2 = \frac{t_u L_{sp}}{2}\text{: the maximum force transmitted from the peripheries to the conduit} \tag{3.2.53}$$

where L_{sp} is the length of the pipeline in the soil mass undergoing ground deformation.

3.2.3.2 Landslide Displacement Perpendicular to the Pipe Axis

As shown in Figure 3.2.23, when landslide displacement acts perpendicularly to the conduit, the pipeline is subjected to large bending deformation, but the high bending performance of steel pipes allows them to follow the landslide displacement without leaking to a considerable degree.

In this case, the bending strain occurring at points A, B, and C in Figure 3.2.23 can be calculated using the following equation:

$$\varepsilon_b = \pm\frac{\pi D_o \delta}{W_G^{\,2}}. \tag{3.2.54}$$

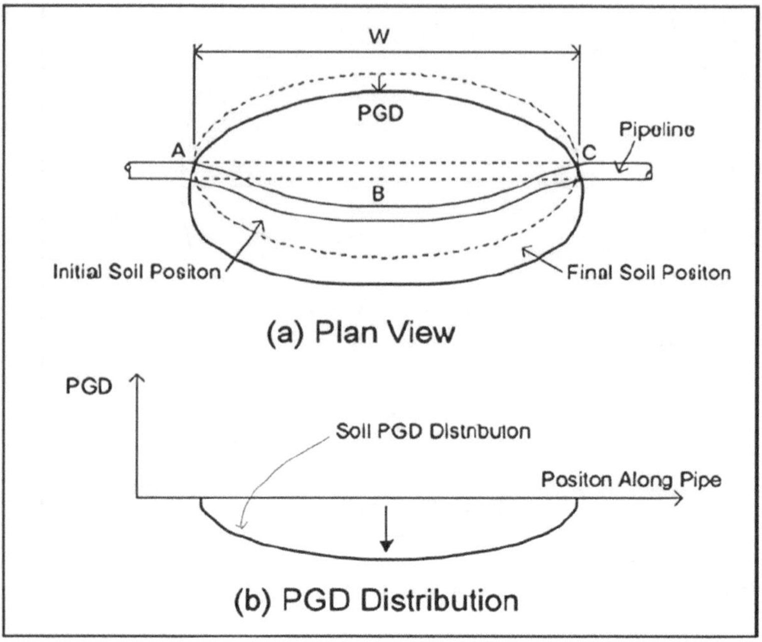

FIGURE 3.2.23 Landslide displacement acting perpendicular to the pipe axis.[30]

where W_G is the landslide width, D_0 is the pipe outer diameter, and δ is the landslide displacement.

If a more accurate evaluation than the calculated bending strain is desired, analysis by FEM is recommended. Since plastic deformation performance of pipelines can be expected to be 4–5% in tensile strain and about 1% in compressive strain, it is desirable to consider plastic deformation performance in the seismic design of pipelines when large ground deformation accompanies the seismic design.

3.2.4 Numerical Methods for Seismic Behavior of Pipelines

Theoretical analysis methods based on elasticity theory or structural analysis theory are very useful for understanding the essential characteristics of a phenomenon, although the structural model under consideration is simple. On the other hand, simplified analysis methods, which are simplified versions of theoretical analysis methods, are useful for design practitioners to understand the general trends of phenomena and to judge whether FEM results are acceptable or not.

However, while the simplified analysis method is effective for models for which the analysis method is easy to apply, when the displacement behavior of a pipeline is in the elastoplastic region and finite displacements must be handled, the FEM analysis method becomes useful in practical terms and in terms of accuracy. In the case of complex pipeline geometries, the FEM model can be used to appropriately model the area to be analyzed and is also useful for understanding the displacement and stress fields in the surrounding area.

However, the following points should be considered when applying FEM analysis methods:

a. Modeling: It is necessary to confirm whether the slip phenomenon between the pipeline and the ground can be correctly modeled by the slip element of the FEM element, or whether the separation between the surrounding soil and the pipe surface can be strictly modeled by the contact element between the buried structure and the surrounding soil.
b. Verifiability: It is necessary to check whether the numerical analysis results by FEM are in good agreement with the actual phenomena. If the numerical model can be easily compared with the actual structure, such as a mechanical product for which the input conditions are easily understood, it will be easy to compare the accuracy of the experimental results with the analytical results. However, it is not easy to verify the consistency between analytical results and actual phenomena for pipeline nonlinear behavior during earthquakes, which occur rarely. Without solving this problem, the validity of the FEM results related to buried pipelines will not be confirmed.
c. Applicability: There may be no other appropriate method than FEM analysis method to determine the location of maximum strain due to local buckling and to check the stress/strain field at the ultimate limit state such as crack initiation in straight and irregularly shaped pipes. However, it is necessary to confirm that the modeling and boundary conditions are appropriate for such local analysis even for the FEM analysis method.
d. Input conditions: In order to correctly apply seismic loads acting on the pipeline, one tries to make the ground model as rigorous as possible. However, various prediction errors must be taken into account, such as errors in the two-dimensional geometry of the ground, errors in the prediction of soil conditions, errors in the estimation of seismic waves incident on the base, etc. In FEM analysis, it is necessary to verify the validity of the analysis results by taking into account various random variables in the input side.

The following are examples of simple numerical models for pipelines in the cases of ground motion, fault crossings, and landslides.

3.2.4.1 Ground Motion Case

Figure 3.2.24 shows an example of a numerical model for the case of ground shaking. In this model, the soil between the buried pipe and the surrounding ground is connected by a nonlinear spring to evaluate the interaction between the pipe and the ground during an earthquake. The mechanical properties of the soil can be characterized by setting specific ground springs to express shear forces along the pipe axis and compressive forces perpendicular to the pipe axis, respectively.

However, when ground shaking becomes large, as in the case of Level 2 ground motion, slip between the pipe and the surrounding ground is expected to occur. To numerically analyze this phenomenon using a FEM model, it is necessary to accurately model the slip behavior between the soil around the pipe and the pipe surface. Seismic ground shaking propagating along the pipe axis causes a section of the pipe

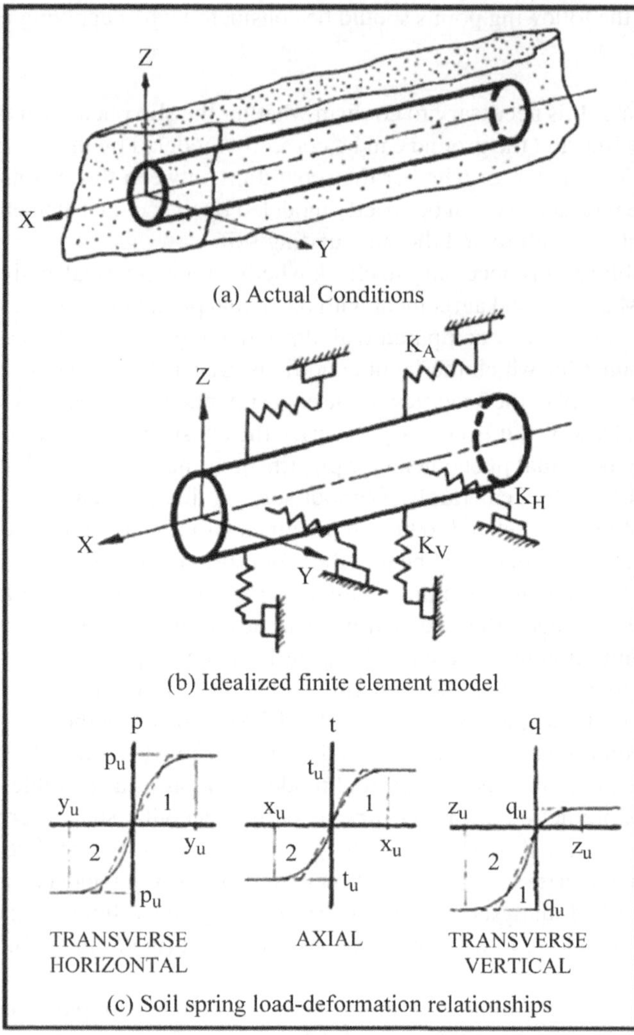

FIGURE 3.2.24 Straight pipe model.[30]

to slip, and the soil returns to a nonslip state in the next instant. This means that it is necessary to assume a considerable load just to perform FEM analysis of seismic slip phenomena in a buried straight pipeline.

3.2.4.2 In the Case of Fault Crossings

In the case of numerical analysis of fault crossing, a simple two-dimensional analytical model as shown in Figure 3.2.25 was used in the seismic design of the Trans-Alaska Crude Oil Pipeline in the 1970s to analyze the pipeline deformation behavior in fault crossing areas. After many studies, simulation analysis using 2D and 3D

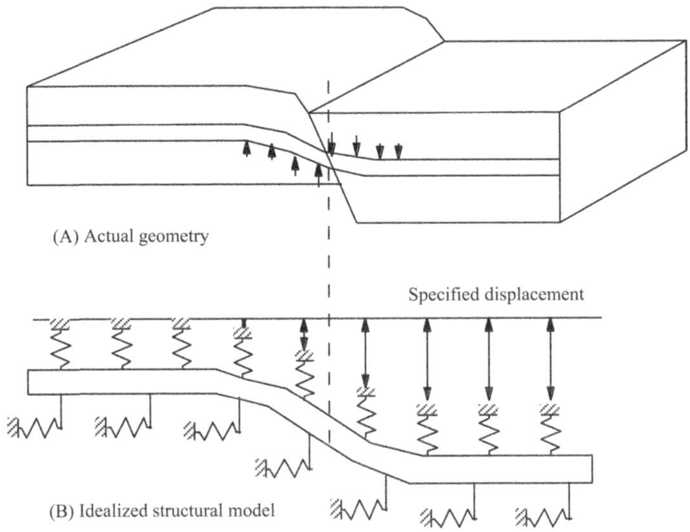

FIGURE 3.2.25 Analytical model of fault crossing.[30]

FEM analytical models as shown in Figures 3.2.17 and 3.2.19 has recently been conducted. Although nonlinear models of soil around pipelines have been refined, the knowledge of seismic loading aspects such as soil pressure distribution around faults has not necessarily progressed that much.

3.2.4.3 Case of Crossing a Landslide Zone

An example of a numerical analysis model for a pipeline crossing a landslide zone is shown in Figure 3.2.26. Landslide models are in the same situation as fault models, and compared to the progress in FEM modeling of the soil around the conduit, knowledge on the distribution of earth pressure during a landslide has not necessarily advanced.

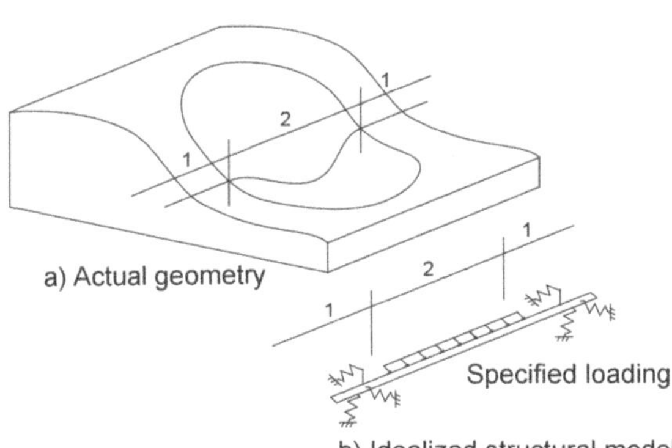

FIGURE 3.2.26 Numerical analysis model of a pipeline crossing landslide zone.[30]

3.3 SEISMIC DESIGN CRITERIA FOR PIPELINES

Pipelines such as water, sewer, gas, electric power, and communication cable networks are smaller in diameter than shield tunnels and are buried underground in shallow areas within a few meters of the ground surface. Seismic design of these underground structures is basically based on the response displacement method.

3.3.1 Domestic Design Standards

Typical domestic standards for pipelines include the *Technical Standard for Petroleum Pipelines*,[32] the *Guidelines for Seismic Design of High-Pressure Gas Pipes*, and the *Guidelines for Seismic Design Methods for Waterworks Facilities*.

The *Technical Standard for Oil Pipelines* was established when oil pipelines were constructed to transport jet fuel to Narita International Airport, and the introduction of the seismic design technique for buried pipes was based on the response displacement method in the standard. The seismic design system based on the response displacement method was followed in the subsequent seismic design guidelines for high-pressure gas pipelines and water main lines.

The first technical standards for oil pipelines were developed based on the design concepts of the 1970s, but many seismic design techniques have been developed since then based on the experience of earthquake damage to many pipelines. The seismic design methods reflecting these results have been reflected in the *Guidelines for Seismic Design of High-Pressure Gas Pipelines* and the *Guidelines for Seismic Design Methods for Waterworks Facilities* and are used to this day. Table 3.3.1 compares these three guidelines for pipeline seismic design.

3.3.2 US Design Criteria

Seismic design methods for pipelines in the United States are established by the ALA,[30] which is affiliated with the Federal Emergency Management Agency (FEMA). For example, for water pipelines:

Seismic Guidelines for Water Pipelines (2005)
Guidelines for the Design of Buried Steel Pipe for Water Pipelines (2001)

For gas and oil pipelines, the American Petroleum Institute (API) has published a number of systematic standards and guidelines for pipeline design technology.

In addition, the Technical Council of Lifeline Earthquake Engineering (TECLEE) is the main organization for research and development of seismic design methods for lifelines, with a focus on these pipelines. TECLEE has published a number of books on seismic design methods in the form of the TECLEE Monograph.

The US seismic design method for lifelines is classified into two categories: segmented pipe and continuous pipe. Mechanical joints are classified as conventional joints without breakaway protection (cast iron pipe with push-on bell-and-spigot gasketed joints). In particular, mechanical joint pipes are classified into two types: conventional cast iron pipe [CIP] with push-on bell-and-spigot gasketed joints, and

TABLE 3.3.1
List of Seismic Design Methods for Underground Pipes

Item	Technical Standard for Petroleum Pipelines	Seismic Design Guideline for Water Supply Facilities	Seismic Design Guideline for High-Pressure Gas Pipes
Ground displacement	Level 1 ground motion $U_h = \frac{2}{\pi^2} S_V T_G K'_{h1} cos\left(\frac{\pi Z}{2H}\right)$ T_G: typical period of surface grounds K'_{h1}: design seismic intensity on the baserock, assuming 0.15 S_V: velocity response spectrum per unit seismic intensity Z: depth (m) H: thickness of surface layer (m)	Level 1 ground motion $U_h = \frac{2}{\pi^2} S_V T_G K'_{h1} cos\left(\frac{\pi Z}{2H}\right)$ S_V: velocity response spectrum per unit seismic intensity Level 2 ground motion $U_h = \frac{2}{\pi^2} S'_V T_G cos\left(\frac{\pi Z}{2H}\right)$ S'_V: velocity response spectrum on the baserock	Level 1 ground motion $U_h = \frac{2}{\pi^2} S_V T_G K'_{h1} cos\left(\frac{\pi Z}{2H}\right)$ S_V: velocity response spectrum per unit seismic intensity Level 2 ground motion $U_h = \frac{2}{\pi^2} S'_V T_G cos\left(\frac{\pi Z}{2H}\right)$ S'_V: velocity response spectrum on the baserock
Typical period of surface ground	$T_G = C \frac{H}{V_{1S}}$ H: ground surface thickness (m) V_{1S}: shear wave velocity (m/s) C: coefficient of soil classification	$T_G = 4 \sum_{i=1}^{n} \frac{H_i}{V_{Si}}$ H_i: i-th surface ground thickness V_{Si}: shear wave velocity of i-th surface ground	$T_G = \frac{4H}{V_S}$ $V_S = \frac{\Sigma V_{Si} H_i}{H}, \ H = \Sigma H_i$
Apparent wave length	$L = \frac{2L_1 L_2}{L_1 + L_2}$ L_1: horizontal propagating wavelength of SH waves in surface layer (m) L_2: horizontal propagating wave length of SH waves in baserock ground (m)	$L = \frac{2L_1 L_2}{L_1 + L_2}$	$L = V \cdot T_G$ V: horizontal propagation velocity of surface waves (m/s) 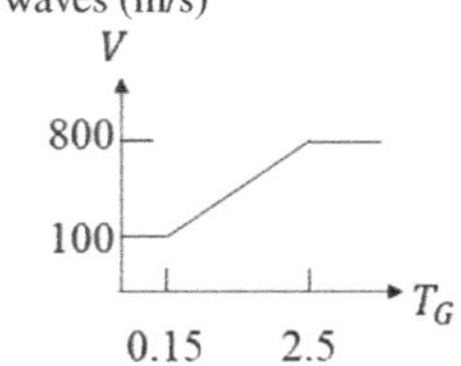

(Continued)

TABLE 3.3.1 *(Continued)*
List of Seismic Design Methods for Underground Pipes

Item	Technical Standard for Petroleum Pipelines	Seismic Design Guideline for Water Supply Facilities	Seismic Design Guideline for High-Pressure Gas Pipes
Response velocity	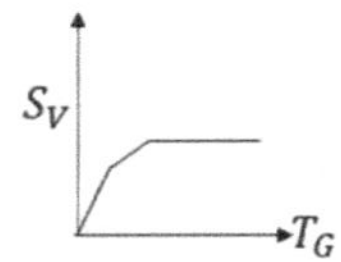	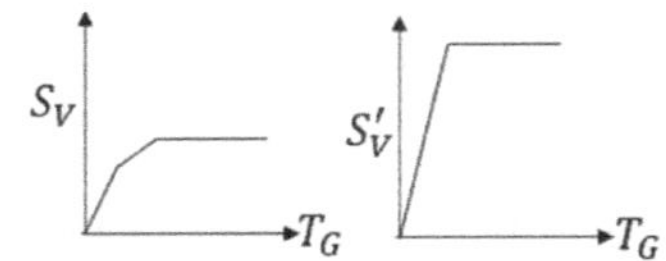	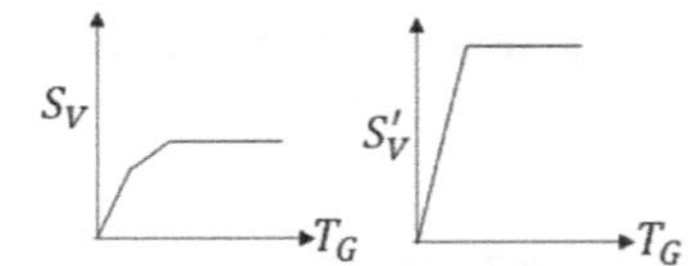
Target pipelines	straight pipe verification point: axial stress in straight pipe for SH waves incident 45°	straight pipe verification point: axial strain in straight pipe for SH waves incident 45°	straight pipe, bend and Tee-junction verification point: axial strain in straight pipe for surface waves
Slippage effect	Slippage effect is not considered	Slippage effect is not considered for Level 1 ground motion Slippage effect is considered for Level 2 ground motion	Slippage effect is considered for Level 1 and Level 2 ground motions
Critical values	Based on allowable stress design method	Critical strain for buckling strength for pipeline with continuous joints	Critical strains 1% for Level 1 ground motion 3% for Level 2 ground motion
Ground displacement	not considered	Assessment of axial strain for straight pipes in liquefied, coastal or landslide sensitive zones	Assessment of deformed angle for straight and bend pipes in liquefied, coastal and landslide sensitive zones
Ground strains	$\varepsilon_G = \frac{2\pi}{\sqrt{2}L} \cdot \frac{U_h}{\sqrt{2}} = \frac{\pi U_h}{L}$ L: wave length of SH waves incident 45°	$\varepsilon_G = \frac{\pi U_h}{L}$ L: wave length of SH waves incident 45°	$\varepsilon_G = \frac{2\pi U_h}{L}$ L: wave length of Rayleigh-type surface wave

Axial strains	Axial strain $$\varepsilon_L = \alpha_1 \varepsilon_G$$ Bend strain $$\varepsilon_B = \alpha_2 \varepsilon_G \frac{2\pi D}{L}$$ Combined strain $$\varepsilon_X = \sqrt{3.12\varepsilon_L^2 + \varepsilon_B^2}$$ where α_1, α_2: conversion factors for axial and perpendicular to the straight pipe $$\alpha_1 = \frac{1}{1 + \left(\frac{2\pi}{\lambda_1 L'}\right)^2}$$ $$\alpha_2 = \frac{1}{1 + \left(\frac{2\pi}{\lambda_2 L}\right)^4}$$ $$L' = \sqrt{2}\,L$$ $$\lambda_1 = \sqrt{\frac{K_1}{EA}},\ \lambda_2 = \sqrt[4]{\frac{K_2}{EI}}$$	Continuous jointing pipelines Axial strain $$\varepsilon_L = q\alpha_1 \varepsilon_G$$ Bending strain $$\varepsilon_B = \alpha_2 \varepsilon_G \frac{2\pi D}{L}$$ Combined strain $$\varepsilon_X = \sqrt{\varepsilon_L^2 + \varepsilon_B^2}$$ where α_1, α_2: conversion factors for axial and perpendicular to the straight pipe q: slippage factor in which $$\tau_G = \frac{2\pi}{L'} Et\alpha_1 \varepsilon_G$$ $$L' = \sqrt{2}\,L$$ Mechanical jointing pipelines Axial deformation and bending angle at joints based on the ground strain (Calculation formula omitted)	Mechanical jointing pipelines Axial strain $$\varepsilon_L = q\alpha_0 \varepsilon_G$$ where $\tau_G \geq \tau_{cr}$ $$q = 1 - cos\,\xi + \Omega\left(\frac{\pi}{2} - \xi\right) sin\,\xi$$ $\tau_G < \tau_{cr}$ $q = 1$ $$\xi = arc\,sin\left(\frac{\tau_{cr}}{\tau_G}\right)$$ in which q: slippage factor Ω: modification factor, 1.5 and $$\tau_G = k_1(1 - \alpha_1)U_h$$ $$\alpha_0 = \frac{1}{1 + \left(\frac{2\pi}{\lambda_1 L}\right)^2}$$

(Continued)

TABLE 3.3.1 *(Continued)*
List of Seismic Design Methods for Underground Pipes

Item	Technical Standard for Petroleum Pipelines	Seismic Design Guideline for Water Supply Facilities	Seismic Design Guideline for High-Pressure Gas Pipes
Pipe strains for bends and Tee-junctions			Bending strains of bends $\varepsilon_B = \beta_B \Delta$ where β_B: conversion factor for bends Bending strains for Tee-junctions branch pipe: $\varepsilon_{T1} = \beta_{T1}\Delta$ main pipe: $\varepsilon_{T2} = \beta_{T2}\Delta$ where $\Delta = \left(1 - q^* \alpha_1\right) U_h$ $\tau_G \geq \tau_{cr}$ $q^* = sin\,\xi \cdot \left(1 + \frac{\pi^2}{8} - \frac{\xi^2}{2}\right) - \xi\, cos\,\xi$ $\tau_G < \tau_{cr}$ $q^* = 1$ in which $\xi = arc\,sin\left(\frac{\tau_{cr}}{\tau_G}\right)$

ductile cast iron pipe [DCIP], the body of which is made by ductile iron. The former corresponds to CIP and DCIP (old type) in Japan, and the latter to DCIP (new type, which is reinforced by mechanical joint). The seismic safety assessment is verified for the former pipes by joint displacement and for the latter pipes by joint tensile strength.

The following is a partial list of the ALA's *Seismic Guidelines for Water Pipelines* as an example.

3.3.2.1 Pipeline Performance Goals

The ALA has identified the following two goals for the seismic design of pipelines:

- Target performance level
- Seismic hazard level

Each pipeline target is set according to the reliability of the operational performance to meet these performance goals. To achieve the performance goals, each pipeline must fulfill its function in the system according to its importance.

3.3.2.2 Classification of Pipeline Functions

Pipelines are classified into the four levels shown in Table 3.3.2 according to the functions they perform.

In the immediate aftermath of an earthquake, pipeline operations should maintain a minimum level of reliability. The reliability of each pipeline classified by function can be achieved through seismic design according to the seismic model (i.e., the

TABLE 3.3.2
Pipe Function Classes

Pipe Function Class	Seismic Importance	Description
I	Very low to none	Pipelines that represent very low hazard to human life in the event of failure. Not needed for past earthquake system performance, response, or recovery. Widespread damage resulting in long restorations times (weeks or longer) will not materially harm the economic well-being of the community.
II	Ordinary	Normal and ordinary pipeline use, common pipelines in most water systems. All pipes not identified as Function I, III, or IV.
III	Critical	Critical pipelines serving large numbers of customers and that present significant economic impact to the community or a substantial hazard to human life and property in the event of failure.
IV	Essential	Essential pipelines required for post-earthquake response and recovery and intended to remain functional and operational during and following a design earthquake.

TABLE 3.3.3
Earthquake Hazard Return Period for Each Pipe Function Class

Pipe Function Class	Probability of Exceedance in 50 Years (P)	Return Period, T (in Years)
I	100%	Undefined
II	10%	475
III	5%	975
IV	2%	2475

probability of exceedance of an earthquake during the 50-year service period and the recurrence period) set for each class. Table 3.3.3 shows the relationship between the functional classification of the pipeline and the seismic model.

3.3.2.3 Seismic Hazard

The seismic hazards for seismic design of pipelines are classified into ground shaking (maximum or peak ground acceleration, or PGA); maximum or peak ground velocity (PGV); maximum or permanent ground displacement, or PGD), and ground deformation (fault displacement, liquefaction, fluidization, landslide, unequal settlement), as shown in Table 3.3.4, in which PSHA is probabilistic seismic hazard analysis.

TABLE 3.3.4
Earthquake Hazards and Parameters Needed for Pipeline Design

Hazard	Earthquake Parameters	Seismic Hazard Analysis Method	Geotechnical Parameters
	Transient ground movement		
General shaking	PGA, PGV spectral response	PSHA	Soil/rock conditions, depth, *Vs*
Near-source directivity	Fault distance	PSHA, fault map	Fault type, orientation, rupture direction
Ground amplification	PGA, PGV spectral response	PSHA	Site soil and rock conditions, *Vs*
	Permanent ground movement		
Faulting	Magnitude, length	De-aggregate PSHA or geologist	Fault type, orientation
Liquefaction	PGA, magnitude	PSHA, de-aggregate	Soil type, relative density, thickness, groundwater
Lateral spread and flow	PGA, magnitude, distance	PSHA, de-aggregate	Topography, soil type, strength, thickness, groundwater
Slope movement, landslide	PGA, acceleration time history	PSHA	Topography, soil type, strength, groundwater
Settlement	PGA	PSHA	Soil type, strength, thickness, groundwater

3.3.2.4 Seismic Design Methods for Each Pipe Type

The following methods are specified for seismic design of pipelines against ground shaking:

$$\text{Ground strain } \varepsilon_G = \frac{PGV}{c}, \tag{3.3.1}$$

where c is the velocity of propagating waves at the location of the pipeline in the ground, and unless otherwise specified 4 km/sec (13,000 feet/sec) unless otherwise specified.

The 4 km/sec value is considered to correspond to the velocity of the entity P wave generated during an earthquake, while in Japan, the propagating wave velocity is based on the surface wave velocity or SH wave velocity in the surface layer, resulting in a value around 400 m/sec, a 10-fold difference between the two countries.

3.3.2.4.1 Continuous Joint Pipelines

$$\text{Pipe strain } \varepsilon_{pipe} = \varepsilon_G. \tag{3.3.2}$$

The fact that the pipe strain is assumed to be equal to the ground strain is considered a safe evaluation based on the assumption that the ground strain is very small and there is no slip between the ground and the pipe.

3.3.2.4.2 Mechanically Fitted Pipelines

A survey of CIP damage experience in the United States indicates that lead-filled joint cast iron pipelines exhibit a large variation in axial restraint, with a tendency for relatively large joint displacements adjacent to joints with strong restraint. For example, a 1% joint exhibits a joint displacement bias that is three times greater than the average value, and a 0.1% joint exhibits a joint displacement bias that is five times greater than the average value. Based on the results of this study and in consideration of safety, ALA[30] recommends that the displacement which the joint must be able to accommodate should be seven times the average joint movement, as shown in the following equation:

$$\text{Joint displacement } \Delta_{\text{joint}} = 7 l_d \varepsilon_G \,(\text{mm}), \tag{3.3.3}$$

where

Δ_{joint}: design joint displacement
l_d: length of the pipe segment
ε_G: ground strain

With this setup, ALA assumes that the probability of pipe failure during an earthquake is less than 10^{-4}.

In the United States, ground strain is estimated to be about 0.0125% based on a ground surface velocity amplitude of 50 cm/s and horizontal propagation velocity of about 4 km/s. The estimated strain is one order of magnitude smaller than that of Japan, where the ground strain against Level 2 ground motion is estimated to be about 0.1%[8] at a ground natural period of around 1 second. As a result, the design joint displacement in the United States is set to 5.25 mm from Equation (3.3.3), which is one order of magnitude smaller than the design joint displacement in Japan[8] of 23 mm on one side (DCIP [old type]) to 54 mm on both amplitudes (DCIP [new type]).

3.3.2.5 Basis of Seismic Design

The safety verification formulas are set as follows depending on the type of joint:

$$\textbf{Continuous joint pipeline } \varepsilon_{pipe} \leq 0.40\frac{F_y}{E}, \tag{3.3.4}$$

where F_y is the yield stress of the pipe and E is the longitudinal modulus of elasticity (Young's modulus) of the pipe material.

3.3.2.4.3 Mechanically Joint Pipelines

In the United States, seismic safety verification is performed using the following equation

$$\Delta_{\text{joint}} \geq \Delta_{\text{seismic}}. \tag{3.3.5}$$

where Δ_{seismic} is the joint displacement during an earthquake.

For the mechanically joint pipeline with detachment protection, assuming that n pipes are connected and move together, the joint displacement to be shared by one joint is $1/n$ of the ground displacement, and n is assumed to be about 3. Therefore, the tensile force (load) of the n-connected pipeline is given by the following equation:

$$F_{stop} = \pi D \tau_G l_d (n+1). \tag{3.3.6}$$

where τ_G are the shear stresses (circumferential friction force per unit area), respectively. Seismic design is performed so that the joint pullout output is within the elastic range:

$$R_y \geq F_{\text{stop}}. \tag{3.3.7}$$

where R_y is the elastic limit of the extraction force.

REFERENCES

1. Toki, K.: *Seismic Analysis of Structures, New System of Civil Engineering* 11, GIHOUDO Publishing, 1981.
2. Takada, J.: *Lifeline Earthquake Engineering*, Kyoritsu Shuppan Co., 1991.
3. T. Seki and T. Nishikawa: Study on the ground surface response of inclined piles by wave line theory considering diffraction waves, *Proceedings of the Architectural Institute of Japan, Structural Engineering*, No. 452, pp.21–30, 1993.

4. Otsuki, A., T. Tazo, and K. Shimizu: Seismic behavior and ground strain of irregularly shaped ground with sloping base, *Proceedings of JSCE*, No. 350/I-2, pp.291–300, 1984.
5. Toki, K. and T. Sato: Seismic response analyses of ground with irregular properties by the boundary element method, *Natural Disaster Science*, Vol. 5, No. 1 pp. 31–52, 1983.
6. Wang, K-H., J. Koizumi, M. Nakano, N. Sugiyama, and H. Tanaka: Experimental study on buckling of an inner tube in a filled conduit, *Journal of JSCE, F1 (Tunnel Engineering)*, Vol. 73, No. 2, pp.59–70, 2017.
7. Mizuno, K., and Koizumi, J.: Fundamental study on seismic design method of shield tunnel in transverse direction, *Journal of JSCE*, No. 687/III-56, pp.107–123, 2001.
8. Japan Gas Association: *Seismic Design Guideline for High-Pressure Gas Pipes,* JGA Finger-206-03, Committee for Investigation of Technical Standards for Gas Facilities, Japan Gas Association, 2004.
9. Koike, T.: Seismic strain evaluation of buried pipelines, *Proceedings of JSCE*, No. 331, pp.13–24, 1983.
10. Markl, R. C.: Piping-flexibility analysis, *SME*, Vol. 77, No. 2, pp. 127–143, 1955.
11. Ogawa, Y., and T. Koike: Structural design of buried pipelines for severe earthquakes, *Soil Dynamics and Earthquake Engineering*, Vol. 21, pp. 199–209, 2001.
12. Koike, T., K. Takahashi, S. Kawaguchi, K. Kiriyama, H. Gohyakuda, T. Yabuguchi, T. Shiozaki, and M. Toyama: Seismic performance of miter-bend for water supply pipelines, *Journal of Structural Engineering*, Vol. 47A, JSCE, pp.685–696, 2001.
13. Japan Water Works Association: *Guidelines for Earthquake-Resistant Construction Methods for Water Supply Facilities*, 2009 edition, Japan Water Works Association, 2009.
14. Japan Water Pipe Association: *Steel Pipe Buckling Test for Water Supply*, WSP060-96, Japan Water Pipe Association, 1996.
15. ASME: *Criterion of the ASME: Boiler and Pressure Vessel Code for Design by Analysis*, ASME Code Section III and VIII, Division 2, White paper, ASME, 1963.
16. JSCE: *Handbook of Urban Lifelines*, Maruzen Corporation, 2009.
17. FEMA: *HAZUS 99 TECHNICAL MANUAL*, Federal Emergency Management Agency (FEMA), USA, 1999.
18. Japan Gas Association: *Seismic Design Guideline for Liquefaction Resistance of High-Pressure Gas Pipes*, Japan Gas Association, 2001.
19. Hamada, M.: *Geotechnical Earthquake Engineering*, Maruzen Publishing, 2012.
20. Bartlet, S., T. L. Youd: Empirical prediction of liquefaction-induced lateral spread, *ASCE, Journal of Geotechnical Engineering*, Vol. 121, No. 4, pp.316–329, 1995.
21. Bardet, J.-P., T. Tobita, N. Mace, and J. Hu: Regional modeling of liquefaction-induced ground deformation, *Earthquake Spectra*, Vol. 18, No. 1, pp. 19–46, 2002.
22. Suzuki, T., Takada, J., Koike, T., Ogawa, Y. and Matsumoto, M.: Plastic Hinge Analysis of Buried Steel Pipes Subjected to Lateral Flow due to Liquefaction, *Journal of JSCE*, No. 710/I-60, pp.79–90, 2002.
23. Mon, K. A., T. Koike, O. Nishikawa, L. E. Garciano, and J. Kiyono: *A Simplified Seismic Design Method of Water Lifelines in Developing Countries.* International Collaboration in Lifeline Earthquake Engineering IRP 1, pp.543–549, 2016.
24. McGuire, R. K.: *Seismic Hazard and Risk Analysis*, MNO-10, EERI, 2004.
25. Uzu, T.: *Seismology*, Kyoritsu Zensho 216, KYORITSU SHUPPAN Publishing, 1984.
26. Irikura, K.: Prediction of strong ground motions due to future earthquakes using active faults as a case study – the case of the Osaka Basin, *Proceedings of the 12th World Congress on Earthquake Engineering* (12WCEE), 3 (CDROM), 2000.
27. Wells, D. L. and K. J. Coppersmith: New empirical relationship among magnitude, rupture length, rupture width, rupture area, and surface displacement, *Bulletin of the Seismological Society of America*, Vol. 84, No. 4, pp. 974–1002, 1994.

28. Bonilla, M. G., R. K. Mark, and J. J. Lienkaemper: Statistical relationship among earthquake magnitude, surface rupture length, and surface fault displacement, *Bulletin of the Seismological Society of America*, Vol. 74, No. 2379, pp.2379–2411, 1984.
29. Hanks, T. C. and W. H. Bakun: A bilinear source-scaling model for M-log A observations of continental earthquakes, *Bulletin of the Seismological Society of America*, Vol. 92, No. 5, pp. 1841–1846, 2002.
30. American Lifelines Alliance: *Seismic Guidelines for Water Pipelines*, American Lifelines Alliance (ALA), FEMA, 2005.
31. Hasegawa, N.: *Research on seismic design methods for fault-crossing pipelines*, Doctoral thesis submitted to Kyoto University, 2016.
32. Ministry of International Trade and Industry, Ministry of Transport, Ministry of Construction and Ministry of Home Affairs: *Technical Standards for Project Facilities of Petroleum Pipeline Projects* (abbreviated as *Technical Standards for Petroleum Pipelines*), Notification No. 1 to Determine Details of the Above Standards of the Ministry of International Trade and Industry, Ministry of Transport, Ministry of Construction and Ministry of Home Affairs, 1973.
33. Miki, C., T. Kobayashi, N. Oguchi, T. Uchida, A. Suganuma, and A. Katoh: Deformation and Fracture Properties of Steel Pipe Bends with Internal Pressure, 12WCEE, No.1814, pp.1–8, 2000.
34. Shimamura K., M. Hamada, S. Yasuda, K. Ohtomo, Y. Fujita, S. Kojima, and Y. Taji: *Load on Pipes Buried in a Non-Liquefaction Layer Due to Liquefaction*. Japan Institute of Public Works, No.1385, pp.1–8, 2000.
35. Japan Water Works Association: *Survey Report on Damage to Water Supply Facilities, 2018 Hokkaido Bold Eastern Earthquake*, Japan Water Works Association, Ministry of Health, Labor and Welfare, Pharmaceuticals, Public Health and Sanitation Water Supply Division, 2018.
36. Kawashima, K.: *Seismic Design of Underground Structures*, Kajima Publishing Inc., 1994.
37. Koike, T.: *Structural Strain Estimation of Segmented Shield Tunnel for Severe Earthquakes*, 12WCEE, No.0292, pp.1–8, 2000.
38. Koike, T.: Longitudinal and transverse strain estimates of shield tunnel under seismic hazard, *JSCE*, No. 759/I-67, pp. 37–52, 2004.

4 Seismic Behavior Analysis of Lifeline Facilities

4.1 SEISMIC RESPONSE ANALYSIS OF STRUCTURES

4.1.1 Inelastic Response Spectrum

4.1.1.1 Definition and Explanation

When the seismic acceleration input to a structure is large, as in the case of Level 2 ground motion, the structural seismic response as a vibration system will result in large response displacements that exceed the yield level of the structure. The inelastic response spectrum is obtained by drawing the maximum value of the seismic response displacement in the elastoplastic state for each eigen-period of the vibration system.

When the design acceleration response spectrum is given, the acceleration spectrum obtained by calculating the yield acceleration level that gives a constant ductility factor μ for each structure eigen-period is called the yield acceleration spectrum.[1,2]

4.1.1.2 Inelastic Response Displacement Based on the Constant Energy Law

When subjected to large seismic motions such as Level 2 ground motion, the response of a structure shows a large response that exceeds the yield level and extends into the plastic region, i.e., nonlinear seismic response.

To simplify the discussion, nonlinear seismic response displacement of one-degree-of-freedom vibration systems is considered. The method based on the constant energy law is used in practice as a simple method to estimate the maximum response value.

The case where the stress–strain relationship of the bilinear system shown in Figure 4.1.1 is established as the response characteristics of a structure are discussed.

The constant energy law is a method for estimating the response displacement δ_{max} plasticized by the elastoplastic response from the maximum response value A_{max} assumed for the elastic response, assuming that the area of the triangle in the diagram is equal to the area of the rectangle.

The response acceleration A_{max}^{L2} and the response displacement δ_{max}^{L2} are produced as the one-degree-of-freedom nonlinear response due to Level 2 ground motion.

By dividing the response acceleration and displacement by the yield acceleration A_y and yield displacement δ_y, the structural characteristic coefficient[2] C_S and ductility factor μ are defined by the following equations:

$$C_S = \frac{A_y}{A_{\max}^{L2}}, \quad \mu = \frac{\delta_{\max}^{L2}}{\delta_y}, \tag{4.1.1}$$

in which

$$\delta_y = \left(\frac{2\pi}{T}\right)^2 A_y. \tag{4.1.2}$$

DOI: 10.1201/9781003655527-4

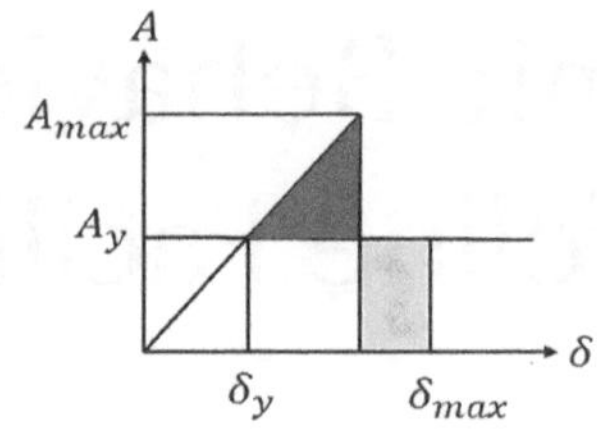

FIGURE 4.1.1 Illustration of the constant energy rule.

According to the constant energy law, they are linked by the following relationship:

$$C_S = \frac{1}{\sqrt{2\mu - 1}} \tag{4.1.3}$$

By the way, regarding the relationship between the nonlinear response displacement and the natural period of a one-degree-of-freedom vibration system, Shibata[3] states that the following relationship is relatively valid in the short-period region due to the constant energy law, while the following relationship is not valid in the long-period region.

$$A_y = C_S \cdot A_{max}^{L2} \quad \text{and} \quad \delta_{max}^{L2} = \mu \cdot \delta_y \tag{4.1.4}$$

It is noted that in the long-period regime (natural period > 1 s), the following equation is relatively valid due to the displacement constant law:

$$A_y = \frac{1}{\mu} A_{\max}^{L2} \quad \text{and} \quad \delta_{max}^{L2} = \mu \cdot \delta_y \tag{4.1.5}$$

4.1.1.3 Handling of Structural Characteristic Coefficient, C_S

The structure characteristic coefficient defined in Equation (4.1.3) has been used in quasi-static design calculations for seismic design against Level 2 ground motion. Safety verification is carried out by multiplying the design acceleration by the structure characteristic coefficient as the load due to Level 2 ground motion. However, as shown in Equation (4.1.5), the designer must accurately understand that "multiplying by the structural characteristic coefficient is a measure based on the assumption that the yield level of the structure to be designed matches this yield acceleration." In some existing design cases, this point has not been clearly confirmed.

When Level 1 and Level 2 ground motions are set as the design acceleration, the yield acceleration exists at the position A_y shown in Figure 4.1.2.

This means that the design conditions required for the structure to be designed are that the maximum acceleration A_{max}^{L1} for Level 1 ground motion must not exceed the yield acceleration A_y, and that the maximum response displacement $\delta_{\max}^{L2}$ due to Level 2 ground motion must not exceed the repair limit displacement $\delta_{cr,2}$.

Therefore, as shown in Figure 4.1.3, the yield acceleration A_y must be located between the Level 1 and Level 2 ground motions. As shown in the same figure, if the yield acceleration is set below 500 gal, the elastoplastic response displacement

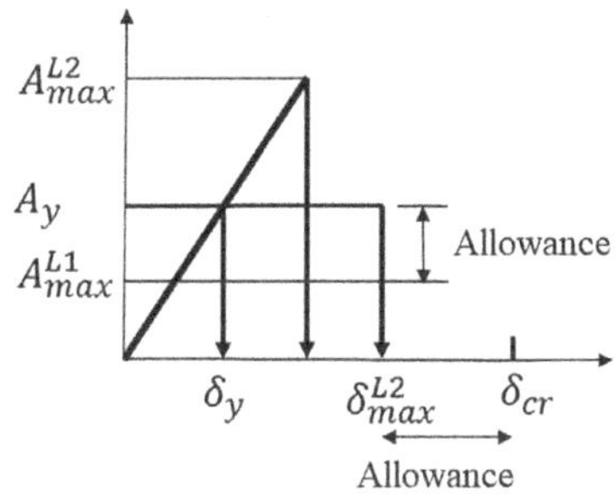

FIGURE 4.1.2 Elastoplastic relations of structures according to the constant energy rule.

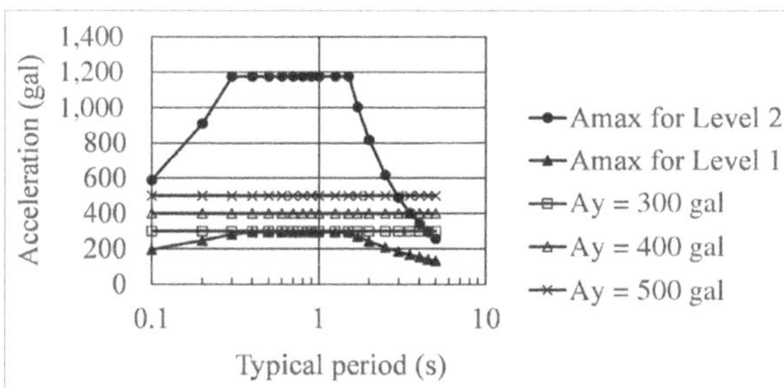

FIGURE 4.1.3 Relationship between input acceleration and yield acceleration (Soil Class III ground).

increases due to a large deviation from the Level 2 ground motion design acceleration response spectrum, while if the yield acceleration is set above 500 gal, the elastoplastic response displacement gradually tends to approach the yield displacement. Specific features are detailed in Figure 4.1.4.

The results of a comparison between the maximum response displacement (δ_{max}^{L2}) obtained simply by applying the hybrid method, which changes the applicable equation around the natural period of 1 second, and the maximum response displacement (u_{max}) obtained from a dynamic earthquake response analysis using the earthquake motion matching the design response acceleration spectrum as input acceleration, as suggested by Shibata.[3] are shown in Figure 4.1.4[1, 22] shows the results of the comparison. Both values show a generally good correspondence over the entire natural period range, confirming the accuracy of the application of the simplified formula.

4.1.1.4 Yield Acceleration Spectrum Satisfying the Ductility Factor

The yield acceleration spectrum can be drawn as shown in Figure 4.1.5 with the ductility factor as a parameter. Figure 4.1.5(a) shows the yield acceleration when the phase of the seismic wave is random and Figure 4.1.5(b) compares the yield acceleration spectrum when real seismic waves, which reflect a specific phase, are used. They are mutually similar, indicating that the yield acceleration spectra are not so affected by the phase difference. As shown in Figure 4.1.6, given the natural period T of the target structure (one-degree-of-freedom vibration system) and its

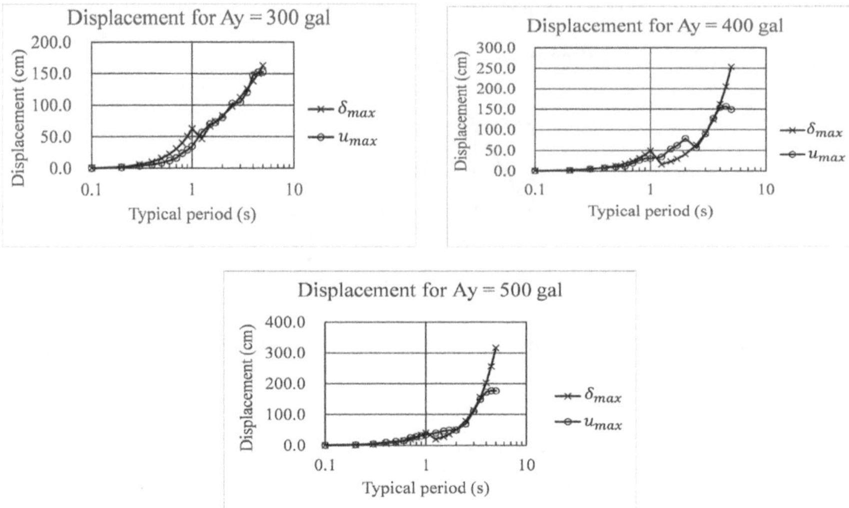

FIGURE 4.1.4 Verification of applicability of displacement prediction equations.

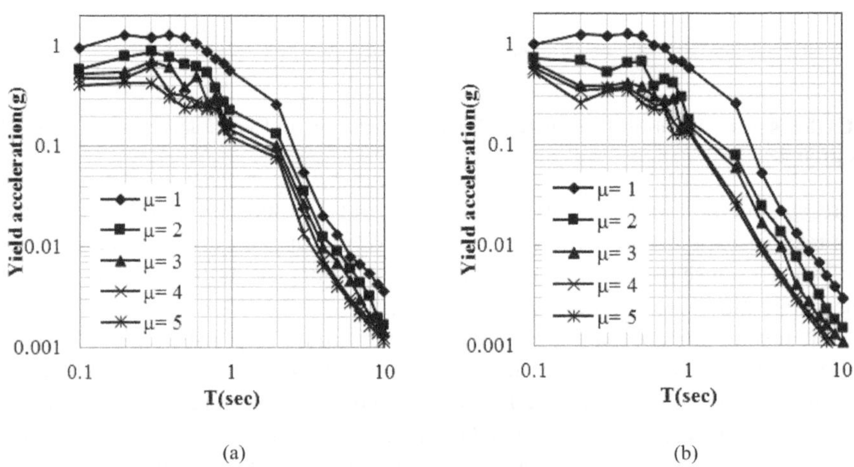

FIGURE 4.1.5 Examples of yield acceleration spectra: (a) for random phase; (b) for real seismic phase.

yield acceleration spectrum $S_y(T)$, the plasticity factor μ that satisfies both can be estimated from Figure 4.1.6.

4.1.2 Overview of Seismic Behavior Analysis Methods for Underground Structures

Underground conduits are often connected to underground structures at their start and end points or at river crossings along the way. Buried pipelines have relatively

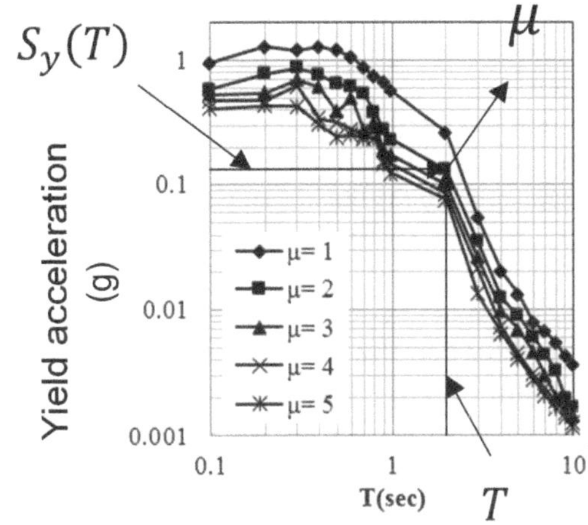

FIGURE 4.1.6 Method for determining the ductility factor from the natural period and yield acceleration.

small cross sections among underground structures, and seismic design methods based on the response displacement method are applied on the assumption that the seismic behavior follows the surrounding ground. On the other hand, the seismic behavior of large underground structures, unlike that of buried pipelines, does not necessarily follow the surrounding ground. Therefore, where buried pipelines connect to relatively large underground structures, there will be a difference in seismic behavior between the two, and care is taken in design to absorb this difference so that seismic damage does not occur.

In general, seismic design methods for underground structures can be broadly classified into static and dynamic analysis methods, with static analysis methods,[4] including seismic intensity, response displacement, and response seismic intensity methods. Dynamic analysis methods include the response spectrum method, the modal analysis method, and the time history response analysis method.

The response seismic method is an analysis method used for seismic response analysis of underground structures modeled by FEM. The ground is modeled using plane strain elements, and seismic ground displacement is simulated by applying seismic ground inertia forces to each element of the analysis model. When the response displacement method is applied to the FEM model, the seismic ground displacement is simulated by applying nodal external forces equivalent to the ground displacement to each node of the analytical model.

4.2 SEISMIC DAMAGE ANALYSIS OF EXPANSION FLEXIBLE JOINTS

The Great East Japan Earthquake of March 11, 2011, caused a lot of seismic damage[5] to water supply lifelines. In this earthquake, leakage damage[6,7] due to de-piping of expansion flexible joints installed in steel pipelines near concrete structures occurred in several locations. The first case was a 1,200 mm diameter welded steel pipe for water supply, which leaked from an expansion flexible joint installed between a shaft and a valve pit, and the second case was a 2,400 mm diameter welded steel pipe for water supply, which leaked from an expansion flexible joint installed at both ends of a concrete roll-up around a river crossing. Other cases of damage included pipe breakage from expansion flexible joints in liquefied ground and damage to the pipe body of the expansion flexible joint tie-rod attachment section due to ground deformation. This section presents the results of detailed investigations of these two cases.

4.2.1 Vicinity of Shafts

4.2.1.1 Relationship Between the Seismic Behavior of the Shaft and Leakage Damage

The water supply line under investigation was a water pipe made of welded steel pipes with a diameter of 1,200 mm and an operating water pressure of 1.2 MPa, which is relatively high for a water pipeline.

The closest seismometer installed in the vicinity was MYG016 (Shiraishi), which was 1.53 km from the accident site, the main earthquake on March 11 and the aftershocks on April 7, when damage to lifelines was relatively large, caused tremors of intensity 6 or lower near the damaged site. The expansion and contraction flexible joint used here is called a closer type[8] and has a watertight structure with a housing and rubber ring as shown in Figure 4.2.1.

As shown in Figure 4.2.2(a) and (b), wear has occurred in approximately one quarter of the top of the housing, and some bolts and nuts in particular show signs of wear to the extent that their function has been lost.

4.2.1.2 Expansion Flexible Joint Damage Mechanism

Figure 4.2.3 shows a CAD diagram of the worn section (yellow section), the lost volume of which corresponds to approximately 6.3% of the original volume of the section in question. There was no leakage in this section at the time of the main earthquake on March 11 that overflowed onto the ground surface, and a large amount of water leakage occurred during the aftershock on April 7. This event can be interpreted by assuming the following damage process. It is thought that the main earthquake on March 11 loosened the housing of the expansion flexible joint, causing a small amount of water leakage, and that the high-water pressure in the water pipe

FIGURE 4.2.1 Expansion flexible joint – closer type.[11]

(a) (b)

FIGURE 4.2.2 Wear damage on the telescopic flexible joint: (a) actual damage scene; (b) worn parts of the telescopic flexible joint.

caused sandblasting involving backfilling sand, which gradually wore away the steel pipe surface coating and housing. Furthermore, it is assumed that the external force of the aftershock on April 7 triggered the loosening of the pipe and the de-piping of the telescopic flexible joint that lost the water sealing function of the housing that was partially cut off, leading to a massive leakage.

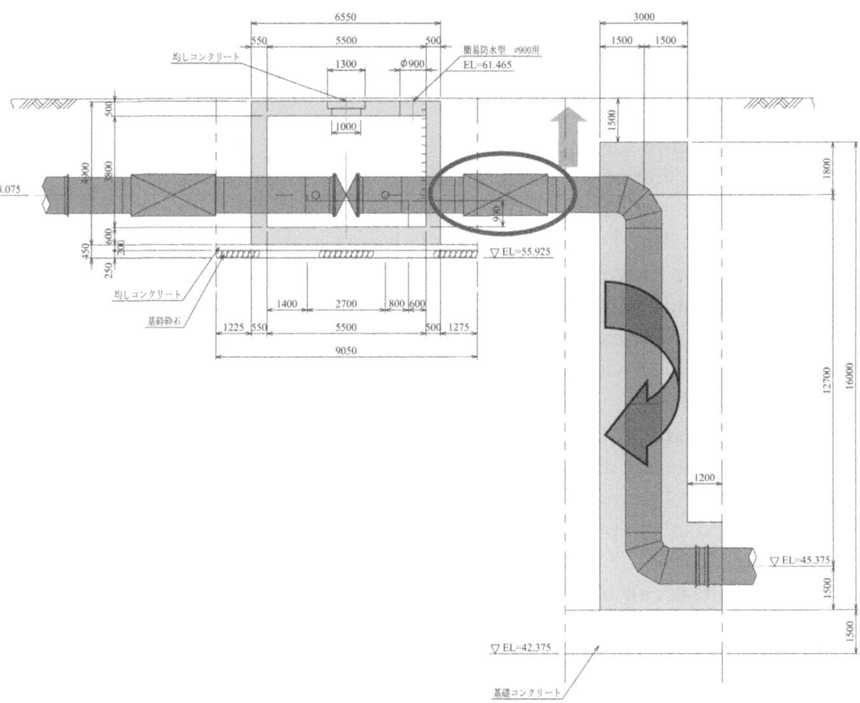

FIGURE 4.2.3 Expansion flexible joint arrangement around valve chambers and shafts.

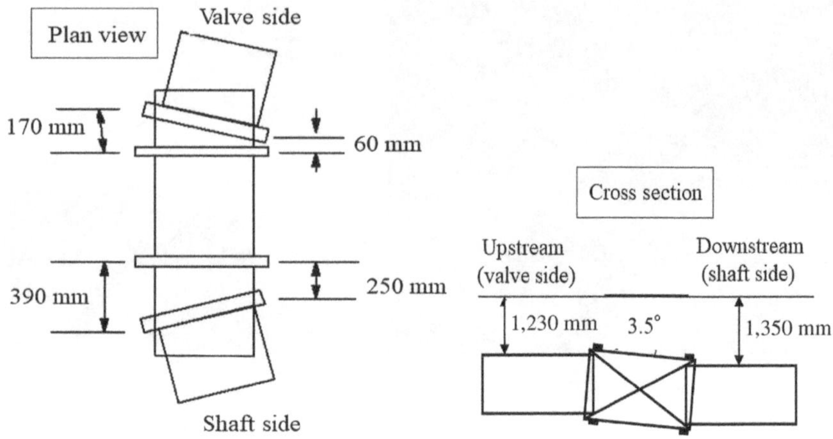

FIGURE 4.2.4 Arrangement of expansion flexible joints after the earthquake.

There have been several reports[5,6] of leakage accidents caused by sandblasting in water pipelines in the past. In particular, a case[5] of abrasion damage to the housing of an expansion flexible pipe caused by high-pressure water of 1.4 MPa, which is almost equal to the water pressure at the accident site, has also been reported, and the damage mechanism described above is considered sufficiently valid.

Figure 4.2.4 shows the deformation state of an expansion flexible joint.

4.2.1.3 Failure Mechanisms of Valve Chamber and Shaft Systems

4.2.1.3.1 Analysis Model

As shown in Figure 4.2.4, expansion flexible joints[5] are installed at both ends 1.2 m apart on the outside of the valve pit. These expansion flexible joints are generally required in water pipeline design guidelines to absorb the relative displacement between the pipeline connected to the valve pit and the buried pipeline that is not connected to the valve pit. However, here the shaft is located 5 m to the right of the valve pit, and the possibility of additional shaft settlement must be taken into account, but here the expansion flexible joints are assumed to have been installed as usual in accordance with the technical document[8] issued by Japan Water Steel Pipe Association (WSP).

Figure 4.2.5 shows the measured relative displacement of the expansion flexible joint before and after the earthquake at the time of the leakage incident on April 7. As shown in the figure, it appears that at least 120 mm of vertical relative displacement occurred in the pipeline between the valve pit and the shaft due to the settlement of the shaft.

As a result, it can be inferred that the telescopic flexible joints had an angular change of at least 3.5° before the earthquake.

Figure 4.2.6 shows the ground characteristics in the vicinity of the shaft. The liquefaction hazard of the ground in question for the earthquake motion has a *PL* value of about 27, which indicates the possibility of liquefaction occurring, but since the occurrence of liquefaction damage could not be confirmed during the leakage

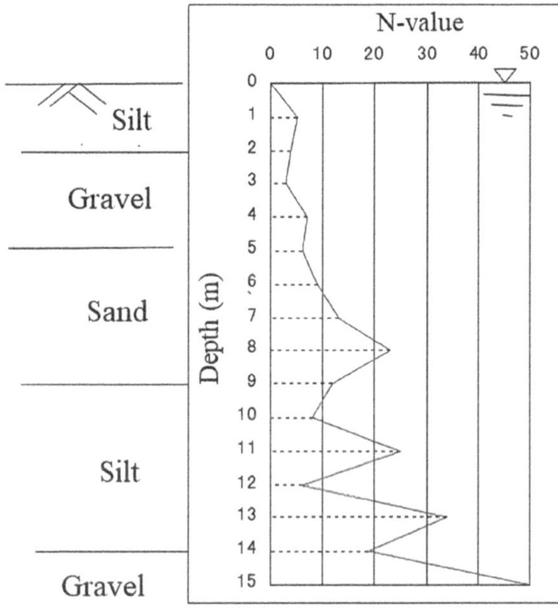

FIGURE 4.2.5 N-value distribution of the subject ground.

accident restoration work, the following discussion is based on the assumption that liquefaction did not occur.

4.2.1.3.2 Occurrence of Leakage during the Main Shock

In order to elucidate the mechanism by which the above leakage damage occurred after the mainshock and aftershocks, the validity of the failure mechanism of the expansion flexible joints due to the seismic response of the shafts is examined.

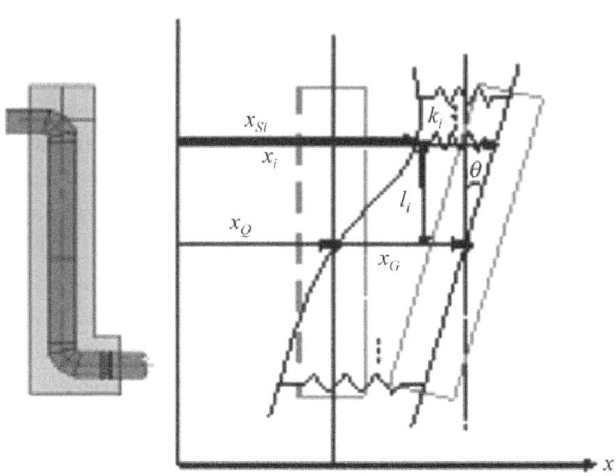

FIGURE 4.2.6 Coordinate system around the shaft.

The seismic response of the shaft was calculated from the seismic records of MYG016 (Shiraishi) using SHAKE (one-dimensional ground response analysis program), and the seismic response of the shaft was calculated using these records as external forces.

The seismic response of the shaft is calculated from the horizontal displacement response of the shaft center of gravity and the rotational response, assuming the shaft to be a rigid body for simplicity of analysis. The ground springs supporting the shaft are calculated based on the road bridge design guideline.[31]

With the preceding analytical model, the shaft motion can be described by the following equation:

$$M\left(\ddot{x}_Q+\ddot{x}_G\right)+\sum_{i=1}^{N}k_i\left(x_i-x_{si}\right)=0, \tag{4.2.1}$$

$$I_G\ddot{\theta}=-k_R\theta-\sum_{i=1}^{N}k_i\left(x_i-x_{si}\right)l_i. \tag{4.2.2}$$

where x_Q, x_G, x_{Si} are the base displacement, the horizontal shaft gravity center displacement and the ground displacement, respectively, as shown in Figure 4.2.6. The rotation angle around the shaft gravity center and the ground displacement are related by the following Equation (4.2.3)

$$x_i=x_G+l_i\theta+x_Q. \tag{4.2.3}$$

Figure 4.2.7 shows the horizontal and vertical displacement response of the pipeline at the shaft rotation angle and the position of the telescopic flexible joint. According to the figure, the maximum value of the vertical displacement response of the pipeline is approximately 11 mm, and it can be inferred that this forced displacement deformed the telescopic flexible joint and induced a small leakage. In other words, when a forced displacement of 11 mm was applied in the vertical direction to this expansion flexible joint, the rotation angle was approximately 0.21° with a push ring spacing of 1960 mm, and when the prior bending angle of 3.5° was added, an angular deformation of approximately 3.7° occurred.

However, since the allowable bending angle of the telescopic flexible joint shown in Figure 4.2.8 is 6°, it is assumed that an angle of at least 4–5° is required for small leakage to occur from the telescopic flexible joint at this point, so it is estimated that the actual bending angle before the earthquake reached around 4.5°. It is also assumed that the housing of the telescopic flexible coupling became worn and damaged due to leakage under high pressure, resulting in the loss of the mechanism that secures the push ring. In this state, the aftershock of April 7 was felt, and it is thought that the impact force of the earthquake destroyed the water sealing function of the damaged expansion flexible joints, causing a large amount of water leakage. Incidentally, as shown in Figure 4.2.9, the seismic response at this time was about half the maximum acceleration during the main earthquake, and the vertical displacement at the location of the expansion flexible joint was about 2 mm.

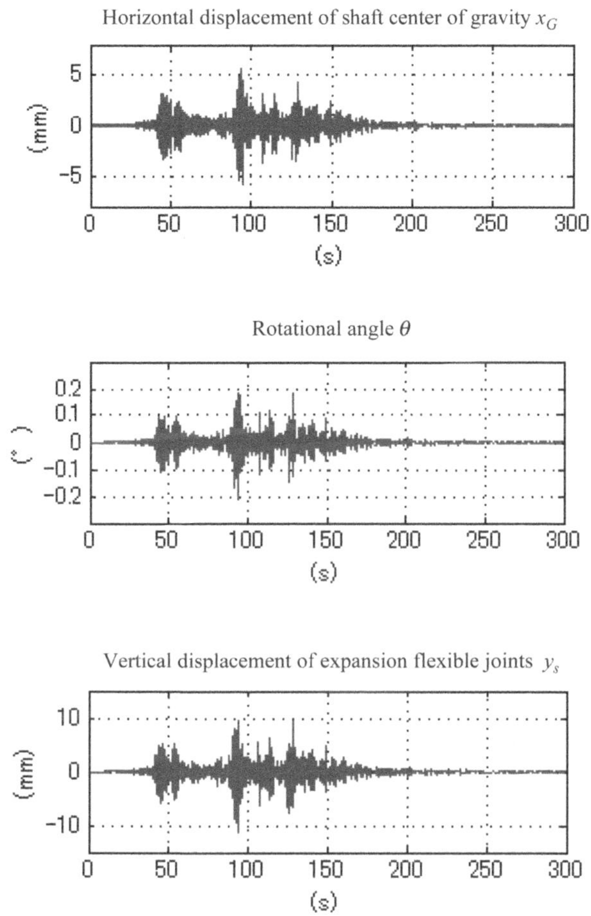

FIGURE 4.2.7 Seismic response of the shaft during the main shock.

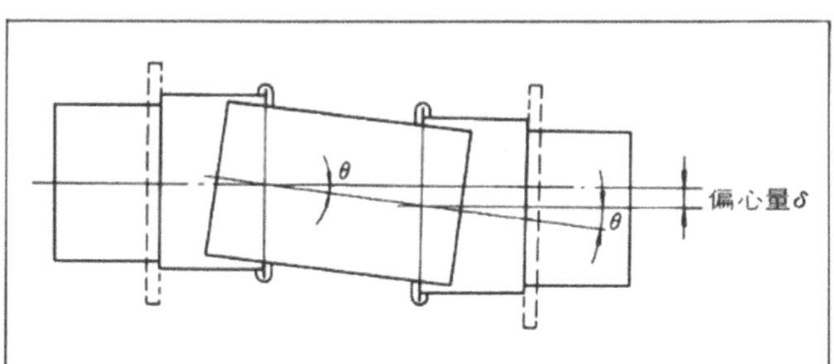

FIGURE 4.2.8 Deformation modes of expansion flexible joints.

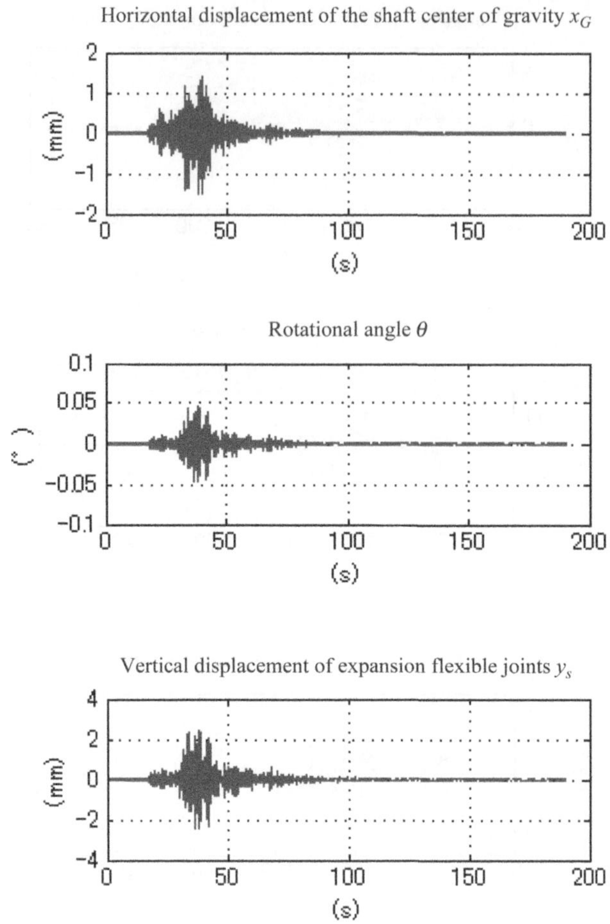

FIGURE 4.2.9 Seismic response of shafts during aftershocks.

4.2.1.4 Seismic Measures for Expansion Flexible Joints

The cause of the pipe breakout of the expansion flexible joint in the 2011 Great East Japan Earthquake was that a considerable degree of relative displacement had occurred in the expansion flexible joint before the earthquake due to the settlement of the shaft installed near the expansion flexible joint, and that the shaft earthquake response under these conditions caused forced displacement in the expansion flexible joint, which in turn caused a small leak. Furthermore, the high-water pressure of the water pipe caused the sealing fittings of the expansion flexible joints to be worn and damaged by the sandblasting phenomenon involving backfilled sand, resulting in the loss of the water sealing function, and the pipe is thought to have been easily dislodged by the seismic impact force of the aftershocks.

Therefore, as future seismic countermeasures, the layout of concrete structures and shafts before and after the installation of expansion flexible joints should be carefully assessed, and where expansion flexible joints may be subjected to forced displacement during an earthquake, the installation position of the expansion flexible joint should be changed to eliminate the effect or to cope with the forcing forces. It is necessary to implement some kind of seismic reinforcement measures in advance, for example, by implementing reinforcement measures such as anti-detachment devices that can cope with the forcing force.

4.2.2 Around the Inverted Siphon

4.2.2.1 Relationship Between Seismic Behavior of Inverted Siphon and Leakage Damage

The second case[8] is a water pipe with a diameter of 2,400 mm, also in Shiroishi City, Miyagi Prefecture, where the expansion flexible joints on both sides of the concrete-wrapped section of a river overbank were simultaneously deformed by the main earthquake. The damage caused by the leakage is shown in Figure 4.2.10, and the field diagram of 2,400 mm diameter pipe and modeling of concrete sections[1] is also shown in Figure 4.2.11.

4.2.2.2 Expansion Flexible Joint Damage Mechanism[5]

It is presumed that the rotational movement of the concrete rolls caused relative displacement at the position of the expansion flexible joint, resulting in de-pipe. The concrete section is assumed to be a rigid body and formulated as a plane motion in

FIGURE 4.2.10 Leakage damage from expansion flexible joints due to deformation of the concrete-wrapped inverted siphon.

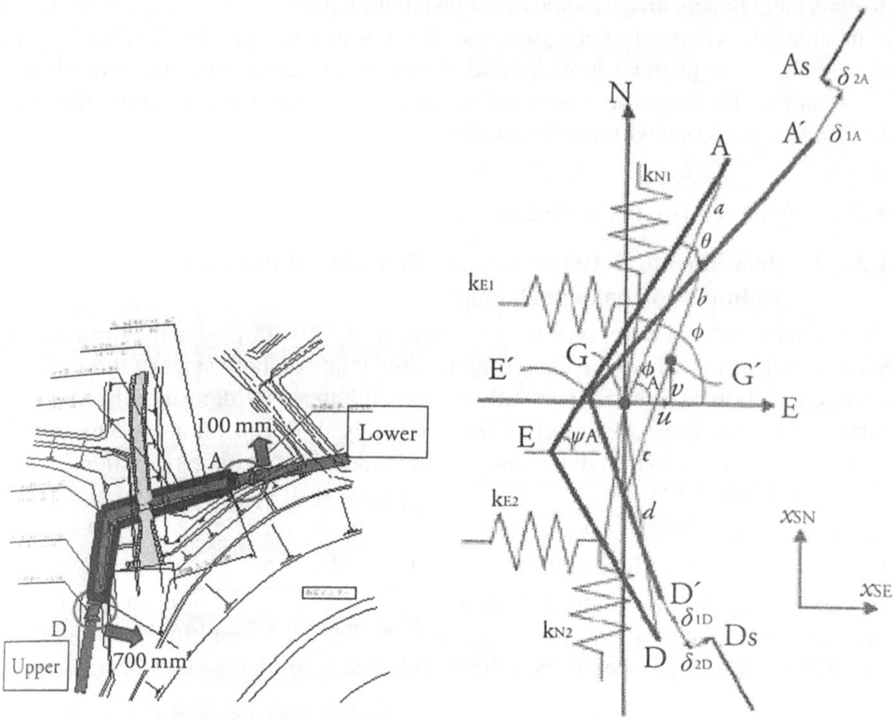

FIGURE 4.2.11 Field diagram of 2,400 mm diameter pipe and modeling of concrete sections.[1]

the east-west, north-south, and rotational directions, and seismic response calculations are carried out.

$$\mathbf{M\ddot{u}} + \mathbf{Ku} = \mathbf{Kx_S} \tag{4.2.4}$$

$$\mathbf{M} = \begin{pmatrix} M & 0 & 0 \\ 0 & M & 0 \\ 0 & 0 & I_G \end{pmatrix} \tag{4.2.5}$$

$$\mathbf{K} = \begin{pmatrix} k_{E1} + k_{E2} & 0 & (k_{E1}b - k_{E2}c)\sin\varphi \\ 0 & k_{N1} + k_{N2} & (k_{N2}c - k_{N1}b)\cos\varphi \\ (k_{E1}b - k_{E2}c)\sin\varphi & (k_{N2}c - k_{N1}b)\cos\varphi & \begin{matrix} k_{E1}b^2\sin^2\varphi + k_{N1}b^2\cos^2\varphi \\ -k_{E2}c^2\sin^2\varphi - k_{N2}c^2\cos^2\varphi \end{matrix} \end{pmatrix} \tag{4.2.6}$$

$$\mathbf{u} = \begin{pmatrix} u \\ v \\ \theta \end{pmatrix} \quad \mathbf{x_S} = \begin{pmatrix} x_{SE} \\ x_{SN} \\ 0 \end{pmatrix}, \tag{4.2.7}$$

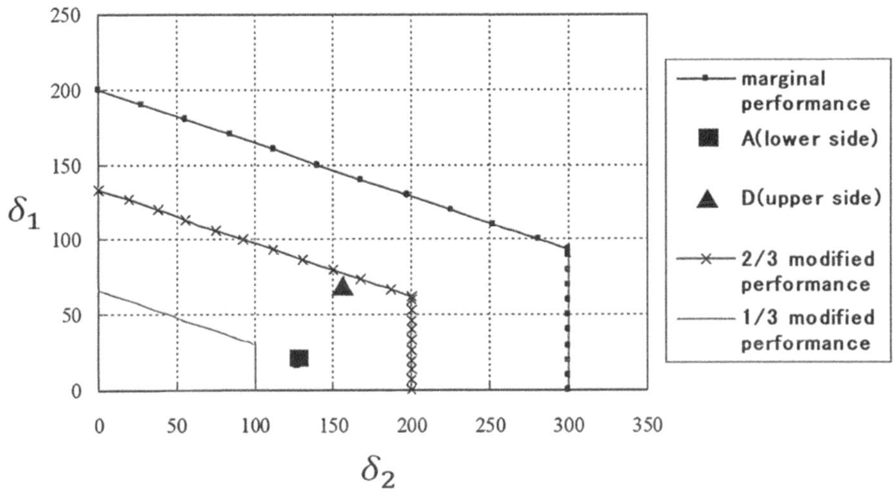

FIGURE 4.2.12 Limit performance and maximum response displacement of 2,400 mm diameter pipe.

where u, v are the displacements in the east-west and north-south directions, θ is the rotational displacement, and x_{SE}, x_{SN} are the ground displacements in the east-west and north-south directions.

The limit performance of the expansion flexible joint is determined from the axial displacement δ_1 and the shear displacement δ_2, but it is possible that the performance was reduced due to unequal settlement before the earthquake. From the response calculation results (triangles and circles) in Figure 4.2.12, it can be said that if there was performance degradation, the rotational movement of the concrete section was one of the causes of the de-pipe.

In the past, it is feared that the perspective of how the concrete overhang as a concrete structure and the telescopic flexible joint connected to it behave together during an earthquake, and what magnitude of displacement acts on the telescopic flexible joint as a result, has been out of the scope of seismic design considerations.

As a lesson learned from the aforementioned large-scale leakage accidents from shafts and overhangs, it can be said that "when installing expansion flexible joints, it is necessary to consider not only the standard installation conditions for the joints, but also measures to prevent them from detaching based on the seismic behavior of the structure and joint as an integrated structural system."

4.3 SEISMIC DESIGN METHODS FOR BURIED EXPANSION FLEXIBLE JOINTS

While above-ground pipelines are constantly affected by temperature changes, internal pressure, and vibration, buried pipelines are less affected by such effects and are more affected by displacement perpendicular to the axis due to unequal settlement. Expansion flexible joints have been installed where necessary to eliminate these effects as much as possible. However, in earthquake-prone countries such as Japan, ground shaking and ground deformation act not only on pipelines but also on expansion flexible joints equally when an earthquake occurs, so it is also necessary to ensure the seismic safety performance of expansion flexible joints.

When an earthquake occurs, the propagation of longitudinal and shear waves along the pipe axis is caused by the propagation of body waves and surface waves in the surface soil, and the pipeline expands and contracts relatively significantly in the direction of the pipe axis, but relatively little displacement occurs in the direction perpendicular to the pipe axis unless ground deformation, such as liquefaction, occurs. As a result, during earthquakes, expansion flexible joints are subjected to expansion and contraction displacement in the direction of the pipe axis under conditions where the displacement in the perpendicular direction due to unequal settlement is maintained.

It is an important requirement that expansion flexible joints buried underground have seismic performance that does not rupture or leak during earthquakes under the unequal settlement conditions described above.

In addition, in above-ground pipes installed in aqueducts, the seismic response displacement of each part of the aqueduct must be absorbed by the pipeline displacement, and for this purpose expansion flexible joints are inserted at the necessary points. Even in this case, if the support columns of the water pipeline bridge experience larger-than-expected displacements and exceed the limits of the expansion flexible joints, joint damage and water leakage will occur. In other words, it must be considered that water pipeline damage and leaks during earthquakes not only occur due to damage to the pipe body and pipe joints, but also due to damage to the expansion flexible joints installed at the connection between the pipeline and the facility.

In the past few decades, many cases of water leakage from expansion flexible joints have been reported in water pipelines during earthquake disasters. For example:

a. Unexpected seismic displacements acting on expansion flexible joints connected to underground structures such as shafts and overhangs, resulting in joint failure[9]
b. Unexpected seismic displacement of the main body of an aqueduct, resulting in joint failure[10]
c. Cases where expansion flexible joints failed due to seismic motion when the seismic deformation capacity had already been reduced due to unequal settlement before the earthquake[11]

In order to prevent seismic failure of expansion flexible joints, it is necessary to clarify the required seismic performance of expansion flexible joints and develop a new seismic design method for expansion flexible joints, based on an analysis of the failure mechanisms of existing accident cases. However, existing expansion flexible joints have been installed for many years without sufficient experimental and design confirmation of their seismic safety performance with respect to the failure limit state for unequal settlement and displacement perpendicular to the pipe axis during earthquakes, i.e., without confirmation of whether their seismic resistance is qualified or not.

In this chapter, the seismic safety performance of expansion flexible joints with respect to the failure limit state for unequal settlement and displacement in the direction perpendicular to the pipe axis during earthquakes is experimentally and design-wise confirmed, and a new seismic design method for expansion flexible joints with qualified seismic performance is proposed based on the performance design method.

4.3.1 Seismic Performance of Expansion Flexible Joints[23]

4.3.1.1 Basic Features of Expansion Flexible Joints

Expansion flexible joints are basically classified into sliding, bellows, and rubber types. All types of joints are installed to absorb pipe displacement caused by temperature changes, internal pressure, and unequal settlement at all times, but seismic performance is also required to prevent expansion flexible joint failure due to large displacements during earthquakes.

The seismic performance of bellows-type[12] expansion flexible joints is discussed in the following section as an example.

4.3.1.2 Experiments to Confirm the Seismic Performance of Bellows-Type Expansion Flexible Joints

WSP 024-2013[13] states that the seismic performance required for bellows-type expansion flexible joints[12] installed in water pipelines "shall take into account the combined displacement of the most severe displacement situations, including both normal and earthquake conditions." However, bellows-type expansion flexible joints have conventionally been designed for a calculated fatigue life of 102–106 cycles subject to temperature and pressure variations, and there has been insufficient discussion of safety against the large displacements that may occur during earthquakes. Therefore, the bellows were subjected to large displacement loading tests assuming ground shaking caused by Level 2 ground motion, and the repair limit and ultimate limit of the bellows were investigated.

The seismic safety verification method for bellows-type expansion flexible joints against Level 2 ground motion is specified in the *Guidelines for Seismic Design of Production Facilities.*[26] It requires that the (apparent) total stress amplitude generated in the bellows is less than the limit total stress amplitude corresponding to 50 and 10 cycles in the design fatigue curve described in JIS B8266 Annex 8.[14]

When Level 2 ground motion is used for safety verification of the repair limit state and the ultimate limit state, the respective design velocity response spectra shall be denoted by $S_V^L(T)$ and $S_V^U(T)$ according to reference [25] in order to distinguish between the two.

4.3.1.3 Deformation Behavior Characteristics of a Single Bellows Pile

4.3.1.3.1 Repairable Limit and Ultimate Limit Displacement for Level 2 Ground Motion

According to Utoguchi's work on low-cycle fatigue,[15] the ASME design fatigue curves cited by JIS B8266 Annex 8 are based on the design fatigue curve concept proposed by Langer,[16] which is directly adopted in the ASME Code, Section III.

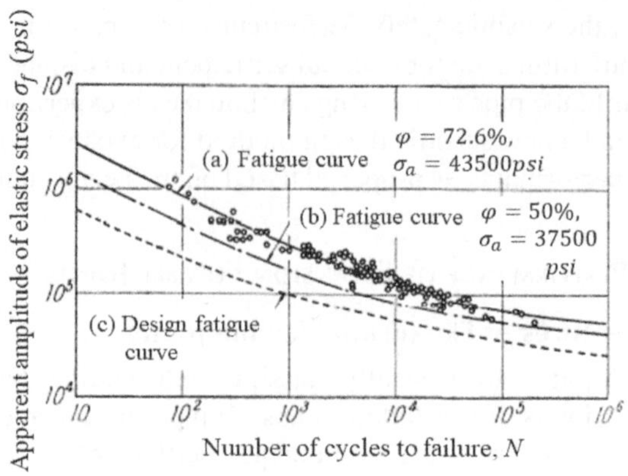

(a)

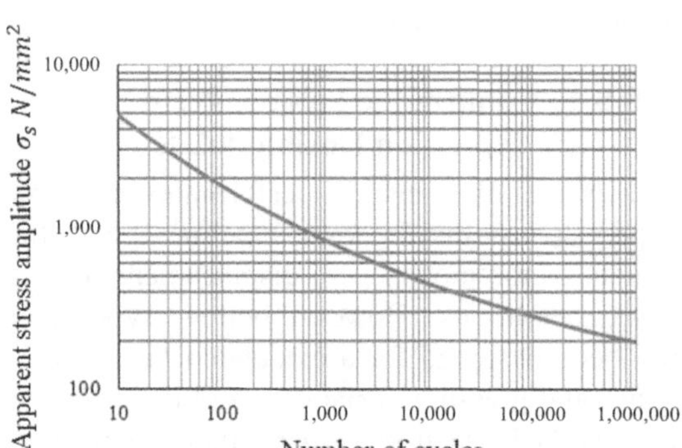

(b)

FIGURE 4.3.1 Fatigue curves for austenitic stainless steels: (a) Langer's fatigue curves;[15,16] (b) ASME design fatigue curves.[6]

Langer proposed that the design fatigue curve for austenitic stainless steel should be the lower limit of the relationship between fatigue life and apparent elastic stress (fragment) amplitude as shown in Figure 4.3.1(a), with the factor of safety set at two times on the stress side and 20 times on the fatigue life side. Figure 4.3.1(b) shows the ASME design fatigue curve obtained in this way. Here, the apparent total stress amplitude corresponds to $2\sigma_f$ in Figure 4.3.2. The strain amplitude can be obtained from the apparent total stress amplitude by using the material Young's modulus $E = 1{,}95{,}000\,N/mm^2$ at 20°C for austenitic stainless steels and $\varepsilon_t = 2\sigma_f/E$.

A bellows-type flexible joint consists of several bellows (mountain type), and the overall expansion and contraction is expressed as several times the bellows displacement of one mountain. Therefore, the basic deformation characteristics of a bellows-type flexible joint can be determined by investigating the deformation characteristics of the bellows of one pile.

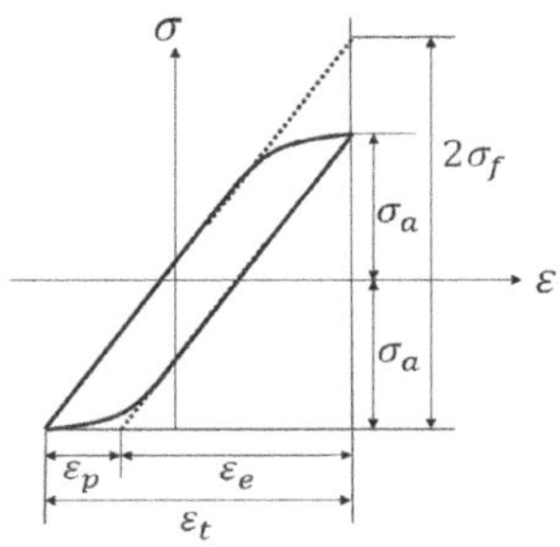

FIGURE 4.3.2 Stress–strain cycles.

TABLE 4.3.1
Test Specimen Dimensions

Item	Symbol	Unit	Value
Size of specimen			300A
Material			SUS304
Internal diameter	D_i	mm	298.5
Height	H	mm	30
Width	W	mm	34
Thickness	t	mm	1.5
Number of elbows	n		1
Young's modulus	E	N/mm²	195000

In a full-scale experiment,[17] a single-pile bellows with the dimensions shown in Table 4.3.1 was used to investigate the acting load/deformation relationship by applying a tensile load to the bellows as shown in Figure 4.3.3. Strain gauges were attached to the inside and outside surfaces of the bellows at the top of the bellows and at both bases to measure the strain distribution in each part of the bellows.

Figure 4.3.4 shows the strain distribution inside the bellows in the elastic limit, obtained by FEM numerical analysis. According to this figure, the maximum strain was greater at the apex than at the base of the bellows, which was used as the focus point for fatigue crack initiation during the experiment.

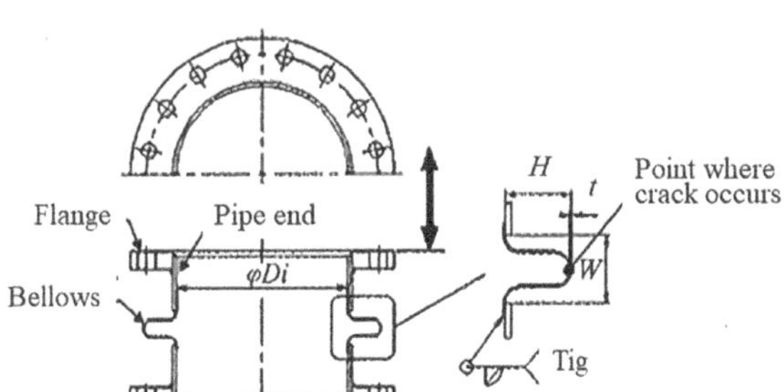

FIGURE 4.3.3 Specimen.

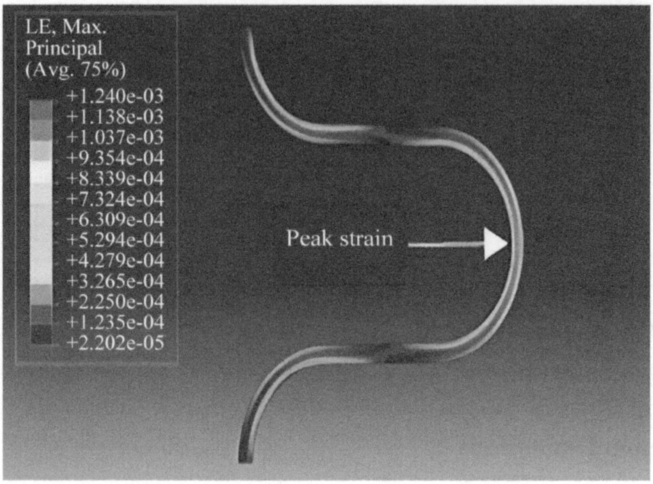

FIGURE 4.3.4 Strain distribution in bellows.

4.3.1.3.2 Hysteresis Curves and Repairable/Ultimate Limit Displacement

Repeated loading experiments were conducted to confirm the limit displacement amplitude at which a stable hysteresis curve could be drawn for 50 and 10 repetition loops, respectively, and the results shown in Figure 4.3.5 were obtained. The hysteresis curve corresponding to the repairable limit was obtained as the maximum repetition amplitude of 25 mm (upper end value of the green load) for a group of curves that could stably ensure 50 repetition loops. The hysteresis curve corresponding to the ultimate limit was obtained as the maximum repetition amplitude of 35 mm (upper end value of the red load) for a group of curves that can stably ensure 10 repetition loops.

The design fatigue curve in Annex 8 of JIS B8266 (quoted from the ASME design fatigue curve), which is required by the seismic design guidelines for high-pressure

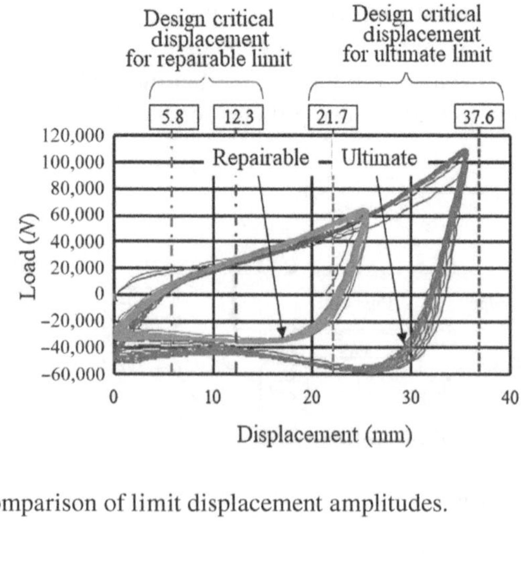

FIGURE 4.3.5 Comparison of limit displacement amplitudes.

gas facilities, includes the aforementioned safety factor, so the limit displacement is evaluated using the original fatigue curve by Langer shown in Figure 4.3.1(a). The restoration limit displacement is 21.7 mm and the ultimate limit displacement is 37.6 mm. As shown in Figure 4.3.5, the limit displacement amplitudes obtained from the cyclic loading tests are generally close to the limit displacement amplitudes calculated by Langer's original fatigue curve.

On the other hand, the limit displacement amplitude (nominal limit value) obtained from the design fatigue curve including the safety factor in JIS B8266 Annex 8, which is specified by the seismic design guideline for high-pressure gas facilities, is 5.8 mm for the repairable limit displacement and 12.3 mm for the ultimate limit displacement, but these values deviate significantly from the safety factor in the ASME fatigue curve, as shown in Figure 4.3.5. The values deviate significantly from the experimental results in Figure 4.3.5 due to the influence of the safety factor of the ASME fatigue curves. In other words, the restoration and ultimate limit displacements of the expansion flexible joints given by the current seismic design guidelines (*Guidelines for Seismic Design of High-Pressure Gas Facilities*) have a considerable safety margin compared to the experimental results.

Therefore, the following procedure for determining the limit displacement of bellows can be proposed as a new seismic design method. The limit displacement $\delta_{cr,joint}$ of the one-pile bellows is estimated as the mean value μ based on Langar's original fatigue curve, and its standard deviation σ is estimated from the variation of the limit displacement data obtained by accumulating the experimental results shown in Figure 4.3.5. This allows the marginal displacement $\delta_{cr,joint}$ of a single mountain bellows to be estimated probabilistically in the form of the following equation:

$$\delta_{cr,joint} = f_{\delta_{cr,joint}}(\mu, \sigma), \tag{4.3.1}$$

in which

$f_{\delta_{cr,joint}}$: function giving the probability statistics of the single mountain bellows limit displacement.

4.3.1.3.3 Critical Displacement Performance as Bellows-Type Expansion Flexible Joints

A standard bellows-type expansion flexible joint consists of a single joint by arranging several single bellows in series. If the maximum axial displacement is given as the design value, the number of bellows peaks, n, of the corresponding expansion flexible joint is calculated. This gives the limit displacement $\Delta_{cr,joint}$ of the expansion flexible joint as

$$\Delta_{cr,joint} = n \cdot \delta_{cr,joint}. \tag{4.3.2}$$

4.3.1.3.4 Effect of Unequal Settlement on Joint Seismic Displacement Performance

The displacement performance of standard expansion flexible joints against unequal settlement differs according to their type. In general, within a small range of unequal settlement of a few centimeters, there is a range where the deflection deformation of

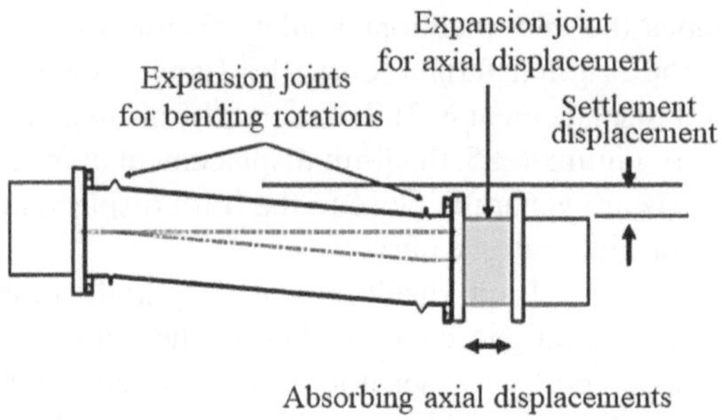

FIGURE 4.3.6 Model of an expansion flexible joint system to absorb unequal settlement.

the pipeline itself can be followed. Beyond that range, the unequal settlement must be absorbed by the displacement performance of the expansion flexible joint in the direction perpendicular to the axis.

However, if unequal settlement is to be dealt with by a single expansion flexible joint, the axial expansion displacement must be absorbed during an earthquake after settlement has been absorbed within the joint. It should be noted that the seismic displacement absorption performance may be slightly reduced due to the effect of unequal settlement.

If a single expansion flexible joint cannot absorb the amount of unequal settlement, a combination of several joints can cope with larger unequal settlements. For example, considering the bending direction rotation performance of bellows, a bellows-type[12] expansion flexible joint can be combined with an axial expansion joint and a bending direction rotation joint function to form a joint system that can cope with a wide range of unequal settlement displacements. Figure 4.3.6 shows an example of this, where a joint dedicated to rotation in the bending direction absorbs unequal settlement and a joint dedicated to axial expansion absorbs seismic movement in the axial direction.

For example, a joint dedicated to rotation in the bending direction can be designed to physically allow only rotation displacement in the bending direction by installing a hinge mechanism and a pin structure, while a joint dedicated to axial expansion and contraction can be designed to absorb the amount of expansion and contraction in the axial direction exclusively.

4.3.2 New Seismic Design Methods for Expansion Flexible Joints

4.3.2.1 Joint Deformation for Normal and Earthquake Conditions

The expansion joint shown in Figure 4.3.7 is a bellows-type expansion flexible joint that has several bellows in series. The maximum elongation characteristics of an expansion joint is given by $\Delta_{cr,joint}$. When an expansion joint is installed for the release of thermal effect in a piping system of engineering plants, the elongation of an expansion joint is not so large, and an appropriate number of bellows are selected to meet the elongation requirement. If an expansion joint is used for severe ground motion by an earthquake, on the other hand, the elongation might need many

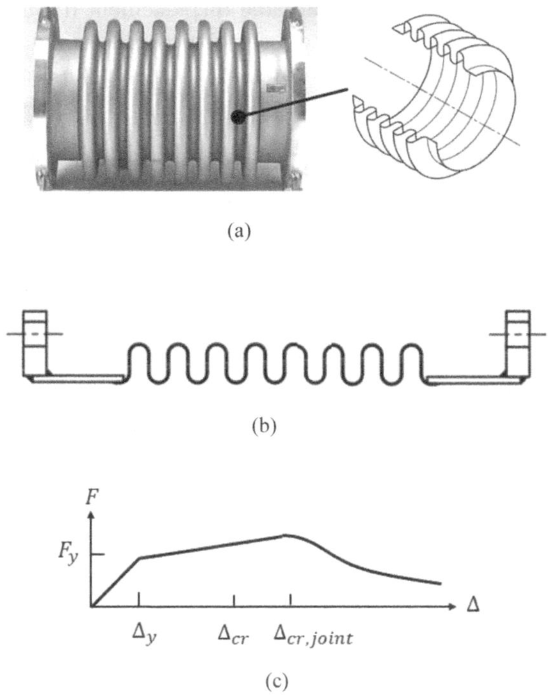

FIGURE 4.3.7 Schematic illustration of a bellows-type expansion joint: (a) overall view of an expansion flexible joint; (b) example of a cross-sectional view of an expansion flexible joint; (c) mechanical property of an expansion flexible joint.

bellows to absorb a large axial deformation due to earthquake ground motions. In this situation, some bellows might exceed the yield level for the extremely large axial deformation.

From the engineering point of view, the limit displacement Δ_{cr} for verification of the expansion flexible joint is given by

$$\Delta_{cr} = \gamma \cdot \Delta_{cr,joint}, \tag{4.3.3}$$

where

γ: limit displacement reduction factor for expansion flexible joints, the magnitude of which is basically $\gamma \leq 1$.

By the way, when the limit displacements for the experimental results of the restoration limit and the ultimate limit in Figure 4.3.5 are used separately, they are denoted as Δ_{cr}^R and Δ_{cr}^U, respectively.

When seismic displacement acts on an elastic flexible joint with the deformation suppression mechanism described above, if the magnitude of the displacement is within the displacement limit of the deformation suppression mechanism, the elastic flexible joint has adequately absorbed the seismic displacement. On the other hand, when the seismic displacement exceeds the displacement limit of the deformation restraining mechanism, the remaining displacement that could not be absorbed by the joint will be absorbed by the deformation of the surrounding water pipeline or underground structures such as valve chambers.

4.3.2.2 Seismic Safety Verification Method for Expansion Flexible Joints

The seismic safety of an expansion flexible joint is verified as follows, depending on whether the seismic displacement stays within the limit displacement of the deformation restraining mechanism or not.

When $\Delta \le \Delta_{cr}$, the expansion flexible joint can absorb the seismic displacement, and the safety of the joint can be confirmed.

When $\Delta > \Delta_{cr}$, the expansion flexible joint cannot absorb the seismic displacement, and the remaining displacement $(\Delta - \Delta_{cr})$ is absorbed by the surrounding buried conduits and underground structures. It is therefore necessary to check whether or not the residual displacement can be absorbed,
where

Δ: seismic axial displacement acting on the expansion flexible joint
Δ_{cr}: maximum deformation of the deformation restraining mechanism

4.3.3 Seismic Safety Verification Methods for Buried Expansion Flexible Joints[23]

4.3.3.1 Effect of Unequal Settlement

Underground structures such as the lower part of aqueduct abutments and valve chambers are basically supported by piles to prevent settlement, but the pipelines connected to them are buried directly in the ground. As a result, expansion flexible joints have been installed to absorb the displacement due to concerns about unequal settlement between the underground structure and the buried pipeline.

Usually, unequal settlement begins immediately after the underground structure, expansion flexible joints, and buried pipelines are installed, and is completed before the occurrence of an earthquake. On the other hand, expansion flexible joints installed as a countermeasure against pipe lifting and settlement due to ground liquefaction are required to be able to follow large displacements in the direction perpendicular to the axis, which are different in scale from normal unequal settlement.

Therefore, it is necessary to adopt an expansion flexible joint of a type that can absorb the assumed unequal settlement displacement for the expansion flexible joint selected for unequal settlement countermeasures, and to check whether the axial expansion absorption performance against ground shaking and ground deformation is affected by unequal settlement.

4.3.3.2 Seismic Displacement Acting on Buried Expansion Flexible Joints

If the valve body is fixedly supported as shown in Figure 4.3.8(a), the valve chamber-side expansion flexible joint will show the same response displacement as the ground, while the buried pipeline-side expansion flexible joint will show the seismic

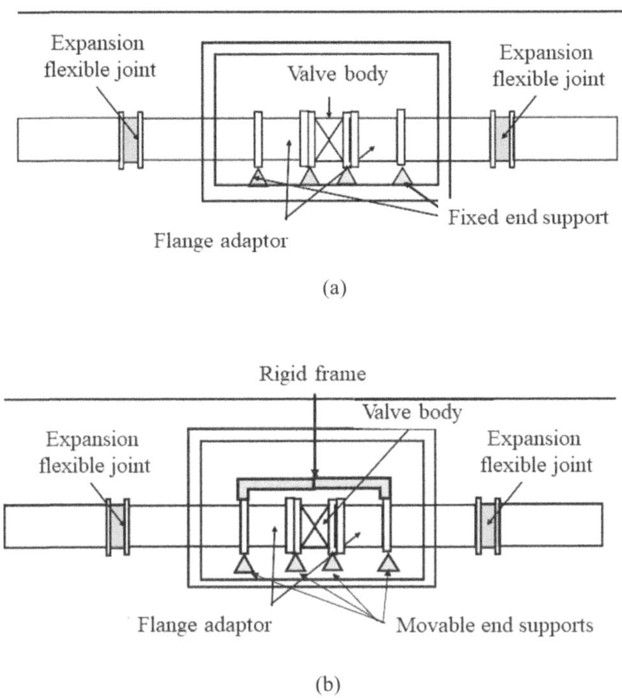

FIGURE 4.3.8 Reinforcement method of the valve body with expansion flexible joint deformation control: (a) reinforcement methods for valve body fixed end supports; (b) reinforcement method for valve body movable end support.

extension displacement of the buried pipeline. In this case, the seismic displacement[23] acting on the expansion flexible joint can be calculated as the relative displacement between ground and pipe[18,19] using the following equation.

Relative displacement Δ^{L1} for Level 1 ground motion:

$$\Delta^{L1} = \left(1 - q^{*}\alpha_S\right)U_h^{L1}, \tag{4.3.4}$$

Relative displacement Δ^{L2} for Level 2 ground motion:

$$\Delta^{L2} = \left(1 - q^{*}\alpha_S\right)U_h^{L2}. \tag{4.3.5}$$

On the other hand, if the valve body is movably supported, as shown in Figure 4.3.8(b), both the valve chamber side and the buried pipeline side show the same response displacement as the pipeline if there is no expansion flexible joint. As a result of the installation of the expansion flexible joints, the pipeline expansion and contraction are interrupted, and the seismic expansion and contraction displacements of the buried pipeline act on both ends of the expansion flexible joints. In this case, the seismic displacement acting on the expansion flexible joint can be

TABLE 4.3.2
Design Conditions for Pipelines

Element	Item	Symbol	Unit	Value
Pipe	Outer diameter	D	m	0.7
	Thickness	t	m	0.0015
	Young's modulus	E	kN/m^2	205,000,000

calculated as the expansion and contraction displacement of the buried pipeline[18] using the following equation:

Expansion and contraction displacement $\Delta^{*,L1}$ for Level 1 ground motion:

$$\Delta^{*,L1} = q^{*}\alpha_S U_h^{L1}. \tag{4.3.6}$$

Expansion and contraction displacement $\Delta^{*,L2}$ for Level 2 ground motion:

$$\Delta^{*,L2} = q^{*}\alpha_S U_h^{L2}, \tag{4.3.7}$$

where

α_S: Transformation coefficient
q^{*}: slip coefficient for calculating relative displacements[19]
U_h^{L1}: Ground-shaking displacement for Level 1 ground motion
U_h^{L2}: Ground-shaking displacement for Level 2 ground motion

- U_h^{L2L}: Ground motion displacement for limit state verification of restoration limit from Level 2 ground motion
- U_h^{L2U}: Ground motion displacement for limit state verification of ultimate limit from Level 2 ground motion

4.3.3.2.1 Relative Displacement Between Ground and Pipe

The design conditions for a water pipeline adjacent to an underground structure are set as shown in Table 4.3.2. The relative displacement between the conduit and the ground and the elongation displacement of the conduit itself are shown in Figure 4.3.9. The figure shows the relationship between the displacements and the ground prevailing period in Equations (4.3.4)–(4.3.7). Figure 4.3.9(a) shows that the relative displacement corresponding to Level 2U ground motion exceeds 100 mm in the region where the ground prevailing period is 1.0 s or more. On the other hand, the maximum displacement of pipe expansion and contraction in Figure 4.3.9(b) shows a similar trend for all three ground motions but a relatively large value exceeding 100 mm for Level 2L ground motion.

4.3.3.3 Pipe Strain around Expansion Flexible Joints

4.3.3.3.1 Pipe Strain in Case of Valve Body Fixed End Support

1. When the relative displacement of the ground and buried pipe does not exceed the limit displacement of the expansion flexible joint:

$$\Delta \le \Delta_{cr}, \quad \varepsilon_S = q\alpha_S\varepsilon_G, \tag{4.3.8}$$

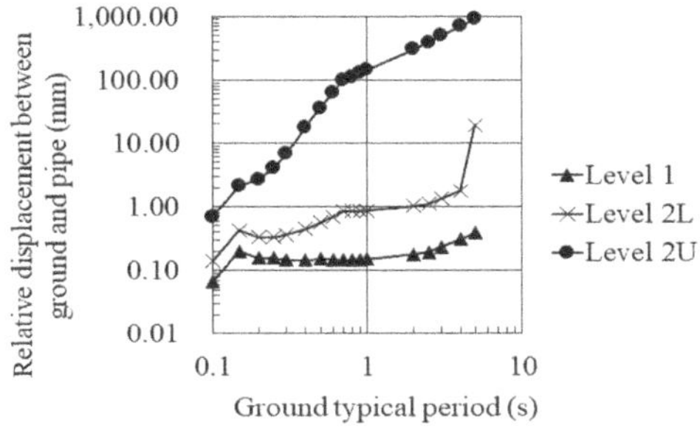

(a)

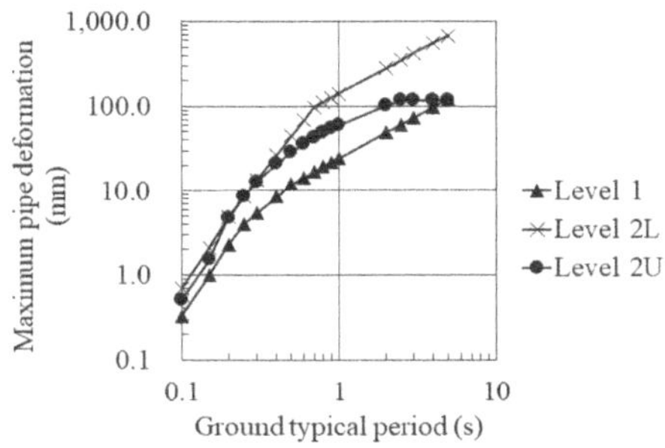

(b)

FIGURE 4.3.9 Axial displacements to be absorbed by expansion flexible joints: (a) relative displacement between ground and pipe; (b) maximum pipe expansion and contraction displacement.

where

ε_G: ground strain
ε_S: seismic pipe strain in the direction of the pipe axis[18]
Δ: relative displacement between ground and pipe, given by $\Delta = \Delta^{L1}, \Delta = \Delta^{L2}$ for Level 1 and Level 2 ground motions, respectively
Δ_{cr}: limit displacement of the expansion flexible joint
q: slippage factor for pipe strain calculation

Since the flexible expansion joint absorbs all displacements between the ground and the pipeline, no tensile or compressive forces due to the expansion flexible joint are generated at the fixed end of the structure.

2. When the relative displacement of the ground and buried pipe exceeds the expansion flexible joint limit displacement:

$$\Delta > \Delta_{cr},\ \ \varepsilon_S = q\alpha_S\varepsilon_G + \frac{\Delta - \Delta_{cr}}{L/4}. \tag{4.3.9}$$

When the maximum relative displacement is applied by seismic waves, the tensile and compressive stresses are zero at the point of action. On the other hand, when it is necessary to absorb the remaining displacement, as in the case of Figure 4.3.8(a), an additional force in either the tensile or compressive direction is generated at the pipe end. The tensile/compressive force F acting on the fixed end of the structure can be calculated as follows, assuming that the residual displacement is absorbed in the quarter-wave section:

$$F = \pi DtE\frac{\Delta - \Delta_{cr}}{L/4} \leq \pi Dt\sigma_{y,pipe}, \quad (4.3.10)$$

where

D: pipe diameter
t: pipe thickness
E: Young's modulus of the pipe material
L: horizontal propagation wavelength of seismic waves
$\sigma_{y,pipe}$: yield stress of the pipe material

4.3.3.3.2 Pipe Strain in Case of Valve Body Free End Support

1. When the expansion displacement of the buried pipe does not exceed the expansion flexible joint limit displacement:

$$\Delta^* \leq \Delta_{cr}, \quad \varepsilon_S = q\alpha_S\varepsilon_G, \quad (4.3.11)$$

where

ε_G: ground strain
ε_S: seismic pipe strain in the direction of the pipe axis
Δ^*: maximum displacement of pipe expansion and contraction, which is given by
$\Delta^* = q^*\alpha_S U_h$
Δ_{cr}: limit displacement of expansion flexible joint
q: slippage factor for pipe strain calculation
q^*: slippage factor for calculating pipe expansion and contraction displacement

In this case, no tension or compression forces from the conduit act on the rigid parts of the valve body and flange adaptor, as the expansion flexible joint absorbs all the pipe expansion and contraction displacements.

2. When the pipe expansion displacement of the buried pipe exceeds the expansion flexible joint limit displacement:

$$\Delta^* > \Delta_{cr}, \quad \varepsilon_S = q\alpha_S\varepsilon_G + \frac{\Delta^* - \Delta_{cr}}{L/4}. \quad (4.3.12)$$

The residual displacement $\left(\Delta^* - \Delta_{cr}\right)$, which the expansion flexible joint could not absorb, occurs additionally in the pipeline as in the second term on the right-hand side of Equation (4.3.10). On the other hand, additional tensile and compressive forces F^* of Equation (4.3.13) from the pipeline are generated in the rigid parts of the valve body/flange adaptor:

$$F^* = \pi DtE\frac{\Delta^* - \Delta_{cr}}{L/4} \leq \pi Dt\sigma_{y,pipe}. \quad (4.3.13)$$

4.4 SEISMIC SAFETY ASSESSMENT METHODS FOR ABOVE-GROUND EXPANSION FLEXIBLE JOINTS[24]

4.4.1 Estimation of Seismic Displacements Acting on Expansion Flexible Joints

In above-ground pipes installed in water bridges, the seismic response displacement of each section of the water bridge has to be absorbed by the pipe displacement, and for this purpose, expansion flexible joints are inserted at the necessary points. Even in this case, if the support columns of the pipeline are subjected to greater displacement than expected and exceed the limits of the expansion flexible joints, joint damage and leakage will occur. In other words, it must be taken into account that water pipeline damage and leaks during earthquakes not only occur due to damage to the pipe body and pipe joints but also due to damage to the expansion flexible joints installed at the joints between the pipelines and facilities.

4.4.1.1 Water Pipeline Bridges and Surrounding Water Pipeline Facilities

Water pipelines[20] crossing rivers become above-ground pipes on the abutments of pipeline bridges from a buried state and are passed over to the abutment at the other end. In general, above-ground pipes are subject to thrust forces due to temperature changes and internal pressure at all times, so expansion flexible joints are required to absorb the resulting axial displacements. On the other hand, during earthquakes, the relative displacements of the various parts of the pipeline bridge act on the expansion flexible joints, so seismic safety assessment of the joints is required.

In the case of the single-span pipeline bridge shown in Figure 4.4.1(a), expansion flexible joints are placed at the abutments at both ends, the adjacent pipelines are fixed-end supported, and the other pipelines are movably supported. In the case of a multiple-span water pipe bridge, as shown in Figure 4.4.1(b) in the same diagram, expansion flexible joints are placed at the top ends of the intermediate abutments, and the pipelines are installed on fixed-end supports or movable-end supports, depending on their role.

The above-ground part of the pipeline is identical to the displacement of the fixed end of the abutment, while the movable end of the abutment on the other bank does not constrain the displacement in the direction of the pipeline axis. Therefore, a certain amount of relative displacement occurs between the abutment and the pipeline on the free-end side, and an expansion flexible joint[21] is installed to absorb this.

4.4.1.2 Seismic Response Displacement of Abutments and Struts

4.4.1.2.1 Abutment Response Displacements of Single-Span Pipeline Bridges

Abutments can be assumed to be rigid because they are generally sufficiently stiff. In this case, the seismic displacement behavior of the fixed bearing of the pipeline on the abutment follows the seismic response displacement of the bottom of the abutment foundation.

When subjected to ground shaking, the response correlation of a pair of abutments with a large separation distance is not significant, but when the separation distance is small, the response phases of the two abutments are considered to be close. Therefore, the displacement of the abutment at the other end can be regarded

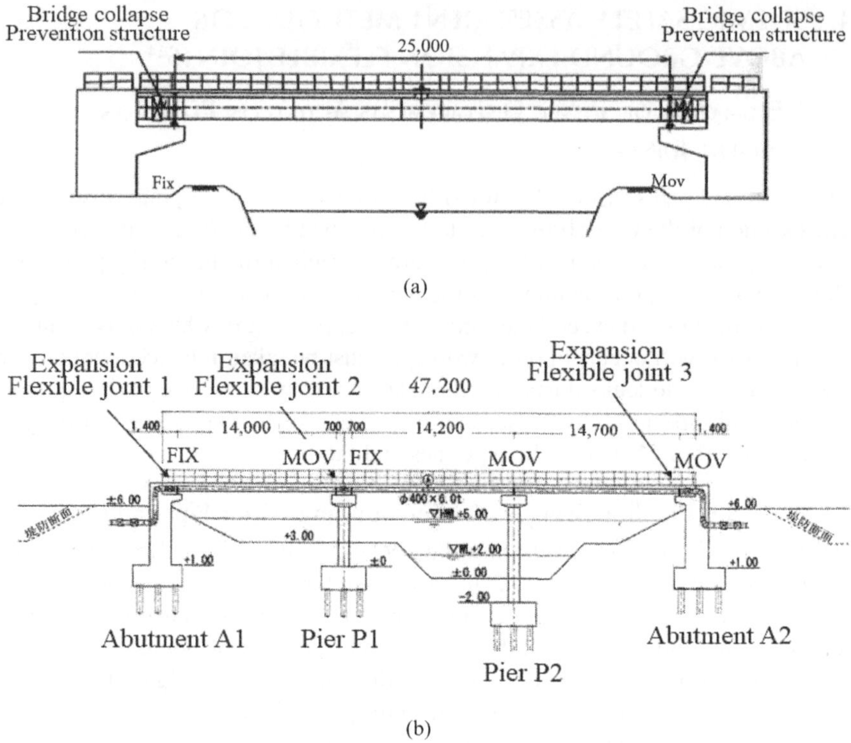

FIGURE 4.4.1 Single-span and multiple-span pipeline bridge models:[20] (a) single-span pipeline bridge; (b) multi-span pipeline bridge.

as causing a difference in response displacement corresponding to the seismic phase difference of the separation distance between abutments. Considering these factors, the response displacements at both ends of the abutment can be estimated by the following equation. The abutment seismic response displacement on one side, δ_{A1}, is given by Equation (4.4.1) as corresponding to the ground response displacement at the bottom of the abutment foundation if the abutment is rigid enough. On the other hand, the opposite abutment seismic response displacement δ_{A2} can be estimated using Equation (4.4.2), taking into account the phase difference of the separation distance between abutments.

$$\delta_{A1}^{L2} = \frac{2\cos\theta}{\pi^2} S_V^{L2}\left(T_{A1}\right) T_{A1} \cdot C_{A1} \cdot \cos\left(\frac{2\pi}{L'} x_0\right), \tag{4.4.1}$$

$$\delta_{A2}^{L2} = \frac{2\cos\theta}{\pi^2} S_V^{L2}\left(T_{A2}\right) T_{A2} \cdot C_{A2} \cdot \cos\left\{\frac{2\pi}{L'}\left(x_0 + W\right)\right\}. \tag{4.4.2}$$

where

S_V^{L1}: design velocity response spectrum for Level 1 ground motion
S_V^{L2}: design velocity response spectrum for Level 2 ground motion
h_{A1}: depth to base of abutment 1

h_{A2}: depth to base of abutment 2
W: separation distance between abutments
x_0: reference point of the horizontal propagation waveform
L: horizontal directional propagation wavelength
L': apparent wavelength of seismic wave incident to the pipeline with an incident angle of θ to be given by

$$L' = L/\cos\theta$$

θ: incident angle
T_{A1}: typical period of the ground at abutment A1
T_{A2}: typical period of the ground at abutment A2
C_{Ai}: adjustment factor to be applied in too shallow ground comparing to the depth of an abutment A_i, which is given by

$$C_{Ai} = \begin{cases} \cos\left(\dfrac{\pi}{2H_i} h_i\right) & for\, h_i \le \dfrac{H_i}{2} \\ \cos\left(\dfrac{\pi}{4}\right) & for\, h_i > \dfrac{H_i}{2} \end{cases}$$

H_i: depth of surface ground to the base-rock at the abutment A_i

4.4.1.3 Abutment and Strut Response Displacements of Multi-Span Pipeline Bridges

4.4.1.3.1 In the Case of Level 2 Ground Motion

As shown in Figure 4.4.2, the depth from the ground surface to the bottom of the foundation of each abutment and pillar in a multi-span bridge depends on the ground conditions.

The seismic displacement behavior of the pipe bearing on the abutment follows the seismic response displacement of the bottom of the abutment foundation as in the case of a single-span bridge.

The seismic response displacement δ_{P1}^{L2} at the top of the pillar is obtained as the sum of the seismic response displacement δ_{P1B}^{L2} at the bottom of the pillar foundation and the seismic response displacement δ_{P1T}^{L2} at the top position of the pillar using the following formula given by

$$\delta_{P1}^{L2} = \delta_{P1B}^{L2} + \delta_{P1T}^{L2}. \qquad (4.4.3)$$

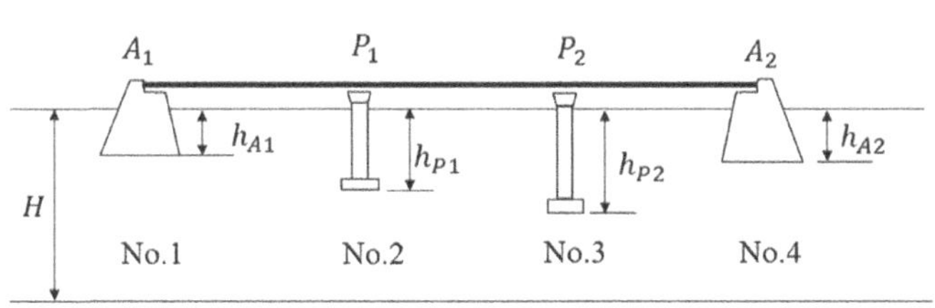

FIGURE 4.4.2 Depth of abutment foundation base.

Here, as a simplified response analysis method for pipeline bridges, it is assumed that the abutments and pillars respond according to their respective typical periods. In this case, the seismic response displacement of the bottom surface of the pillar foundations under Level 2 ground motion is expressed by the following equation:

$$\delta_{P1B}^{L2} = \frac{2\cos\theta}{\pi^2} S_V^{L2}\left(T_{P1}\right) T_{P1} C_{P1}. \tag{4.4.4}$$

where

h_{P1}: depth to the bottom of the foundation of pillar P1
h_{P2}: depth to the bottom of the foundation of pillar P2
θ: incident angle
T_{P1}: typical period of pillar P1
C_{P1}: adjustment factor to be applied in very shallow ground comparing to the depth of the pillar P_1 which is given in the same way as in the abutment A_i

If an elastoplastic zone occurs at the base of the pillar due to Level 2 ground motion, the seismic response displacement δ_{PiT}^{L2} at the top end of the pillar can be estimated using the following equation:[22]

$$\delta_{PiT}^{L2} = \mu_{Pi} \cdot \delta_{y,PiT} \quad (i = 1, 2), \tag{4.4.5}$$

where

$\delta_{y,PiT}$: yield displacement at the top end of pillar P_i
μ_{Pi}: ductility factor of pillar P_i

4.4.1.3.2 In the Case of Level 1 Earthquake Motion

By changing the design earthquake acceleration spectrum and velocity spectrum from Level 2 to Level 1, the preceding equations can be applied as they are. However, if the design for Level 1 ground motion is an elastic design, it must be confirmed that the earthquake response results are basically within the elastic range.

4.4.2 Seismic Safety Assessment of Above-Ground Expansion Flexible Joints

4.4.2.1 For Single-Span Pipeline Bridges

In above-ground pipelines, axial strain due to temperature changes and, depending on piping conditions, axial strain due to internal pressure occurs at all times.

Before the earthquake, the pipeline is subjected to axial strain equivalent to the Poisson's ratio of axial strain due to temperature change and circumferential strain due to internal pressure.

During the earthquake, the pipeline undergoes the same displacement as the abutment displacement through the fixed bearings on the abutment. If the seismic wave propagates horizontally with a wavelength L, the relative displacements between the two abutments are produced by the phase difference. The relative displacement is absorbed by the expansion flexible joints installed adjacent to the free-end bearings.

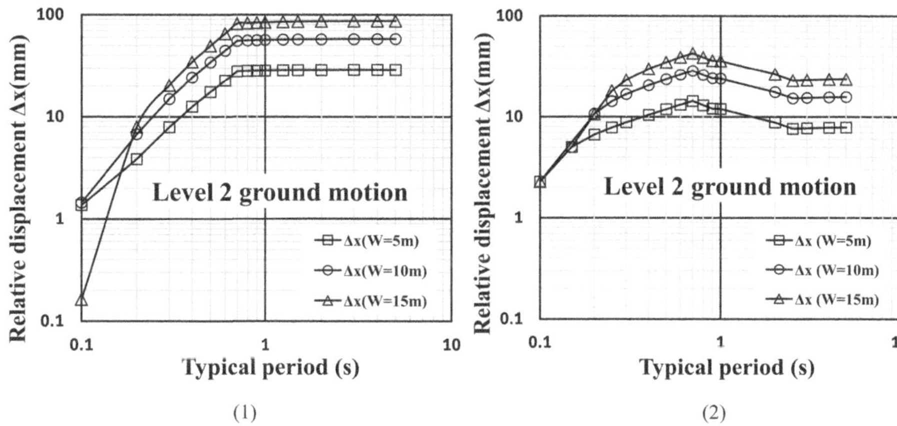

FIGURE 4.4.3 Relative displacement between abutments for Level 2 ground motion on a single-span pipeline bridge: (1) the case of JWWA; (2) the case of JGA.

Here, the relative displacement between abutments A1 and A2 is determined by the following equation:

$$\Delta = \delta_{A1} - \delta_{A2}. \tag{4.4.6}$$

Figure 4.4.3 shows the relative displacement of the single-span pipeline shown in Figure 4.4.1(a) for Level 2 ground motion for the cases of JWWA and JGA. The figure suggests that, in the case of JWWA, relative displacements are less than 100 mm and at the typical period more than 0.7 s, relative displacements are constant. In the case of JGA, on the other hand, relative displacements are comparatively lower than those in the case of JWWA, and the maximum value is located at 0.7 sec.

When the abutment oscillates in the direction perpendicular to the pipe axis, relative displacements occur in the same direction. If the free-end bearing can be displaced to some extent in the perpendicular direction, this relative displacement will be absorbed by the perpendicular relative angle of the expansion flexible joint. The relative angle can be evaluated on the safe side using the following equation:

$$\Theta = tan^{-1}\left\{\frac{|\delta_{A1}|+|\delta_{A2}|}{W}\right\}. \tag{4.4.7}$$

4.4.2.2 Case of Multi-Span Pipeline Bridges

The case of the multi-span pipeline bridge shown in Figure 4.4.1(b) is considered. As shown in Figure 4.4.2, in this pipeline bridge, abutment A_1 (No. 1) has fixed-end supports, while the other abutment A_2 (No. 4) has free-end supports. Expansion flexible joints are installed in each abutment. An expansion flexible joint is installed at the top end of pier P_1 (No. 2) between the free-end support and the fixed-end support. On the other hand, the top end of pier P_2 (No. 3) only has a free-end support, but no expansion flexible joint.

Relative displacements acting on the expansion flexible joints will occur between the fixed end supports and the concrete blocks on abutment A_1 (1,1), between structural abutment A_1 and pillar P_1 (1,2) and between pier P_1 and abutment A_2 (2,4). Pier P_2, on the other hand, is a free-end support and therefore has no effect on the seismic horizontal movement of the pipe.

4.4.2.2.1 Relative Displacements Occurring in Pipeline Bridges

Relative displacements occur in the sections (1,1), (1,2), and (2,4), using the numbers 1, 2, 3, and 4. In other words, the relative axial displacement and the relative angle perpendicular to the axis can be determined respectively by the following equations:

$$\begin{aligned}\Delta_{11} &= \delta_{A1} - \delta_{A1} = 0\\ \Delta_{12} &= \left|\delta_{A1} - \delta_{P1}\right|\\ \Delta_{24} &= \delta_{P1} - \delta_{A2}\end{aligned} \tag{4.4.8}$$

$$\begin{aligned}\Theta_{12} &= tan^{-1}\left\{\frac{\left|\delta_{A1}\right| + \left|\delta_{P1}\right|}{W_{12}}\right\}\\ \Theta_{34} &= tan^{-1}\left\{\frac{\left|\delta_{P2}\right| + \left|\delta_{A2}\right|}{W_{34}}\right\}\\ \Theta_{24} &= tan^{-1}\left\{\frac{\left|\delta_{P1}\right| + \left|\delta_{A2}\right|}{W_{23} + W_{34}}\right\}\end{aligned} \tag{4.4.9}$$

4.4.2.2.2 Comparison of Simplified Analysis Results for Relative Displacements with Previous Numerical Analysis Results

In the 2009 edition of the *Guidelines and Commentary on Seismic Engineering Methods for Waterworks Pipelines*[21] *Water Pipe Bridges (Substructure)*, the design conditions for the multi-span pipeline bridge model shown in Figure 4.4.4 are summarized

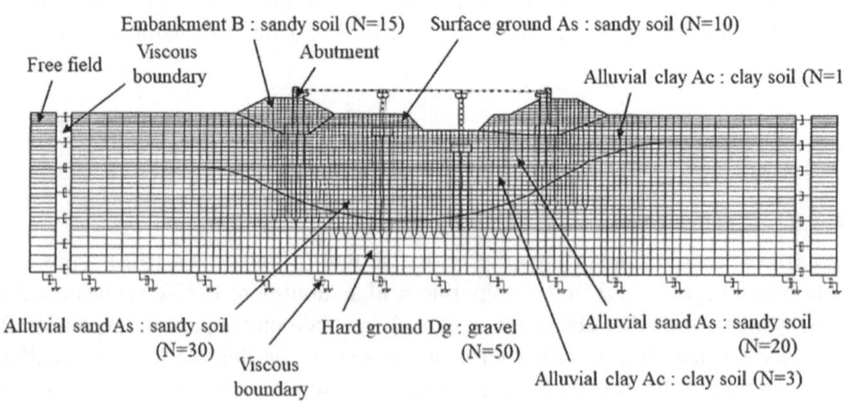

FIGURE 4.4.4 Water pipe bridge FEM analysis model.

TABLE 4.4.1
Design Conditions for Water Pipes and Ground for the Pipeline Bridge

Element	Item	Symbol	Unit	Value
Pipe	Outer diameter	D	m	0.4064
	Thickness	t	m	0.006
	Young's modulus	E	kN/m^2	205,000,000

TABLE 4.4.2
Characteristic Period, Yield Displacement, and Ultimate Displacement of Abutments and Struts

				Ground	Abutment 1	Abutment 2	Pier 1	Pier 2
Item		**Symbol**	**Unit**	**G**	**A1**	**A2**	**P1**	**P2**
ID number					No.1	No.4	No.2	No.3
Depth to the Bedrock		h	m	40	6.75	6.75	7.75	8.75
Ground Typical Period		T	s	0.68	0.822	0.822	0.417	0.58
Pillar	Yield Displacement	δ_y	mm				36.34	113.06
	Ultimate Displacement	δ_U	mm				325.53	371.29
	Yield Acceleration	Sy	m/s^2				5.02	5.24
	Ductility Factor	μ	-				2	2

TABLE 4.4.3
Comparison of Simplified Analysis Results with Previous Numerical Analysis Results[24]

		Maximum Relative Displacement (mm)		
Calculation Method	**Direction**	**1~1 span**	**1~2 span**	**2~4 span**
Simplified Method	Axial Direction	0	73	102
FEM Analysis	Compression	0.54	6	91
	Extension	0.4	16	104

in Table 4.4.1. In addition, the characteristic period, yield displacement, ultimate displacement, yield acceleration, and plastic modulus of abutment and strut are summarized in Table 4.4.2 based on the *Specification and Commentary for Road Bridges (V Seismic Design Edition).* The 2D FEM analysis results for this pipeline bridge model subjected to Level 2 ground motion are then obtained as follows. However, in the FEM analysis results, a distinction is made between the push-in and pull-out directions in the direction of the pipe axis, whereas in the simplified Equation (4.4.17), the two are calculated as the same without distinction. Table 4.4.3

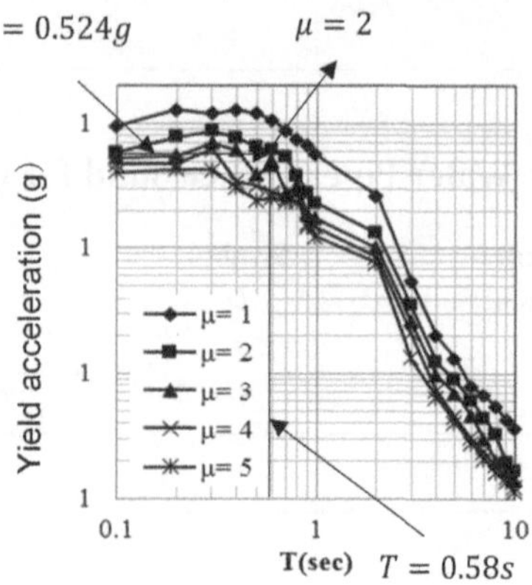

FIGURE 4.4.5 Yield acceleration spectrum diagram of the top end of a bridge support column in the direction of the bridge axis.

compares the results of this numerical analysis with the results of the simplified analysis in the equation.

To obtain the results in the same table, the ductility factor in the elastoplastic response of, for example, a water pipeline bridge strut P2 with characteristic period $T = 0.58(\text{sec})$ and yield acceleration response spectrum[22] $S_y(T) = 0.524g$ is calculated as $\mu = 2.0$, as shown in Figure 4.4.5 according to Equation (4.4.5).

Figure 4.4.6 compares the two results. In the same figure, the relative displacements (1,2), (2,4) are represented in the figure with the ID number (No. 1) of abutment A1 as 1, the ID number (No. 2) of pillar P1 as 2, the ID number (No. 3) of pillar P2 as 3, and the ID number (No. 4) of abutment A2 as 4. The relative displacement between abutment A1 and the fixed end on its abutment (relative displacement between 1 and 1) is zero in the simplified equation and close to zero in the FEM analysis results. Next, the relative displacement (1,2) between abutment A1 and pillar P1 are calculated to be larger in the simplified equation compared to the FEM analysis results. Furthermore, the relative displacement (2, 4) between pillar P1 and abutment A2 shows that the results of the simplified equation and the FEM numerical results are relatively similar.

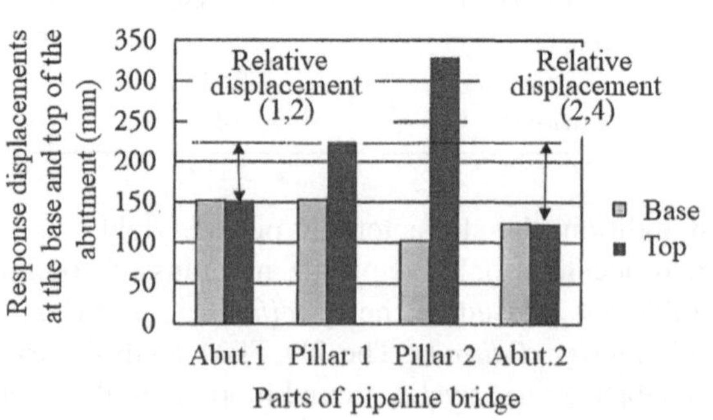

FIGURE 4.4.6 Relative displacement between abutment and strut on multi-span pipeline bridges.

4.5 SOIL-STRUCTURE INTERACTION IN LIFELINE COMPONENTS

In this section, a difference in soil-structural interaction between soil surrounded structure and buried pipelines are discussed and clarified.

An earthquake brings not only ground shaking but also ground collapse such as fault movement, ground liquefaction, and landslide. In the case of ground shaking, the ground is in elastic condition, so that the pipeline motion is proportional to the ground motion. In the case of ground collapse, on the other hand, all the structures to be above-ground and/or underground are brought to unstable conditions or significant damage stages that are caused by extremely large ground movements.

The soil–structural interaction is always observed not only in the above-ground structures to be supported by the ground but also in the buried pipelines surrounded by the free field ground.

In this section, the discussion will be focused on the effect of soil–structure interaction from structural differences and also from different ground movements.

4.5.1 The Soil–Structure Interaction in a Buried Pipeline and Its Surrounding Ground

When an earthquake ground motion propagates along a pipeline, the pipe is elongated or compressed in the axial direction. The pipe deformation is forced by shear stress acting on the pipe surface as shown in Figure 4.5.1. This shear stress is produced by the difference of rigidity between a pipe material and its surrounding free field ground. This shear stress is a result produced by the soil–structure interaction in a buried pipeline and its surrounding ground.

In the soil-structure interaction for a buried pipeline, the ground is classified into two layers: a free field ground and a soil surrounding the pipeline. The free field ground is intrinsically elastic, while a surrounded soil can be deformed in elastic-plastic mode. For a strong ground motion, this surrounded soil is deformed to produce a plastic shear strain that appears in a slippage effect along the pipeline as described in Section 3.1.2.

When soil–structure interaction is taken into consideration in a buried pipeline, a pipe stress/strain will be decreased in a straight pipeline as a slippage effect.

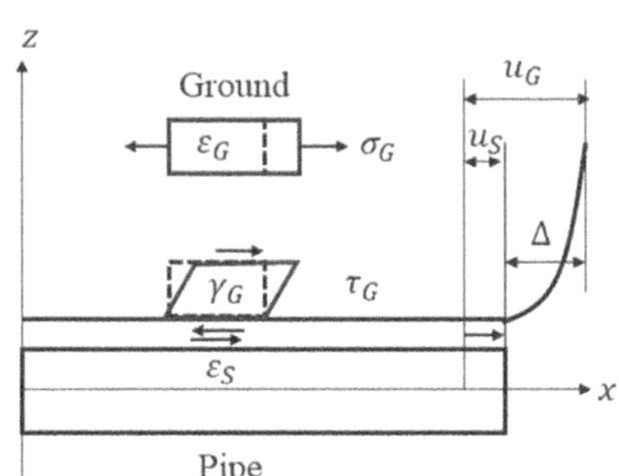

FIGURE 4.5.1 Soil–structure interaction in a buried pipeline.

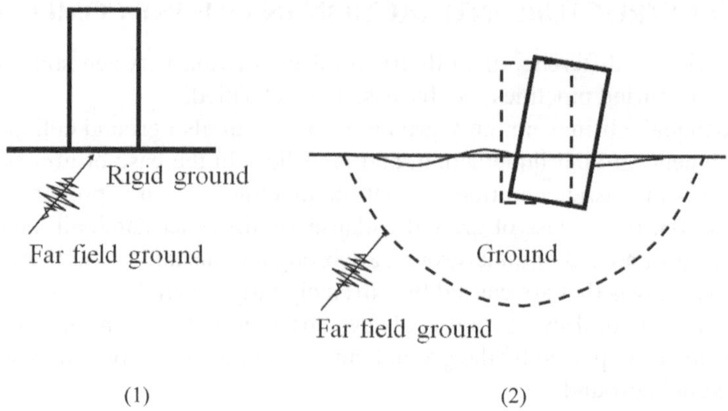

FIGURE 4.5.2 Illustration of a soil–structure interaction: (1) structural response on the rigid ground; (2) structural response interacted with the surrounding ground.

4.5.2 The Soil–Structure Interaction in a Structure-Foundation System and Its Surrounding Ground

When a structure is constructed on a rigid ground, any foundation is not necessary. Then, an earthquake ground motion will be fully transmitted to the structure, in which there is no soil–structure interaction between a structure and its surrounding ground.

When a structure is constructed on an elastic ground, however, any foundation is seismically designed to support the structure. If a soil–structure interaction is taken into consideration, the structural seismic response will be expected to decrease, compared to that in a rigid ground.

In an elastic ground, an input seismic force is classified into effective seismic force, kinematic effect on a geometrical restriction, and radiation effect due to structure-foundation reaction to the ground. These effective seismic forces will be expected to decrease the structural response. Both ground models are compared in Figure 4.5.2.

Currently many researchers,[27, 28, 29, 30] have been discussed on this research theme. So the detail formulations on the soil–structure interaction problems are not discussed in this book.

It is noted that the soil–structure interaction is discussed on the elastic ground condition even in a severe ground motion. If a ground motion is changed into ground collapse such as fault displacement, liquefaction, or landslide, the structures in the ground collapse are unstable and the structural integrities are also lost.

4.5.3 Interaction between the Structure-Foundation System and Its Connected Lifeline System

In general, urban structures are functional under the support of electricity, water, gas, and any other services. Once a structure survives, but these service systems are lost by a severe ground motion, the structure cannot maintain its functional services. That is, under sound functional conditions, the structure can be maintained with all the services to be supplied through conduits, pipeline, optical fiber, and any other

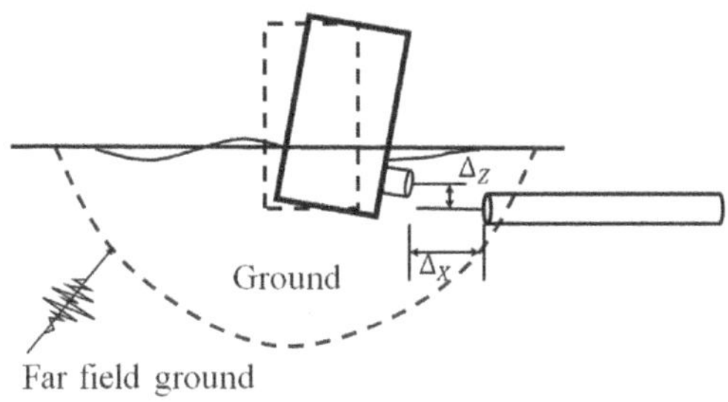

FIGURE 4.5.3 Interaction between a structure and its connected pipeline.

lifelines. So, it is important that the interaction between the structures and their connected lifelines should be recognized as one more intrinsically significant problem.

A simple configuration between the structure and its connected pipeline is shown in Figure 4.5.3. When an earthquake induces a large unstable deformation, buried pipeline produces a large relative displacement between the structure and its connected pipe end such as horizontal displacement of Δ_X and vertical settlement of Δ_Z. If these relative displacements are large enough to be absorbed by a pipeline elongation, the structural wall might be broken or the pipeline will be disconnected. In the actual design stage, an expansion joint is installed to solve this difficulty.

This approach suggests that, in an elastic level of ground motion, a soil–structure interaction should be discussed to obtain more accurate estimation of the structural response, but in the ground collapse condition, a seismic defect of unstable structures might produce unexpected seismic damages between a structure and its connected lifelines.

REFERENCES

1. Lwin, T., T. Koike, and D. Ji: A rational seismic design method for buildings in earthquake-prone developing countries, *ASEAN Engineering Journal*, Vol. 11, No. 4, 2021.
2. T. Koike, H. Watanabe, M. Hamano, N. Hasegawa, H. Onuma, N. Party, and N. Nakagawa: Proposal of a seismic safety performance verification method for aboveground water distribution basins, *Journal of JSCE*, Vol. 77, No. 4, I_24–I_34, 2021.
3. A. Shibata: *Latest Seismic Structure Analysis*, Morikita Publishing, 2014.
4. A. Tateishi: Study on the method of earthquake load action in response displacement method, Journal of JSCE, No. 441, pp. 157–166, 1992.
5. Higashide, T., T. Yabuguchi, T. Imai, and T. Koike: Seismic performance assessment of expansion joints for water lifelines, 15th World Conference of Earthquake Engineering, Lisbon, Portugal, 2012.
6. Kobe University: *Survey of Water Supply Damage and Actual Water Supply and Restoration Support for the Great East Japan Earthquake*, Kobe University Water Supply Earthquake Countermeasures Study Group, Kobe University, 2012.
7. Higashide, T., T. Yabuguchi, T. Imai, and T. Koike: Failure mechanism of expansion flexible pipes for water supply in the Tohoku Earthquake, 2012th JSCE Annual Research Conference at Kansai Division, Section 1, pp. I-55–I-56, 2012.
8. Japan Water Pipe and Steel Pipe Association: *Expansion and Contraction Flexible Pipes in Steel Pipelines for Water Supply* WSP 024-97, Japan Water Pipe and Steel Pipe Association, 1997.

9. Higashide, T., T. Yabuguchi, T. Imai, and T. Koike: Damage mechanism of expansion joints of water lifelines in the 2011 Great East Japan Earthquake. *JSCE, Journal of JSCE A1*, Vol. 68, No. 4, I-800–I-806, 2012.
10. WSP: *Damage Analysis and Proposals for Disaster Mitigation of Steel Water Pipelines in the Kumamoto Earthquake*, WSP, 2016.
11. WSP: *Damage Analysis and Proposals for Restoration and Reconstruction of Steel Water Pipelines in the East Japan Great Earthquake*, 2011.
12. Iwamoto, S.: Neuron Product Catalogue, Nippon Neuron Corporation, 2021.
13. Japan Water Pipe Association: *Expansion Flexible Pipes in Steel Water Pipelines* WSP 024-97, Japan Water Pipe Association, 1997.
14. Japanese Industrial Standards: JIS B8266 Annex 8 Structure of Pressure Vessels – Specific Standards, Japanese Industrial Standards (JIS), 2006.
15. Unoguchi, E.: Low cycle fatigue and thermal fatigue problems of pressure vessels, *High Pressure (JHPI)*, Vol. 3, No. 1, p. 32–46, 1965.
16. Langer, B. F.: Design of pressure vessels for low-cycle fatigue, *Journal of Basic Engineering, Transaction of ASME*, Vol.84, No.3, Ser. D, pp.389–399, 1962.
17. Kanamaru, Y., N. Nishikawa, Y. Nishi, and T. Koike: Investigation of Performance Limits for Bellows-Type Expansion Flexible Joints, 2022 Waterworks Research Meeting, Japan Water Works Association, 2022.
18. Koike, T.: Estimation of buried pipe strains under seismic risk, *Journal of JSCE*, Vol. 331, 1983.
19. Japan Gas Association: *Guidelines for Seismic Design of High-Pressure Gas Pipelines* (JGA Finger-206-03), Japan Gas Association, 2004.
20. Japan Water Works Association: *Design Case Book for Water Pipelines, Seismic Design Guideline for Water Pipelines 2009th Version*, JWWA, 2009.
21. Japan Water Works Association: *Seismic Design Guideline for Water Pipelines*, 2009 edition,
22. Alhourani, A., D. Ji, and T. Koike: A probabilistic seismic design approach for developing countries, *International Symposium on Earthquake Engineering, JAEE*, Vol. 1, pp. 525–532, 2012.
23. Kanamaru, Y., Y. Nishi, and T. Koike: Seismic design method of expansion flexible joints around underground structures targeting performance design, Proceedings of the 42nd Conference on Earthquake Engineering, JSCE, 2022.
24. Nishi, Y., Y. Kanamaru, and T. Koike: Seismic design method of expansion flexible joints installed on water pipe bridges targeting performance design, Proceedings of the 42nd Conference on Earthquake Engineering, JSCE, 2022.
25. Koike, T., M. Hamano, T. Watanabe, and N. Hasegawa: Proposal on performance design of seismic design method for water supply facilities, *Journal of JSCE, A1*, Vol. 76, No. 4, I-196–I-204, 2020.
26. Japan Gas Association: *Guidelines for Seismic Design of Production* Facilities (JGA-101-1), Japan Gas Association, 2001.
27. Kawashima, K.: *Seismic Design of Underground Structures*, Kajima Publishing Company, Tokyo, 1994.
28. Kramer, S.: *Geotechnical Earthquake Engineering*, Prentice Hall, New Jersey, USA, 1996.
29. Wolf, J. P.: *Dynamic Soil–Structure Interaction*, Prentice Hall, New Jersey, USA, 1985.
30. Novak, M., M. Sheta, L, El-Hifnawy, H. El-Marsafawi, and O. Ramadan: DYNA4: A Computer Program for Calculation of Foundation Response to Dynamic Loads, Report GEOP93, Geotechnical Research Centre, The University of West Ontario, Canada, 1993.
31. Japan Highway Association: *Road Bridge Specifications and Commentary I Common Edition,* Japan Highway Association, 2017.

5 Performance Design Method for Lifelines

This chapter describes how to formulate a safety assessment for lifeline seismic design according to a performance design method. This is in response to the performance design of structure methods that are currently being actively developed in Japan and abroad. In addition, as a case study, the seismic safety assessment method for post-seismic operation of lifelines designed under the old seismic design guideline and the normal-time risk mitigation measures to ensure the seismic safety of lifelines are described in detail from a performance design method perspective.

5.1 PERFORMANCE DESIGN OF SEISMIC DESIGN METHODS

5.1.1 Background of Performance Design

5.1.1.1 What Is Performance Design?

To stimulate the global economy, unifying trade regulations and standardizing product quality worldwide are essential steps toward market expansion. The World Trade Organization (WTO) serves as the principal body for establishing these international standards. Recognizing its role as a leading trading nation, Japan has proactively ratified WTO agreements and is actively promoting global standardization across various economic sectors.

In the realm of structural design, numerous standards – including the Japanese Highway Bridges Specification[1] – have been revised to align domestic design methodologies[2,3,4] with international norms. These revisions aim to enhance the reliability and safety of infrastructure, ensuring that design practices meet globally recognized criteria.

Modern cities rely on various lifeline systems – including water and sewage systems, energy systems (electricity and gas), transportation networks (highways and railways), and information and communication systems (telephones, internet, etc.) to maintain functionality and provide residents with a comfortable lifestyle.

The significance of the performance design method is set out here as follows:

a. Establishing a standardized system for global design evaluation is essential. Adopting reliability-based design methods facilitates this by providing a consistent framework for assessing structural performance internationally.
b. Performance-based design empowers designers and manufacturers with greater flexibility to propose innovative structural properties for new components. This approach streamlines the introduction of novel designs, making the process more efficient compared to traditional prescriptive methods.

DOI: 10.1201/9781003655527-5

c. The establishment of common evaluation indicators is crucial for conducting international performance comparisons of water supply facilities. Such standardized metrics enable operators worldwide to assess and benchmark the effectiveness of their systems, fostering improvements and facilitating knowledge exchange.

In Japan, the majority of water utilities have established water supply facilities that deliver a stable supply to local populations. Given this extensive infrastructure, opportunities for constructing new facilities may be limited. However, following the 1995 Hyogo-ken Nanbu Earthquake, the Ministry of Land, Infrastructure, Transport, and Tourism (MLIT) introduced the concept of Level 2 ground motion as a seismic design criterion. Structures built before this update should be evaluated for seismic safety against this new standard.

Furthermore, the absence of cost-effective safety assessment methods, such as simplified formulas, poses challenges in evaluating the seismic adequacy of existing structures. This issue affects numerous entities across the country.

Implementing performance-based design (PBD) in water supply facilities offers several notable advantages. First, PBD facilitates the introduction of new components by allowing designers and manufacturers greater flexibility to propose structural properties that may not be covered by existing codes. Second, PBD emphasizes evaluating the functional performance of the entire water supply system, which is essential for service providers aiming to determine the extent of necessary improvements in systems that are both deteriorating and exposed to seismic hazards. By focusing on these aspects, PBD provides a framework for considering potential hazards and explicitly evaluating the performance capabilities of facilities as a whole. This approach aids in making informed decisions about system enhancements and resource allocation.

Quantifying the functional performance of an entire water supply system is essential for informed decision-making regarding system improvements and resilience strategies. By estimating the probability of water supply availability – from the source to the demand endpoints – while accounting for potential failures in network nodes and links, stakeholders can accurately assess the system's reliability. Employing performance-based design methods facilitates this process, providing the necessary framework to evaluate and enhance system performance effectively.

Modern cities rely on essential lifeline systems – including water supply and sewerage, energy (electricity and gas), transportation (highways and railways), and information and communication (telephones, Internet, etc.), which consist of multiple backbone facilities interconnected by pipes, lines, roads, and communication networks. These facilities have been constructed over extended periods, leading to variations in their construction dates and safety characteristics. Consequently, evaluating the seismic safety of these interconnected network systems requires a comprehensive assessment of their functional performance, considering the diverse safety attributes of each facility within the system.

The safety design of individual structures varies according to each country's safety standards and the specific types of structures involved. A comparison of international and national performance design standards is presented in Table 5.1.1.

TABLE 5.1.1
International Comparison of Performance Design Criteria (Example)

Design Guidelines	ISO2394	EURO	AASHTO/LRFD	JSCE
Area	International	International	International	Japan
Design method	Partial safety factored design method	Partial safety factored design method	Load and resistance factor method	Partial safety factored design method
Target reliability index	3.5	3.8	3.5	None
Life cycle of the design	50~100	50~100	75	50~100
Limit states	Serviceability and ultimate limit state	Serviceability and ultimate limit state	Serviceability, fatigue, ultimate & contingency limit state	Serviceability, repairable and ultimate limit state
Classification	Level I & II reliability-based design method	Level I reliability-based design method	Load and resistance factor method	Level I reliability-based design method
Applicable country	International	EU	USA	Japan

According to this table, design standards in Europe and the United States recommend a safety index[5] of 3.5 to 3.8, whereas Japanese standards often lack such specific guidelines. This discrepancy highlights the importance of clarifying target design levels. Without clear definitions, designers cannot accurately assess the safety of their designs, and operators may face uncertainties in maintenance and management objectives, potentially affecting the long-term reliability and safety of these infrastructures.

5.1.1.2 Performance Design Methods in Practice

Despite the proposal of performance-based seismic design for public facilities several decades ago, its widespread adoption remains limited. One contributing factor is the complexity of integrating performance design methods[1], which are rooted in reliability design principles[6], into standard design practices. This complexity arises from the challenges associated with incorporating uncertain design variables into the design process.

Another possible reason for the limited adoption of performance design is the perceived difficulty in realizing its advantages over conventional design methods. However, there are scenarios where the benefits of performance design become evident. For instance, in international projects where design methodologies differ between countries, performance design facilitates mutual understanding and alignment. Similarly, when introducing new products to the domestic market, comparing the performance of existing and newly proposed products using performance design methods aids in making informed decisions.

In addition, the benefits of performance design methods can be realized by the client or the project operator when the product or service result satisfies the performance target values required by the project. Therefore, when the performance targets to be achieved are clear and the results of performance design are foreseeable, all people involved in the project will consequently realize the benefits of performance design.

When the performance targets of a project are clearly defined and the outcomes of performance design are predictable, clients and project operators can directly observe the benefits of performance design methods. This clarity ensures that all stakeholders involved in the project understand the objectives and expectations, leading to more effective collaboration and successful project outcomes.

Material manufacturers supply products that meet specific performance targets based on their internal standards. However, without established criteria to assess these products' suitability for ensuring the seismic safety of their systems, water utilities face challenges in evaluating their effectiveness. This lack of standardized evaluation metrics can lead to uncertainties in selecting appropriate materials that align with the utilities' seismic safety objectives.

The relationship between water utilities and pipe material manufacturers exemplifies this scenario. To address this issue and fully realize the benefits of performance design for all stakeholders, it is essential for each department to establish quantitative evaluation indicators that align with one another as performance targets.

Establishing standardized Key Performance Indicators (KPIs) allows for consistent performance monitoring and fosters improved communication among stakeholders. These indicators enable the assessment of various aspects of water utility operations, facilitating the evaluation of efficiency and effectiveness in service delivery. Implementing such standardized metrics ensures that all parties involved have a clear understanding of performance expectations and outcomes, thereby enhancing collaboration and the overall effectiveness of performance design initiatives.

Traditionally, water utilities have aimed to achieve Seismic Performance Levels 1–3 in their seismic designs. However, questions arise regarding how these utilities assess whether their water supply systems meet these performance criteria, the certainty with which current systems can achieve Seismic Performance Level 2 under future Level 2 ground motions[3], and the appropriate scales for such assessments. Without a solid technical foundation to address these questions, the advancement of performance-based design[7] cannot proceed effectively. Pipe material manufacturers could demonstrate the seismic performance achievable by the dimensions of their products, but they naturally lacked the means to specify how this performance related to the seismic performance of the entire water supply system.

In addition to the pipelines, water treatment plants, distribution reservoirs, pump rooms, and valve rooms are part of the pipeline network, with support facilities such as energy supply and demand systems, and communication and control systems located outside these facilities. Each of these components has its own seismic performance and vulnerability to seismic forces. Therefore, when assessing the seismic safety of water supply systems, it is necessary to evaluate the overall system, considering both the pipeline network and individual facilities as integral components of the system.

A method to assess the seismic performance of water supply network systems, managed and operated by water utilities, is introduced here. This method, based on the Reliability Design Method[5], is called the Seismic Safety Performance Assessment Method for Water Supply Systems.

5.1.2 Safety Assessment Methods

5.1.2.1 Basic Framework

In the Level I reliability design method, various uncertainties are addressed by incorporating characteristic values for load and material strength variables. Additional uncertainties and the social impact upon reaching the limit state are accounted for by applying specific coefficients in the verification equation. These coefficients are referred to as partial coefficients.

Research and development of design code formats in Europe and the United States commenced in the 1970s, leading to the proposal of safety evaluation formulas in the design guidelines of these regions by the late 1970s. Notably, the European Standards Institute (Comité Européen de Normalisation, CEN) released the CEB-FIP International Recommendations[6] for the Design and Construction of Concrete Structures in 1970, which provided foundational principles and recommendations for concrete structure design.

$$g_R\left(\frac{f_k}{\gamma_{m1}\gamma_{m2}\gamma_m}\right) \geq g_S\left(\gamma_{f1}\gamma_{f2}\gamma_{f3}Q_k\right), \tag{5.1.1}$$

where γ_R *and* γ_S are functions that convert strength and load into their effects. f_k *and* Q_k are characteristic values of strength and load, while γ_X is a partial coefficient[8] of the variable X.

γ_{mi}: the resistance coefficient.

γ_{f1}: a coefficient evaluating the deviation from the characteristic values (load coefficient)

γ_{f2}: a coefficient to evaluate the combined load effect (combination coefficient)

γ_{f3}: a coefficient to assess load imprecision (structural analysis coefficient)

The characteristic values of strength and load mentioned generally mean 95% or 5% non-exceedance probability values.

Furthermore, Ravindra and Galombos[9] use a safety evaluation equation with partial coefficients for nominal and average values of strength, as follows:

$$\phi R_n \geq \sum_{k=1}^{n} \gamma_k S_{km}, \tag{5.1.2}$$

where ϕ and R_n are the resistance coefficient of strength and strength nominal, γ_k is the partial coefficient of load, and S_{km} is the mean value of load.

5.1.2.2 Safety Assessment Methods for Japanese Infrastructure Systems

The safety assessment formulas in Japan generally adhere to the aforementioned methodologies. For instance, the Japanese Highway Bridges Specifications and Commentary[2] employ specific safety evaluation formulas to verify the safety of road and bridge structures. These formulas incorporate partial coefficients that account for various uncertainties, ensuring a comprehensive assessment of structural integrity under different loading scenarios:

$$\xi_1 \cdot \xi_2 \cdot \Phi_R \cdot R(f_c, \Delta_c) \geq \sum S_i(\gamma_{pi} \cdot \gamma_{qi}, \cdot P_i). \tag{5.1.3}$$

where P_i is the characteristic value of the load, S_i is the load effect, R is the resistance characteristic value of the member, etc., f_c is the characteristic value of the material, Δ_c is the characteristic value of the dimension, and the following coefficients are specified:

Φ_R: the resistance coefficient
γ_{qi}: the load coefficient
γ_{pi}: the load combination coefficient
ξ_1: the investigation and analysis coefficient
ξ_2: the component and structural coefficients

As the Japanese Highway Bridges Specification follows the same evaluation formula structure as the CEB, it is assumed that each partial coefficient is evaluated in the same way as in the CEB.

In the Japanese Highway Bridges Specification and Commentary[2] V, Seismic Design Edition, the resistance coefficient R is predefined as 1 for Limit State 1 and 0.65 for Limit States 2 and 3. Similarly, the investigation and analysis coefficient (1) and the member and structural coefficient (2) are also predefined as 1.

The resistance coefficient essentially corresponds to a safety margin set by the designer, based on the reliability design method, to ensure the required level of safety for the client. This takes into account the predictive uncertainty of the resistance characteristic value R and the load effect S. If this margin is specified as a priori by the road bridge specifications, the designer's freedom in the performance design method is restricted.

This differs from the ISO partial factor setting method, which allows the designer to calculate the partial factor by considering variations in each design variable. In this sense, the partial coefficient design method in the Japanese Highway Bridges Specification limits the designer's ability to set partial coefficients independently.

On the other hand, the *Seismic Design Guidelines for Liquefaction Resistance of High-Pressure Gas Pipes*[10], based on the revised road bridge specifications, presents a safety assessment method for gas pipelines buried in liquefied ground. Therefore, the partial coefficients in this guideline are essentially the same as those in the CEB and the Road Bridge Specifications.

While the Specification for Highway Bridges covers combinations of several loads, only the seismic loads are considered here. The safety of the structure is determined by the following equation:

$$S_d \leq R_d, \tag{5.1.4}$$

where S_d is the load for verification and R_d is the strength for verification.

The load for verification and strength for verification are formulated as follows, following the general load-resistance factor design method

$$\begin{aligned} S_d &= \gamma_a \cdot S(\gamma_f \cdot F_k), \\ R_d &= R(f_k/\gamma_m)/\gamma_b, \end{aligned} \tag{5.1.5}$$

where

$S(*)$: load calculation function
F_k: characteristic value of the load
γ_f: load coefficient (variation of the characteristic value of the load)
γ_a: structural analysis coefficient (variation of the load calculation formula)
$R(*)$: resistance calculation function
f_k: characteristic value of resistance
γ_m: material coefficient (variation of the characteristic value of the material)
γ_b: member coefficient (variation of limit value calculation formula)

In the calculation examples of the *Seismic Design Guidelines for Liquefaction Resistance of High-Pressure Gas Pipes*, some cases set the partial factor values as $\gamma_b = 1$, $\gamma_m = 1$, $\gamma_a = 1$, and $\gamma_f = 1$ in the design study. Generally, a value of 1 for the partial coefficients can be interpreted as indicating that the uncertainty of the design factor being treated is zero. For example, $\gamma_b = 1$ would imply that the limit strength formula for the pipe is accurate, with no prediction errors.

Furthermore, the resistance values of the pipe material inherently vary with respect to the specimens used in standard tests. The characteristic values are usually influenced by deviations from the standard shape of the machined form of the pipe material, installation conditions, and construction conditions. Therefore, $\gamma_m = 1$ means that there is no deviation between the resistance f_k of the pipe material under actual service conditions and that of the standard test specimen. This implies that $R(f_k/\gamma_m)/\gamma_b = 1$ has no prediction errors except for the intrinsic variation in the resistance of the pipe material, and only this intrinsic variation needs to be evaluated.

A similar argument applies to the load. Specifically, $\gamma_a = 1$ means there are no prediction errors in the load calculation equation, and $\gamma_f = 1$ means there are no errors in how the multiple design variables F_k, which make up the load calculation equation, are set. However, seismic loads and other loads included in the load calculation function are inherently variable and need to be evaluated.

The same safety assessment method, based on the guidelines for seismic design of liquefied high-pressure gas pipes, is used in the *Guidelines and Commentary on Seismic Engineering Methods for Water Supply Facilities* (2009):

$$S_d \leq \frac{R_d}{\gamma_i}, \tag{5.1.6}$$

In the equation, S_d represents the load to be verified, R_d denotes the strength to be verified, and γ_i is the structure coefficient. The structure coefficient i is determined by considering the importance of the structure and the social impact when the limit state is reached. It should generally be between 1.0 and 1.2. In addition, it is also stated that

$$R_d = R/\gamma_b, \qquad S_d = \gamma_a S, \tag{5.1.7}$$

where S is the load characteristic value, R is the strength (resistance) characteristic value, γ_a is the structural analysis coefficient, and γ_b is the member coefficient. The design variables such as resistance (R) and load (S) are treated as random variables, each characterized by a mean value μ_x and a coefficient of variation δ_x, reflecting their inherent uncertainties. This probabilistic approach allows for the assessment of structural performance under varying conditions:

$$R = R(\mu_R, \delta_R), \qquad S = S(\mu_S, \delta_S). \tag{5.1.8}$$

In practical design, all the values of the partial coefficients[11] included in the safety assessment equation are

$$\gamma_i = 1, \qquad \gamma_a = 1,\ \gamma_b = 1.$$

This is often the case. In this case, no variation in the design or prediction equations is interpreted, but the intrinsic variation of the design variables R and S must be evaluated. In other words, the intrinsic variability of R as a strength property value and the intrinsic variability of S, for example, in seismic loading, must be evaluated.

5.1.2.3 Relationship between Partial Coefficients and Safety Indices

As shown in Equation (5.1.8), since the design variables R and S are inherently random variables, safety verification function[12] is defined by the strength variable R and the load variable S as follows:

$$Z = ln(R) - ln(S). \tag{5.1.9}$$

If the design variables R and S for strength and load, respectively, are random variables following a lognormal distribution characterized by the mean value μ_X and the coefficient of variation δ_X, respectively, their safety index[12] can be defined by the following equation:

$$\lambda_X = E[lnX] = ln\left(\frac{\mu_X}{\sqrt{1+\delta_X^2}}\right), \qquad \beta = \frac{\lambda_R - \lambda_S}{\sqrt{\zeta_R^2 + \zeta_S^2}}, \tag{5.1.10}$$

where

$$\lambda_X = \mathrm{E}[lnX] = \ln\left(\frac{\mu_X}{\sqrt{1+\delta_X^2}}\right), \tag{5.1.11}$$

$$\zeta_X^2 = Var[lnX] = ln\left(1+\delta_X^2\right). \tag{5.1.12}$$

In practical design, the characteristic values of resistance R_n and load S_n are adjusted by their respective strength characteristic coefficients (*R*) and load characteristic coefficients (*S*) to determine the design values used for verification. This relationship ensures that the inherent variability of these design variables is appropriately accounted for in the safety assessment equation:

$$R_n = \frac{\mu_R}{\gamma_R}, \qquad R_d = \frac{R_n}{\gamma_b}, \qquad S_n = \gamma_S \mu_S, \qquad S_d = \gamma_a S_n, \tag{5.1.13}$$

in which

$$\gamma_R = 1 + \Phi^{-1}(P_R)\varsigma_R \qquad \gamma_S = 1 + \Phi^{-1}(P_S)\varsigma_S, \tag{5.1.14}$$

where P_R *and* P_S are the nonexceedance probabilities of *R* and *S*, respectively. On the other hand, by using the verification strength R_d and the verification load S_d, the following apparent safety indices can be formally defined:

$$\beta_d = \frac{ln(R_d/\gamma_i) - lnS_d}{\sqrt{\varsigma_R^2 + \varsigma_S^2}}. \tag{5.1.15}$$

The two safety indicators are then linked by the following relationship:

$$\beta = \beta_d = \frac{ln\ \gamma_R + ln\ \gamma_S + ln\ \gamma_a + ln\ \gamma_b + ln\ \gamma_i}{\sqrt{\varsigma_R^2 + \varsigma_S^2}}. \tag{5.1.16}$$

5.1.3 Limit States of Lifelines

5.1.3.1 Definition of Limit State

Pipe materials used for pipelines include ductile cast iron pipes for water supply, polyethylene pipes, and welded steel pipes used for high-pressure gas pipelines, each of which has a unique joint structure. When the limit state of any of these pipe materials is exceeded due to tensile or compressive forces acting in the direction of the pipe axis during an earthquake, the pipe material itself or the joint will fail, causing leakage of the contents.

The limit state of a pipe material is generally as shown in Figure 5.1.1.

1. Serviceability limit state
2. Repairable limit state
3. Ultimate limit state

The serviceability limit corresponds to the yield limit or elastic buckling limit of the pipe material in the case of steel pipes. However, conventional ductile iron pipes do not have force-type limits, such as yield or repair limits, because of the

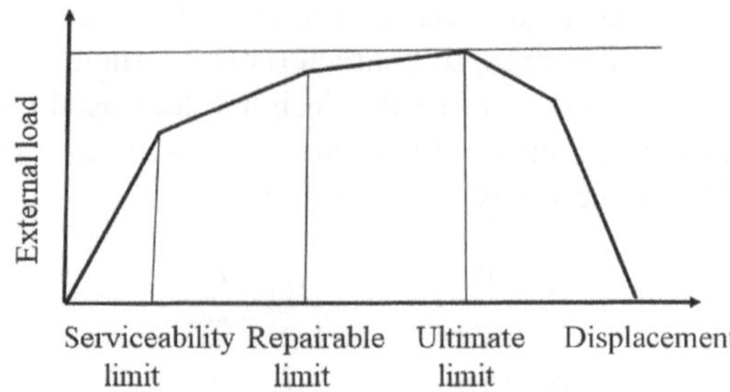

FIGURE 5.1.1 Marginal state models in Japan.

displacement-controlled joint type. The joint will pull out and cause a leakage accident if the displacement exceeds the joint displacement. On the other hand, in ductile iron pipes with protrusions, several single pipes are connected to absorb seismic displacement. Joint failure occurs when the protrusions in the joint eventually shear off. Polyethylene pipes, being viscoelastic, do not show a definite yield point and continue to deform, elongating up to their maximum strength point, after which they continue to elongate to a considerable extent, although at a slightly reduced strength level. However, the same leakage incident occurs in both pipe materials at the point of joint rupture or when a crack occurs in the pipe itself. This rupture limit corresponds to the ultimate limit.

On the other hand, the repairable limit is a limit state established to ensure early recovery after an earthquake and is set as a limit where only minor damage occurs, but no damage requiring days of repair. Therefore, the repairable limit is not a clear-cut limit in terms of structural mechanics, but a limit at which repair work is feasible. It can only be said to lie between the serviceability limit and the ultimate limit.

5.1.3.2 Relationship Between Earthquake Ground Motion and Limit State

In Japan, two types of seismic ground motions, Level 1 and Level 2, are used as seismic loads for the seismic design of structures. On the other hand, the US seismic design guideline[13] sets the magnitude of design acceleration from three types of earthquake models: maximum operational earthquake (MOE), design-based earthquake (DBE), and maximum considered earthquake (MCE).

Table 5.1.2 shows the relationship between design earthquake ground motions and limit states in the United States and Japan. The table classifies Level 2 ground motion into two categories: Level 2L ground motions for the repairable limit state and Level 2U ground motions for the ultimate limit state.

The serviceability limit state means that the response of the structure is considered in the elastic state, if multiple earthquakes (MOE) will occur during the 50-year service period. The repairable limit state is considered in the elastoplastic state, meaning that an earthquake (DBE, Level 2L ground motion) is sufficient to enable early functional recovery through simple restoration. From this point of view, the probability of occurrence is set at around 10%, assuming a few earthquakes in a 50-year period.

TABLE 5.1.2
Correspondence between Limit States and Design Earthquake Ground Motion[11]

		Performance Level of the Structures				Seismic Design Codes	
		S-1	S-2	S-3	S-4		
Probability of Occurrence	**Return Period**	IO		LS	CP	**Japan**	**USA**
50% in 50 yrs	72 yrs	●				Level 1	MOE
20% in 50 yrs	225 yrs	■	●				
10% in 50 yrs	475 yrs		■	●		Level 2L	DBE
5% in 50 yrs	970 yrs						MCE
2% in 50 yrs	2475 yrs			■	●	Level 2U	

Definition of performance

S-1	Serviceability
S-2	Repairable
S-3	Life safety
S-4	Ultimate
●	Standard structures
■	Important structures
Level 2L	70% nonexceedance probability of Level2 ground motion
Level 2U	90% nonexceedance probability of Level2 ground motion

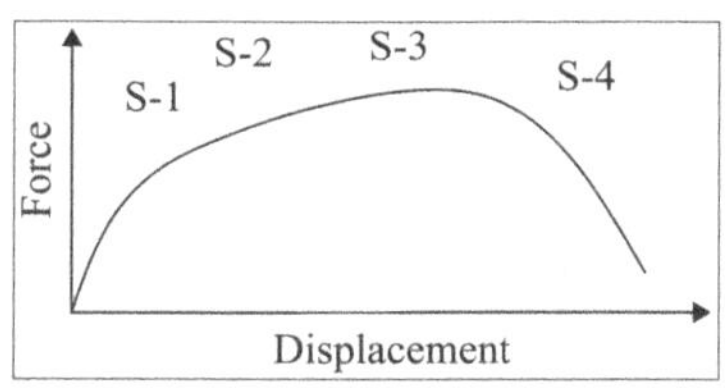

On the other hand, the human safety limit state represents a response level designed to prevent loss of human life, with the probability of a rare earthquake (MCE) occurring within 50 years estimated at approximately 5%. In contrast, the ultimate limit state in Japan assumes that the structure is near collapse, necessitating consideration of scenarios where early repair is challenging. Consequently, strong earthquake ground motions (Level 2U) with a probability of occurrence of 2% or less over a 50-year period are assumed, reflecting their rare occurrence.

5.1.4 Seismic Performance Goals for Lifeline Systems and Pipe Materials

5.1.4.1 Performance Objectives of the System – Water Supply Lifeline Systems Discussed as an Example

The *Guidelines and Commentary on Seismic Engineering Methods for Water Supply Facilities* provide an example of the number of days required for emergency restoration as an evaluation index and set this as the target for earthquake resistance. To achieve this, seismic performance levels 1–3 are presented as goals for seismic design. The problem is that water utilities do not evaluate how confident they are in their ability to meet the emergency recovery time as per the target values. The level of seismic design of individual elements should depend on whether the possibility of achieving the target is 20–30% or 70–80%, but this point has not been addressed. To convert seismic design methods into performance design means evaluating the probability of achieving this seismic target in terms of probabilities, considering the prediction errors of the design variables governing the seismic design. Japanese seismic design methods[1,2,3] generally require that the maximum response value of a structure be within the yield limit for Level 1 ground motion, and that the plastic deformation should be within the limit displacement for Level 2 ground motion.

The current water supply guidelines quantify seismic resistance targets for water supply systems based on the estimation of pipeline damage from past measured data and the estimation of the number of days for emergency restoration. However, if various uncertain variables in the evaluation procedure are not considered, the estimation of the number of days for emergency restoration becomes a fixed value. If all uncertain variables are evaluated as probabilities, probability statistics such as mean values and standard deviations for the number of days for emergency restoration can be obtained, though the calculation becomes enormous. In practice, to reduce design costs, each variable is treated as a deterministic quantity, making the emergency recovery time estimation deterministic. As a result, the accuracy of estimating the number of days for emergency restoration becomes 100%, and it is not possible to move toward performance design. This situation is believed to be the reason why seismic design has not progressed to performance design in the past.

However, how much seismic design benefit would be gained by considering the variation in the number of days for emergency recovery in the performance design method? If the discussion on the adoption or rejection of performance design is based solely on this issue, it becomes difficult to determine the seismic design benefits. However, if the relationship between water supply performance and the seismic performance of the constituent pipe materials and elemental facilities can be quantitatively evaluated, without limiting the seismic resistance target to the number of emergency recovery days, it should be possible to discuss the specific seismic performance required for the pipe materials and elemental facilities to achieve water supply performance. On the other hand, if the probability of achieving water supply performance for specific demand (critical water supply) sites can be determined, the estimation of the number of days for emergency restoration can be treated as a separate task based on these values.

5.1.4.2 Seismic Performance of Pipe Materials

The individual pipe members are structural elements connected in series to form a pipeline. As a result, a section of pipeline is vulnerable in that the rupture of a single pipe joint can lead directly to a failure. However, this vulnerability is mitigated by the redundancy of the network. In other words, the seismic resistance of a water supply system cannot be adequately evaluated unless both the seismic strength of the pipe joints – which are the fundamental components of the system – and the redundancy of the network are considered simultaneously.

The current *Guidelines and Commentary on Seismic Engineering Methods for Water Supply Facilities* only assume a single straight pipeline and assess the seismic safety at its maximum strain generation point. They do not address the seismic performance of the entire water supply system, i.e., the water supply performance from the source to the demand end. Therefore, even if the seismic safety of the straight pipeline is verified, it does not imply that the seismic performance of the water supply system has been confirmed. Ductile iron pipes have traditionally been used as a general-purpose pipe material for water supply systems. However, steel pipes are commonly employed for large-diameter pipes in trunk lines, while polyethylene (PE) pipes[14] have recently gained widespread use for small- and medium-diameter pipes. Consequently, the range of pipe materials used is becoming increasingly diversified. Additionally, seismic-resistant pipes have been developed for ductile iron pipes[15] with seismic safety in mind. A system of protrusions inside the joints has been introduced to prevent the joints from coming loose during an earthquake. This allows for the inner dimensions of individual pipe materials to be designed to account for manufacturing errors in the joint's internal dimensions, as well as the strength of the displacement-restricting protrusions inside the joint. However, existing joints are manufactured without the necessary seismic target information from the water supply system, meaning they do not guarantee the seismic performance of the section pipeline or the water supply system comprising these pipe materials. There has, therefore, been a lack of an integrated approach between the pipe materials and the water supply system in terms of seismic performance. The current guidelines for water supply systems do not specify the probability statistics of design variables, meaning that pipe material manufacturers have not been able to provide such data from their own companies to external parties. However, for seismic design methods for water supply systems to evolve into performance-based design methods in the future, the *Guidelines and Commentary on Seismic Design Methods for Water Supply Facilities* should include the probability statistics of the design variables that are required from pipe material manufacturers.

5.2 FUNCTIONAL PERFORMANCE ANALYSIS AND PERFORMANCE DESIGN

5.2.1 Connectivity Analysis[11]

In this section, the water supply system is used as an example. Figure 5.2.1 presents a simplified model of a water supply network, featuring one water supply end (water treatment plant) labeled as N1. Facilities such as distribution reservoirs and demand ends (nodes) are represented by zeros in the diagram, encompassing seven facilities in total. Here, node N7 is designated as an important base (A1). The data corresponding to these nodes are detailed in Tables 5.2.1 and 5.2.2.

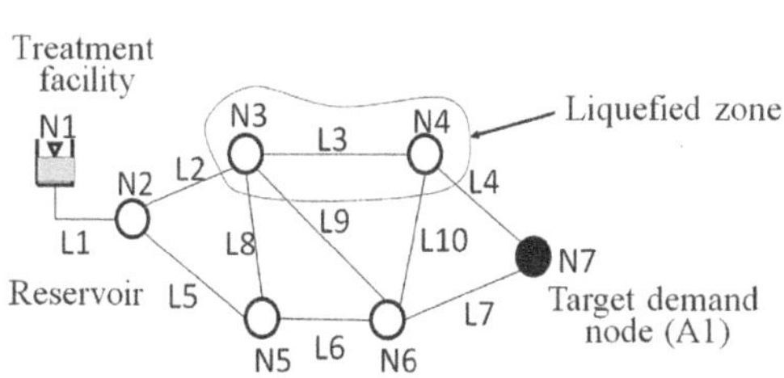

FIGURE 5.2.1 Simplified model of a water supply network.

TABLE 5.2.1
Pipeline Data

ID Link	Pipe Elongation Km
L1	10
L2	5
L3	9
L4	3
L5	6
L6	4
L7	2
L8	8
L9	10
L10	9

TABLE 5.2.2
Facility Data

ID Node	Safety Index
N1	4
N2	3.5
N3	3
N4	3
N5	3.5
N6	3.5
N7	3.75

Pipelines connect the nodes, with 10 pipelines installed in the diagram. These pipelines correspond to the links in a network. The data for each pipeline are provided in Table 5.2.1. For seismic safety assessment, the seismic probability of failure for each base facility has been predetermined and is presented in Table 5.2.2 as a safety index. It is assumed that all pipelines are ductile iron pipes equipped with projecting seismic pipes. The pipelines between nodes (links) consist of single pipes connected in series, and the probability of failure, $p_{f,link}$, is calculated based on the probability of failure $p_{f,joint}$ of a single pipe joint and the number of pipes, L/ℓ, using the following equation:

$$p_{f,link} = 1 - \left(1 - p_{f,joint}\right)^{L/l}, \tag{5.2.1}$$

where L is the pipe length between nodes and l is the length of the single pipe.

Furthermore, the joint safety index β_{joint} is assumed to be connected to the fracture probability $p_{f,link}$ of the single pipe joint by the following equation:

$$\beta_{joint} = -\Phi^{-1}\left[p_{f,joint}\right]. \tag{5.2.2}$$

To understand the relationship between the seismic performance of pipe joints and the functional performance of the water supply system, it is assumed that all pipe joints have the same seismic performance. In other words, the seismic performance of the joints is expressed using the probability of failure p_f or the safety index β. In the liquefaction zone, as shown in Figure 5.2.1, it is assumed that the pipelines (links) and base facilities (nodes) within the zone are located on liquefied ground and are subject to seismic liquefaction hazards. The ability to supply water from the water supply end (N1) to the demand end (N7) can be assessed by determining whether the water can reach the demand end via an unbroken pipeline. The proportion of this reachability depends on the probability of failure of each pipeline and the probability of failure of the base facilities. The reachability of the water supply from the supply end to the demand end of the network – referred to as linkage performance – can be

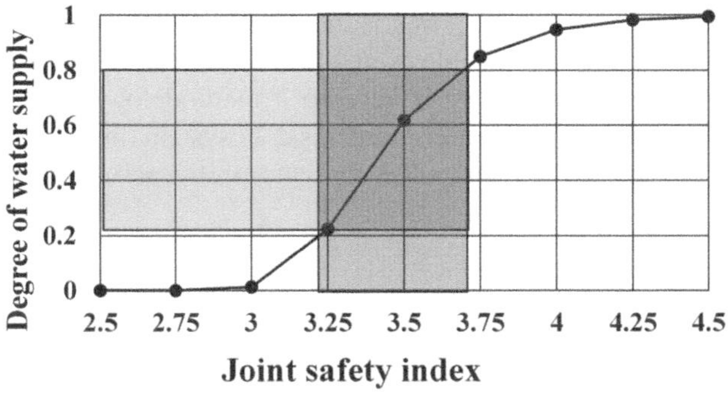

FIGURE 5.2.2 Reachability of water supply to critical locations.

obtained by using Monte Carlo simulation, as shown in Figure A5.1 in Appendix A5. This linkage performance can be evaluated in terms of the probability of successfully supplying water from the source to the target demand end, and the linkage probability (water supply reachability) is defined by the following equation:

$$p_C = \frac{M}{MCS}, \tag{5.2.3}$$

where

M: number of times a connecting route from the water supply end (water treatment plant) to the key location has been secured.

MCS: number of Monte Carlo simulations

In the case of this water supply system, the system will meet the required seismic safety performance if the cross-sectional dimensions and material strength of the pipe joints are manufactured to satisfy the values specified by the joint safety index. Figure 5.2.2 illustrates the relationship between the degree of water supply to reach critical points and the safety index, which serves as an indicator of the required safety of the pipe joints. The figure shows that, within the range of the joint safety index, water supply reachability falls between 0.2 and 0.8:

$$\beta_{joint} = 3.2 \sim 3.7. \tag{5.2.4}$$

5.2.1.1 Relationship Between Functional Serviceability and Physical Connectivity in a Network System

In the previous section, the state of connectivity of the network is used as a measure of the system's performance. Connectivity is defined as the capability of the network to transmit water from at least one of the supply stations to the demand nodes in the systems. However, the connectivity thus defined does not necessarily reflect the functional serviceability of the system. For example, even when a transmission path remains open between one of the supply stations and one of the nodes following an earthquake, the water pressure and flow rate at that node may drop, due to possible earthquake damage, below the level required for effective post-firefighting. Hence, if the capability of effective firefighting is the criterion for system serviceability, the system can be deemed unserviceable even while its connectivity till remains intact.

The purpose of this section is to develop a methodology for the seismic risk assessment[37] of underground water transmission system in a seismically active areas in terms of the degree of possible unserviceability resulting from destructive earthquakes.

5.2.1.1.1 Network Model

A water transmission system adapted from the Los Angeles City system as shown in Figure 5.2.3 is used for the analysis. The data to be used for the analysis including its

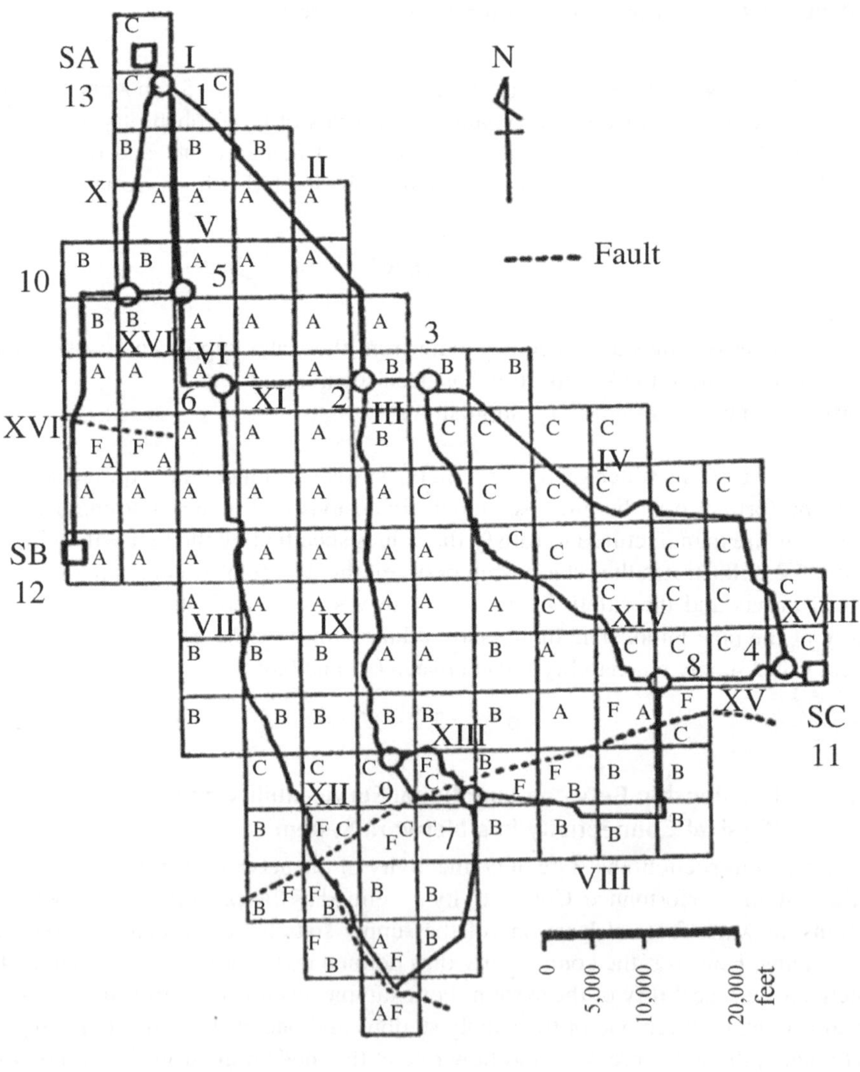

FIGURE 5.2.3 The water transmission network in the city of Los Angeles.

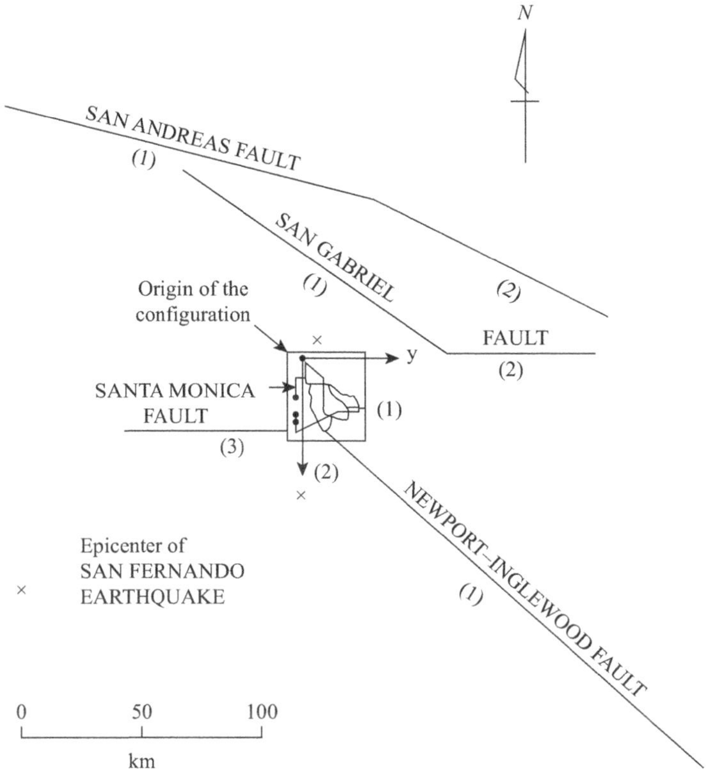

FIGURE 5.2.4 Major faults surrounding the city of Los Angeles.

topological and structural characteristics, conditions of the soil and ground in which it is imbedded, and the locations of major faults surrounding and penetrating it, as shown in Figure 5.2.4, are described in detail hereunder.

Several symbols and numbers are shown in Figure 5.2.3:

SA, SB, and SC: the water source nodes
1 to 13: the node numbers
I to XVIII: the name of links
F: fault zone
A, B, and C: soil classifications, in which
- A = consolidated to poorly consolidated sediments
- B = semi-consolidated to moderately consolidated sediments
- C = very dense igneous and metamorphic rocks

Table 5.2.3 shows the detail soil properties for each ground, A, B, and C.

TABLE 5.2.3
Soil Conditions

Local Soil Properties	Symbol	Unit	Ground Surface A		B		C	Base Rock
Thickness	H	m	30		30		30	∞
Mass density of soil	ρ_G	kN/m^3	15		17		19	2.65
Shear wave velocity	V_S	m/s	150		300		500	692.3
Dilatational wave velocity	V_C	m/s	780		870		1200	1695.7
Shear modulus	G	MPa	34.44		156.1		484.7	1296
Lamé constant	λ	MPa	862.3		1001		1822	5183
Poisson ratio	ν		0.481		0.433		0.396	0.4
Coefficient of soil reaction/volume	k_0	kN/m^3	0.069	*or*	0.1	*or*	0.228	
Critical shear strain*	γ_{cr}		10^{-5}	*or*	10^{-4}	*or*	10^{-3}	

* slippage takes place beyond this value.

5.2.1.1.2 Definition of Damage States

The structural damage states for lifeline network are defined as

a. Major damage
   A structural element in the network is in the ultimate limit state.
b. Moderate damage
   A structural element in the network is not in the ultimate limit state nor in the serviceability limit state.
c. Minor damage
   A structural element in the network is in the serviceability limit state.

The lifeline network system includes supply stations, transmission lines, substations, and service networks to demands. Link elements are a basic unit for transmission lines and distribution networks.

5.2.1.1.3 Link Element

Limit state for major damage mode of a link element:

$$Z_i^{major} = \delta_{cr}^{major} - \delta_i, \tag{5.2.5}$$

Limit state for minor damage mode of a link element:

$$Z_i^{minor} = \delta_{cr}^{minor} - \delta_i, \tag{5.2.6}$$

Link element can be modeled as a poly-line passing through many meshes which belong to various soil conditions. One segment located in a mesh is defined as

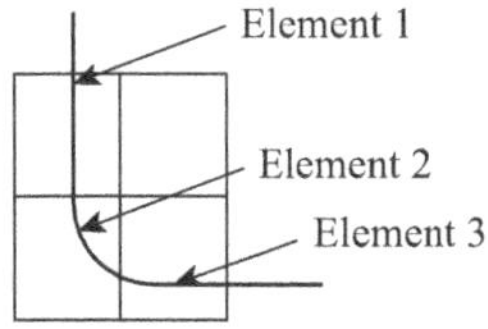

FIGURE 5.2.5 An element model of a passing through many meshes.

an element. In Figure 5.2.5, elements 1, 2, and 3 are located at their own meshes. Elements 1 and 2 are connected at the mesh boundary.

5.2.1.1.4 Link Damage State and Its Probability

It should be noted that a link is a series system of several line elements (i.e., NL_k) connecting neighboring nodes.

Major damage of the kth link:

$$E^k_{major} = \bigcup_{i=1}^{NL_k} E\left[Z_i^{major} < 0\right], \tag{5.2.7}$$

Moderate damage of the kth link:

$$E^k_{moderate} = \overline{E}^k_{major} \bigcap \overline{E}^k_{minor}, \tag{5.2.8}$$

Minor damage of the kth link:

$$E^k_{minor} = \bigcap_{i=1}^{NL_k} E\left[Z_i^{minor} > 0\right] \tag{5.2.9}$$

Once an earthquake occurs, the probability for the kth link damage is evaluated as $P\left[E^k_{major}\right]$, $P\left[E^k_{moderate}\right]$, and $P\left[E^k_{minor}\right]$, respectively.

5.2.1.1.5 Criteria on Loss of Connectivity and Unserviceability

It is noted that a pipeline is a series system of unit pipes and is connected with the starting node and the end node. Each pipeline is defined as a link, and all the links constitute one network system.

If any links in the water network system are physically damaged by an earthquake, and water supply to the target demand node is suspended, the serviceability to the target demand node is in a significant damage. Simply, this unserviceability state can be called damage of connectivity.

The connectivity criterion of the network system is defined in the following way:

1. Criterion for connectivity under the major damage criterion for links is that some routes from the source nodes to the target demand node are connected in the major damaged network.
2. Criterion for connectivity under the moderate damage criterion for links is that some routes from the source nodes to the target demand node are connected in the moderate damaged network.

The probability of a link failure in the network is given in Equation (5.2.1). The probability of connectivity under the major damage criterion for links is given the same way in Equation (5.2.3).

$$P\left[C_{major}\right]=1-\frac{M_{major}}{MCS},$$
$$P\left[C_{moderate}\right]=1-\frac{M_{moderate}}{MCS}, \tag{5.2.10}$$

in which

C_{major}: the major loss of connectivity

$C_{moderate}$: the moderate loss of connectivity

M_{major}: numbers of success of route connection from the source nodes to the target demand nodes in the major damaged network in which all the links are checked by major damage criterion with E^k_{major}

$M_{moderate}$: numbers of success of route connection from the source nodes to the target demand nodes in the moderate damaged network in which all the links is checked by moderate damage criterion with $E^k_{moderate}$

MCS: number of Monte Carlo simulations

The serviceability criterion[38] is defined in the following way.

For the system to be serviceable under normal conditions, it has to be capable, for example, of continuously responding to a minimum level of domestic and industrial water demands. This section, however, disregards this and other serviceability considerations and concentrates on system serviceability under emergency conditions, in which the adequacy of the water pressure and flow rates for firefighting in post-earthquake environments serve as a measure of the system serviceability.

Three possible states of serviceability (adequacy) or equivalently unserviceability (inadequacy) are defined at each demand node.

1. A state of major unserviceability corresponds to the condition that at least either the water pressure or the flow rate is less than the respective minimum requirements (E_{min} for pressure and Q_{min} for flow rate) for firefighting.
2. A state of moderate unserviceability represents the condition under which both the pressure and the flow rate satisfy the minimum requirements, but neither of them exceeds the design allowable amount associated with the undamaged original network (E_{mod} for pressure and Q_{mod} for flow rate).
3. A state of minor unserviceability is then defined as that in which the pressure and flow rate exceed E_{mod} and Q_{mod}, respectively, and therefore implies a state of normal serviceability.

Figure 5.2.6 shows these serviceability criteria schematically.

Based on the serviceability criteria, the probability of unserviceability to the target demand node k is formulated as follows.

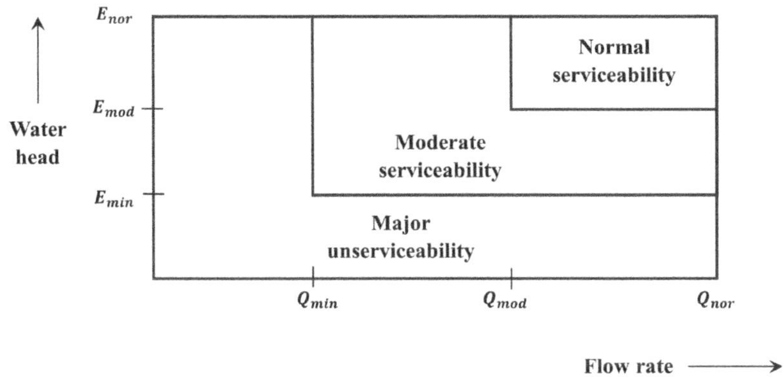

FIGURE 5.2.6 Serviceability criteria.

$$P\left[S^k_{major}\right] = P\left[\left(H_k \le H_{cr}\right) \cup \left(Q_k \le Q_{cr}\right)\right]$$
$$P\left[S^k_{minor}\right] = P\left[\left(e_0 H^N_k < H_k\right) \cap \left(f_0 Q^N_k < Q_k\right)\right] \quad (5.2.11)$$
$$P\left[S^k_{moderate}\right] = 1 - P\left[S^k_{major}\right] - P\left[S^k_{minor}\right],$$

in which

- S^k_{major}: system unserviceability at the target demand node k in the major damage state in which either water head or flow rate is less than the minimum requirement, H_{cr} or $Q_{cr,}$ under locally disconnected network due to major damage links
- S^k_{minor}: system unserviceability at the target demand node k in the minor damage state in which both of water head and flow rate are more than the relieved flow conditions, $e_0 H^N_k$ and $f_0 Q^N_k$, which are relieved from the normal requirements H^N_k *and* Q^N_k with the relieving rate of e_0 *and* f_0 under locally disconnected network due to major damage links
- $S^k_{moderate}$: system unserviceability at the target demand node k in the moderate damage state, that is, a state that is not classified as major or minor
- H_k: water head at the target node k in the damaged network
- Q_k: flow rate at the target demand node in the damaged network

5.2.1.1.6 *Flow Analysis*[39]

The hydraulic conditions at the nodes, i and j, are shown in Figure 5.2.7. At a non-isolated demand node i, the continuity condition requires that

$$\sum_j Q_{ij} + Q_{ir} + \ell_i = 0, \quad (5.2.12)$$

where Q_{ij} is the flow rate in link ij, Q_{ir} is the rate at which the water is extracted from node for a specified purpose (firefighting, in this case), and ℓ_i is the rate of leakage at node i.

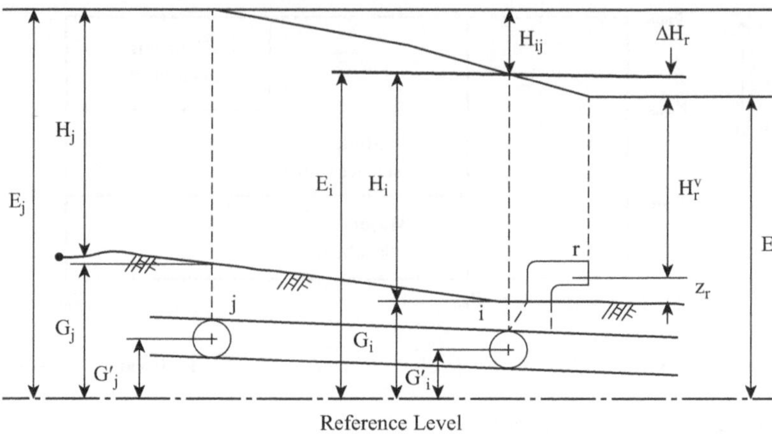

FIGURE 5.2.7 Configuration of buried pipe for hydraulic analysis.

With respect to the relationship between the head loss and the rate of flow in each link, the present flow analysis uses the Hazen–Williams equation since it is valid for a range of surface roughness appropriate for the steel pipes under consideration here:

$$\begin{aligned} Q_{ij} &= R_{ij} \left| H_{ij} \right|^{-0.46} H_{ij} \\ R_{ij} &= 0.27853 D_{ij}^{2.63} L_{ij}^{-0.54}, \end{aligned} \tag{5.2.13}$$

with

$$H_{ij} = E_i - E_j, \tag{5.2.14}$$

where

H_{ij}: the head loss associated with link i, j
E_i: the water head at node i
D_{ij}: the diameter of the pipe used for link i, j
L_{ij}: the length of link i, j

The flow rate Q_{ir} is equal to Q_{nor} if the water head relative to ground level at node i is at least equal to a specified value H_{min}. On the other hand,

$$Q_{ir} = Q_{nor} \left(E_i - G_i \right)^{1/2} H_{min}^{-1/2} \tag{5.2.15}$$

if the relative water head is less than H_{min} but larger than zero. For those cases where $E_i \le G_i$, Q_{ir} is taken as zero.

The rate of leakage ℓ_i at node i may be estimated as

$$\begin{aligned} \ell_i &= d_i \left(E_i - G_i' \right)^{\kappa} && \text{for } E_i > G_i' \\ &= 0 && \text{for } E_i \le G_i', \end{aligned} \tag{5.2.16}$$

where

d_i: a leakage coefficient with appropriate dimensions

G_i': the level at which the center of the pipe for link ij is located at node i and $\kappa = 1.15$

The water transmission network to be considered herein is a simplified version of the Los Angeles system and consists of 3 supply stations, 10 demand nodes, and 18 links, as shown in Figure 5.2.3. It is assumed that $Q_{min} = 0.63Q^3/sec$, $Q_{mod} = 1.87Q^3/sec$, $Q_{nor} = 2.4Q^3/sec$, $H_{min} = 15m$, $G_i = 10m$ *and* $G_i' = 8m$. The minimum water head requirement E_{min} at each demand node is also taken as 15 m, while the design maximum water head E_{nor} is obtained as the result of a flow analysis performed on the original undamaged system when the output flow rate Q_{ir} is equal to the specified normal flow rat $Q_{nor} = 2.4m^3/sec$ at each demand node. Such an analysis is performed under the assumption that the water heads at the three supply stations are all 55 m and the diameter of the steel pipes is 60″ for all links.

5.2.1.1.7 Monte Carlo Simulation

A straightforward Monte Carlo simulation can be performed to determine whether link ij has failed or survived by generating a random number uniformly distributed between zero and unity. The link is considered to have failed (or survived) if the realization of the random number is less than or equal to (or larger than) the probability of link ij in the major damage: $P\left[L_{ij,major}\right]$. If a link is judged to have failed, it is removed from the network. Furthermore, if a node becomes isolated due to the removal of all the links terminating at that node, it is also removed from the network. A flow analysis is then performed on the network system consisting of the remaining nodes and links.

The numerical observation with respect to the correlation between the unserviceability and the loss of connectivity seems to be in order. In fact, Figure 5.2.8 plots the probabilities of a moderate loss of connectivity (line A), major unserviceability (line B), and major loss of connectivity (line C) at demand nodes 1~10 when an earthquake of magnitude 7 occurs with its epicenter at the nearest point from the network. At all the nodes considered, the probabilities of major unserviceability are much larger than the probabilities of a major loss of connectivity (indeed, the latter are practically zero in this case), although the probabilities of a moderate loss of connectivity are larger than those of major unserviceability. The way in which these probabilities depend on the demand nodes is similar both for major unserviceability and for a moderate loss of connectivity. The trend, indeed, is consistent with the variability of the average ratios (line D) of the water heads at these nodes after an earthquake of magnitude 7 to the corresponding values associated with the original undamaged system. Similar ratios for the output flow rate are also indicated by line E. A similar value of these ratios indicates a greater loss of the system performance capability either in terms of water pressure or outflow rate.

The probabilities of unserviceability and loss of connectivity are pronouncedly higher at demand nodes 2, 6, 7, and 9, reflecting the fact that these nodes are farthest from the supply node in the topological sense. In contrast, the probabilities are much smaller at nodes 1, 4, and 10 due to their topological proximities to the source nodes.

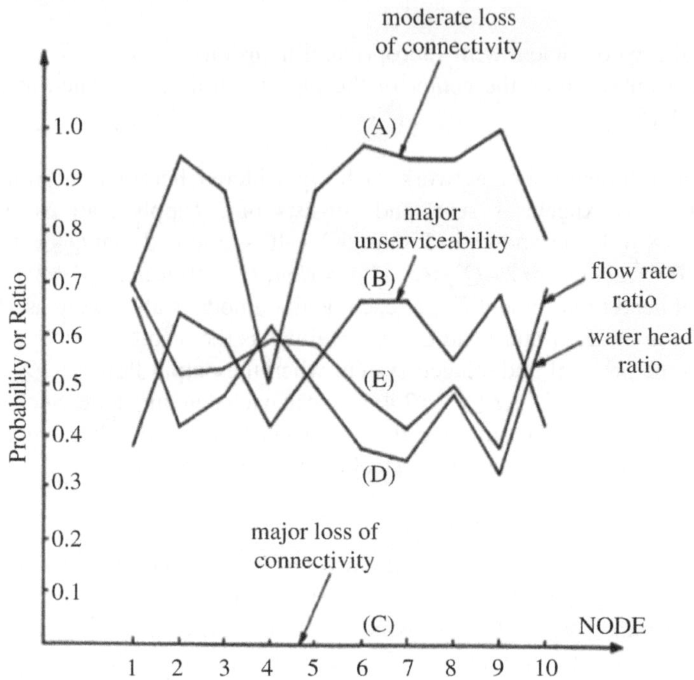

FIGURE 5.2.8 Flow characteristics and correlation between loss of connectivity and unserviceability.

5.2.1.1.8 System Serviceability and Connectivity by an Earthquake from Major Faults[37]

In Figure 5.2.9, the relationship between system serviceability and connectivity is compared in terms of the probabilities of system failure for the four faults. In this figure, closed curves are shown for each fault. Each closed curve is enclosed with the target points, which correspond to the probability of unserviceability or loss of connectivity. The typical characteristics of these curves are observed in the following way:

1. Closed area of the NEWPORT-INGLEWOOD Fault is different from that of the other three faults.
2. These other three faults show a small, closed area such as a rectangle, and their occupied areas are in close proximity.
3. Closed areas of the SAN ANDREAS Fault and SAN GABRIEL Fault are also similar, while the probability of unserviceability for the SAN ANDREAS Fault is relatively small.
4. Closed area of the SANTA MONICA fault is similar to the closed areas of (3), but its size is larger than that of (3).
5. Common characteristics of these four faults are that the probability of system failure at the target demand nodes of $k = 1, 2, 5, 6, 7, 9$ in Figure 5.2.9 is located along line AB, and the other nodes of $k = 3, 4, 8, 10$ are located

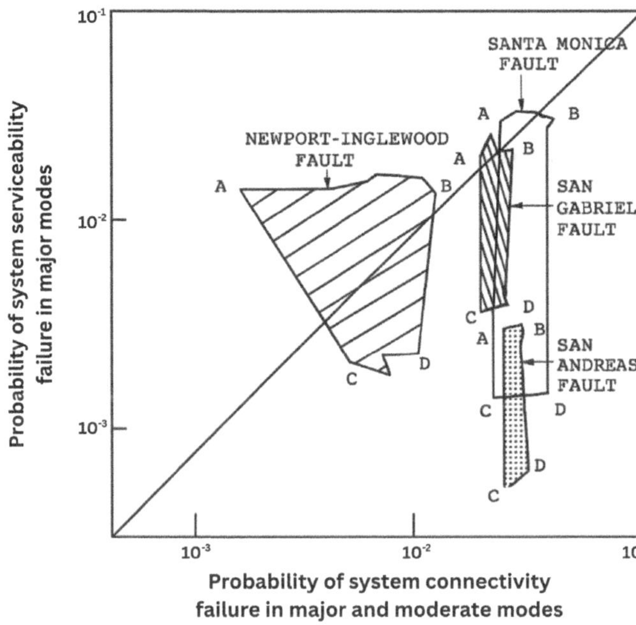

FIGURE 5.2.9 Comparison between system serviceability and connectivity in the probabilities of system failure of the four faults.

along line CD. The length of the lines AB and CD are relatively longer than that of lines AC and BD. This result suggests that the variance of the probability of serviceability is larger than that of the probability of connectivity. On the other hand, it is noted that the nodes of k = 3, 4, 8, 10 along line CD always show a relatively small probability of serviceability for all these faults.

6. In the SAN ANDREAS fault, the probability of connectivity is larger than that of serviceability.

Those observations can be understood by the geological configuration of these four faults as follows:

1. The NEWPORT-INGLEWOOD fault is directed and partially arrives to the network, so that this fault is characterized with (1).
2. Parallel location of the SAN ANDREAS and SAN GABRIEL faults show similar characteristics (3) of the probability of system failures. Smaller probability (6) in the SAN ANDREAS Fault suggests the effect of location farther from the network.
3. The SANTA MONICA Fault passes through the network, so that the epicentral distance for each demand node is clearly different and shows the different probabilities among them. This trend leads to a larger closed area, which is pointed out in (4).

It is confirmed, from the aforementioned observation, that fault location can affect to the system failure of the network, and this is an intrinsic characteristic of spatially extended systems.

5.2.2 Performance Design of Pipe Joints

5.2.2.1 Formulation

5.2.2.1.1 Number of Single Pipes Required to Absorb Ground Displacements U_h

The number of single pipes required, m, can be calculated by the following formula:

$$m = \frac{U_h}{\delta}, \tag{5.2.17}$$

where δ is allowable displacement of single pipe joint.

It is important to note that variations in pipe joint displacement can occur due to fabrication errors in the pipe material, as well as variations in joint positions during the buried construction of individual pipe joints and during seismic events

5.2.2.1.2 Maximum Number of Single Pipe Connections up to the Breaking Limit

The joint tensile strength of earthquake-resistant ductile cast iron pipes is defined as the strength at which the locking ring in the joint shears the insertion butt. This strength is approximately $3D$ (kN), where D is the diameter of the pipe, expressed in millimeters. A rupture of jointed pipes occurs when the joint tension, caused by the shearing force acting along the circumference of the connected pipes, exceeds the joint tensile strength of a single pipe. The number of connected pipes at this point – the number required to cause rupture – is expressed by the following equation:

$$m_{cr} = \frac{3D}{\pi D \ell \tau_{cr}}, \tag{5.2.18}$$

where

τ_{cr}: the maximum shear stress of the soil around the pipe for the slip initiation
l: length of a single pipe

$3D$ and τ_{cr}, which make up this equation, are the limiting strengths of the cast iron pipe material and the surrounding soil and are essentially random quantities; to clarify the randomness of $3D$, it is replaced by the symbol F_3 instead of 3. The random quantity F_3 can therefore be expressed as follows:

$$\text{Mean: } E[F_3], \text{ Variance: } VAR[F_3].$$

The critical shear strength of the soil at slip initiation is determined through a shear experiment using soil surrounding a buried pipe with a soil cover of 1.8 m.

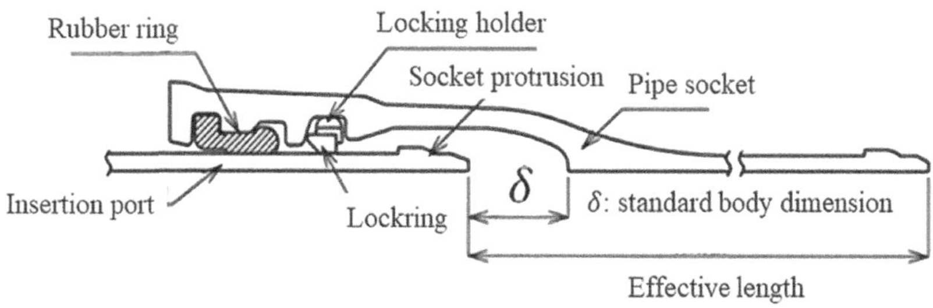

FIGURE 5.2.10 Displacement δ of ductile iron pipe seismic pipe joint.[15]

This strength is essentially a random variable. Therefore, its probability statistics can be expressed as follows:

$$\text{Mean: } E[\tau_{cr}], \text{ Variance: } VAR[\tau_{cr}].$$

5.2.2.1.3 Prediction of Pipe Joint Rupture

A rupture of coupled pipes occurs when the number of pipes, m, required to absorb ground displacement exceeds the number of pipes, m_{cr}, required for rupture. Therefore, the probability of rupture of a pipe joint can be formulated as follows:

$$p_{f,joint} = P[m_{cr} - m < 0]. \tag{5.2.19}$$

5.2.2.2 Performance Design of Pipe Joints

The allowable displacement dimension for pipe joints currently in use is set to 60 mm, as shown in Figure 5.2.10. The validity of this value can be assessed through the joint displacement performance verification outlined in this section. By rewriting the fracture condition equation in Equation (5.2.20) in natural logarithm form, the following fracture condition equation is obtained:

$$Z = ln\left(\frac{F_3 \cdot \delta}{\pi\tau_{cr}\ell}\right) - ln(U_h). \tag{5.2.20}$$

By transforming the random variable Z into the standard normal variable z, Equation (5.2.21) can be expressed as

$$p_{f,joint} = P\left[z < -ln\left\{E[\delta] \Big/ \left(\frac{\pi E[\tau_{cr}]\ell E[U_h]}{E[F_3]}\right)\right\} \Big/ \sigma_z\right], \tag{5.2.21}$$

where the mean of the standard normal variable z is 0 and the standard deviation is 1 and the standard deviation of the random variable Z is given by

$$\sigma_Z = \sqrt{\delta_{U_h}^2 + \delta_{\tau_{cr}}^2 + \delta_{F_3}^2 + \delta_\delta^2}. \tag{5.2.22}$$

TABLE 5.2.4
Analytical Conditions

	T	S_V	U_h	F_3	τ_{cr}	ℓ	σ_Z	δ
	sec	m/s	m	kN/mm	kN/mm²	m		m
Mean	1	1	0.2026	3	12.5	6	*	0.06
Cov	0	0	0.3	0.05	0.1	0	0.324	0.05

However, the δ used for δ_X^2 is the symbol for the coefficient of variation of the variable X.

Equation (5.2.21) can therefore be expressed using the standard normal distribution function Φ as follows:

$$p_{f,joint} = \Phi\left[-ln\left\{E[\delta] \Big/ \left(\frac{\pi E[\tau_{cr}]\ell E[U_h]}{E[F_3]}\right)\right\} \Big/ \sigma_Z\right]. \tag{5.2.23}$$

From Equation (5.2.23), the mean required displacement of the joint can be obtained by the following equation:

$$E[\delta] = \frac{\pi E[\tau_{cr}]\ell E[U_h]}{E[F_3]} exp\left(-\sigma_Z \cdot \Phi^{-1}[p_{f,joint}]\right). \tag{5.2.24}$$

The values of each design variable are provided in Table 5.2.4, assuming a buried pipe soil cover of 1.5 m. The velocity response spectrum S_V is calculated in accordance with the domestic seismic design guidelines. For the subsequent numerical calculations, S_V is assumed to adhere to these guidelines. Since S_V is given as a specific value, the ground displacement U_h is also determined. However, from a performance design perspective, it is important to account for potential prediction errors of U_h. Therefore, a coefficient of variation of 0.3 is assumed a priori.

Figure 5.2.4 illustrates the relationship between the required joint displacement shown in Figure 5.2.3 and the joint safety index. As indicated in Figure 5.2.3, the current allowable joint displacement is 60 mm. However, Figure 5.2.11 compares

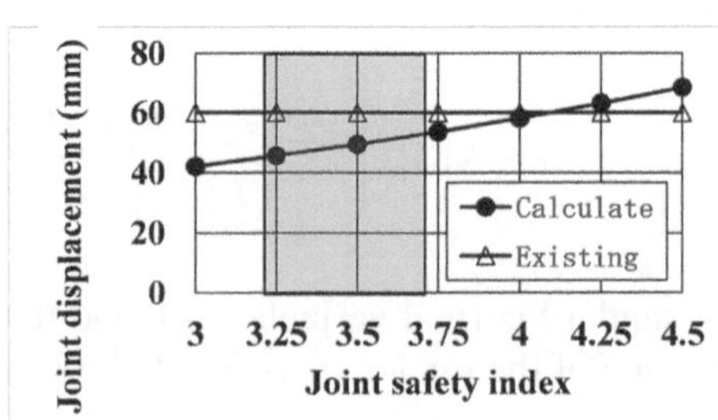

FIGURE 5.2.11 Relationship between joint safety index and joint displacement.

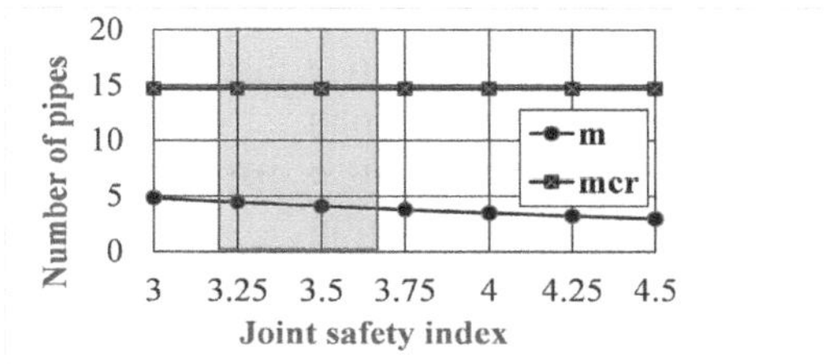

FIGURE 5.2.12 Required number of pipes for the joint safety index.

the design value from Equation (5.2.24) with the current value. The figure shows that the current joint displacements exceed the design values within the joint safety index range of 3.0–4.1, confirming the seismic performance of the seismic joint in the blue section (3.2–3.7) of the figure. However, if the water utility requires a joint safety index of 4.1 or higher, this suggests a need to review the current allowable joint displacements.

Figure 5.2.12 shows the relationship between the joint safety index and the number of pipes required when the joint displacement is defined by Equation (5.2.24). The figure indicates that the number of pipes required to break, m_{cr}, exceeds the number of pipes required to absorb ground displacement m, by approximately three times.

5.2.3 Performance Design of Buried Expansion Flexible Joints

If the safety verification function is now defined as

$$Z = ln\Delta_{cr}^{R} - ln\Delta^{R}, \tag{5.2.25}$$

The corresponding apparent safety index is given by the following equation using the verification values of the design variables:

$$\beta_d = \frac{ln\left\{\left(\Delta_{cr}^{R}/\gamma_i\right)_d\right\} - ln\left\{\left(\Delta^{R}\right)_d\right\}}{\sigma_Z}, \tag{5.2.26}$$

where

$\left(\frac{\Delta_{cr}^{R}}{\gamma_i}\right)_d$: verification value of the limit relative displacement to the repair limit state

$\left(\Delta^{R}\right)_d$: verification value of the relative displacement for the repairable limit state

σ_Z: standard deviation of the safety verification function Z, given by

$$\sigma_Z = \sqrt{\varsigma_{\Delta_{cr}^{R}}^{2} + \varsigma_{\Delta^{R}}^{2}}. \tag{5.2.27}$$

Therefore, given the assumed maximum unequal settlement the safety index can be calculated as

$$\beta\left(\Delta_v^{*}\right) = \beta_d\left(\Delta_v^{*}\right) + \frac{ln\gamma_{\Delta_{cr}^{R}} + ln\gamma_{\Delta^{R}} + ln\gamma_a + ln\gamma_b + ln\gamma_i}{\sqrt{\varsigma_{\Delta_{cr}^{R}}^{2} + \varsigma_{\Delta^{R}}^{2}}}, \tag{5.2.28}$$

where

$\gamma_{\Delta_{cr}^R}$: partial coefficient for the limit relative displacement Δ_{cr}^R to the repairable limit state of the expansion flexible joint

γ_{Δ^R}: partial coefficient for the limit relative displacement Δ_{cr}^R to the repairable limit state of the expansion flexible joint

By back calculating Equation (5.2.28), the performance design value $\left(\frac{\Delta_{cr}^R}{\gamma_i}\right)$ of an expansion flexible joint for the reparable limit state can be obtained using the target safety index β^{Target} by the following equation:

$$\left(\frac{\Delta_{cr}^R}{\gamma_i}\right) = \Delta^R exp\left(\beta^{Target}\sqrt{\varsigma_{\Delta_{cr}^R}^2 + \varsigma_{\Delta^R}^2}\right), \tag{5.2.29}$$

where

$$\beta^{Target} \geq \beta\left(\Delta_v^*\right) \tag{5.2.30}$$

5.2.4 Performance Design of Expansion Flexible Joints for Water Pipeline Bridge

5.2.4.1 Target Safety Performance of the Expansion Flexible Joints of Water Pipeline Bridges

Expansion flexible joints around a water pipeline bridge include both above-ground expansion flexible joints installed on abutments and pillars, and buried expansion flexible joints installed between the middle of the abutment and the buried pipelines. These expansion flexible joints are connected in series to form a group of expansion flexible joints around the water pipeline bridge. When assessing the seismic safety of the expansion flexible joints, it is essential to evaluate the seismic safety of these joints as a whole.

Let us assume that the safety index β_{cr}^{Group} represents the required seismic safety performance for the group of expansion flexible joints. The corresponding probability of failure is assumed to be given by

$$p_f^{Group} = 1 - \Phi\left[\beta_{cr}^{Group}\right]. \tag{5.2.31}$$

5.2.4.2 Seismic Safety Assessment of Expansion Flexible Joints

5.2.4.2.1 Safety Verification of Above-Ground Type Expansion Flexible Joints against Level 2 Ground Motion

$$p_{f,(i,j)} = P\left[\Delta_{cr} \leq \Delta_{(i,j)}\right]$$
$$for\ (i,j) = (1,2)\ or\ (2,4) \tag{5.2.32}$$

$$p_{f,(k,l)} = P\left[\Theta_{cr} \le \Theta_{(k,l)}\right]$$
$$for\ (k,l)=(1,2)\ or\ (2,4)\ or\ (3,4), \tag{5.2.33}$$

where

$\Theta_{(i,j)}$: relative angle between abutment and strut (i,j) perpendicular to the bridge axis

$\Theta_{(1,2)}$: relative angle in the direction perpendicular to the bridge axis between abutment A1 and pillar P1

$\Theta_{(2,4)}$: relative angle in the direction perpendicular to the bridge axis between strut P1 and abutment A2

$\Theta_{(3,4)}$: relative angle in the direction perpendicular to the bridge axis between pillar P2 and abutment A2

Therefore, the probability of failure of a group of expansion flexible joints can be calculated as follows, assuming that the individual failure events are independent of each other:

$$p_f^{Group} = 1 - \left\{\prod_{(i,j)}\left(1 - p_{f,(i,j)}\right)\right\} \cdot \left\{\prod_{(k,l)}\left(1 - p_{f,(k,l)}\right)\right\}. \tag{5.2.34}$$

Assuming now that the fracture probabilities of the m expansion flexible joints are equally $p_{f,joint}$, the fracture probability p_f^{Group} of the multispan pipeline bridge model is roughly related to the following relationship:

$$p_{f,joint} = 1 - \left(1 - p_f^{Group}\right)^{\frac{1}{m}} \cong \frac{1}{m} p_f^{Group}. \tag{5.2.35}$$

5.3 SEISMIC SAFETY ASSESSMENT METHOD FOR POST-QUAKE OPERATION OF LIFELINES DESIGNED UNDER THE OLD GUIDELINES[16]

The 9.0 magnitude earthquake that occurred on March 11, 2011, along with the tsunami it triggered, caused extensive human and material damage across many areas on the Pacific coast of the Tohoku region. In particular, the tsunami severely impacted sewage treatment plants, power stations, and various manufacturing facilities in the coastal areas, leading to the cessation of many lifelines. As a result, it took considerable time for residents to resume their daily activities and for urban functions to recover.

In Sendai City, gas supply was suspended immediately after the earthquake due to tsunami damage to gas plants located in the coastal area. However, the long-distance, high-pressure trunk gas pipeline, which crosses the Japanese archipelago from the Sea of Japan to the Pacific Ocean, sustained minimal damage. As a result, gas supply was restored to the city within a short period.

This section examines why the long-distance, high-pressure gas pipeline demonstrated robust seismic performance during the wide-area earthquake, considering both seismic damage and seismic design aspects.

The gas pipeline consists of the main pipeline and ancillary facilities such as delivery stations, valve stations, supervisory control and data acquisition (SCADA) systems, and an operation and monitoring center. The pipeline is operated by S Company. While the pipeline experienced seismic shaking and tsunami effects, several facilities along the coast were severely damaged by the tsunami. Despite this, the main pipeline itself remained intact, even in areas affected by permanent ground displacement in mountainous regions. Due to the early restoration of the pipelines and related facilities, gas supply to the city network in Sendai was restored just two weeks after the earthquake.

Seismic design standards for high-pressure trunk gas pipelines have been revised based on damage surveys from previous earthquakes, mandating high seismic performance for pipeline materials and systems. However, Japan has had limited opportunities to experience major earthquakes that would allow for verification of the seismic safety of long-distance high-pressure gas pipelines. The Tohoku Earthquake provided a valuable opportunity to assess the seismic performance of these pipelines, which are subjected to rigorous quality control.

This section presents:

1. A seismic safety assessment of trunk pipelines that experienced various types of seismic damage in mountainous, plain, and coastal areas
2. A quantitative evaluation of the seismic performance of trunk pipelines designed according to seismic design standards established before the 1995 Hyogo-ken Nanbu Earthquake, in the context of a Level 2 class giant earthquake

5.3.1 Seismic Damage Along the Long-Distance Gas Pipelines

5.3.1.1 Characteristics of the Pipeline Route and Seismic Intensity Distribution

The seismic intensity distribution along the Niigata-Sendai gas pipeline route and the areas it traverses in Miyagi Prefecture during the Tohoku–Pacific Ocean Earthquake is shown in Figure 5.3.1. This seismic intensity map is based on the detailed distribution estimated by the Japan Meteorological Agency.[17,18] According to the velocity response spectrum of the seismic motion record measured in Zao Town (near VS23), one of the transit areas shown in Figure 5.3.2, values around 100 cm/sec are observed in regions where the excitation period is 0.4 seconds or longer. This is nearly identical to the design velocity response spectrum[19] upper limit for Level 2 ground motion, as specified in the *High-Pressure Gas Pipe Seismic Design Guidelines*.[20] This suggests that the seismic ground motion experienced was nearly equivalent to the upper limit of the response spectrum for Level 2 ground motion outlined in the seismic design guidelines for high-pressure gas pipelines.

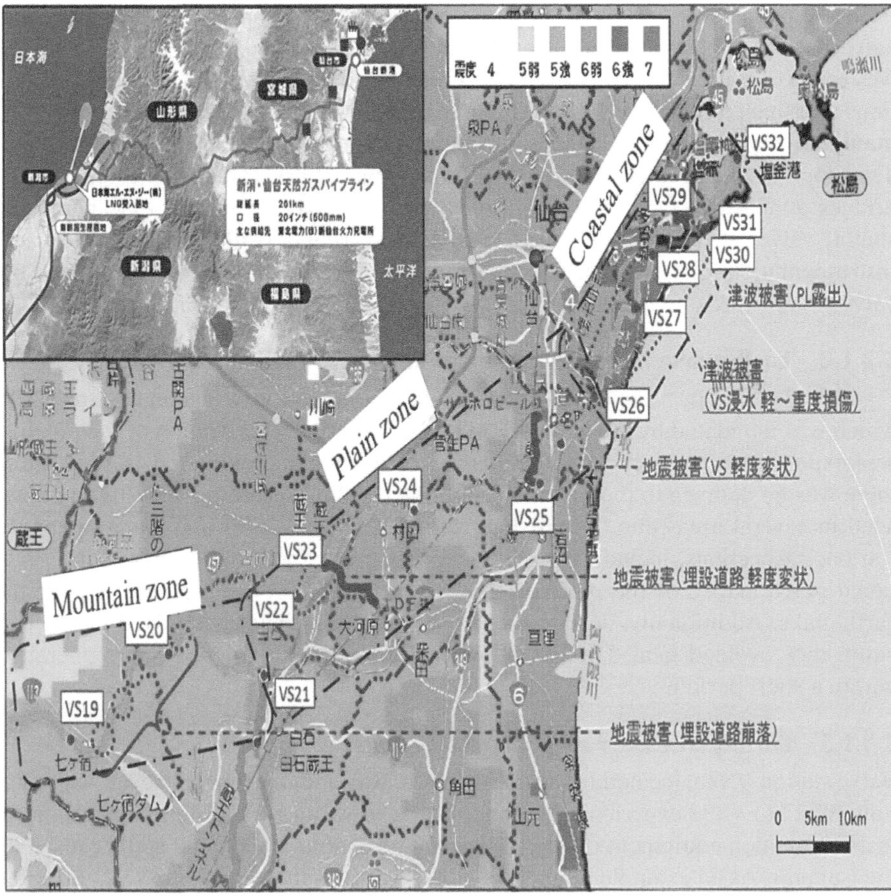

FIGURE 5.3.1 Distribution of seismic intensity along the pipeline route (thick line).

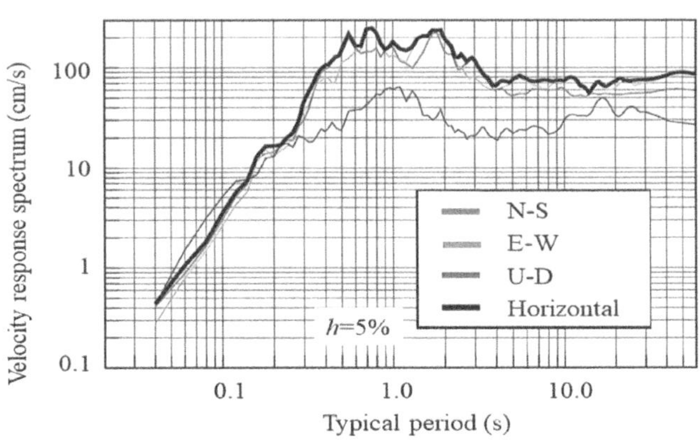

FIGURE 5.3.2 Velocity response spectra measured in Zao Town, Miyagi Prefecture.

The pipeline section shown in Figure 5.3.1 is divided into 14 sections, each with valve stations at both ends, covering a total distance of approximately 112.7 km. The westernmost three sections are located in mountainous areas, followed by five sections crossing plains, with the remaining sections running through coastal areas. The main delivery facility is located at the New Port of Sendai, near the city of Sendai. The pipelines are primarily buried, with independent dedicated bridges or pipeline bridges attached to existing road bridges used for river crossings. In general, the shutoff valves at valve stations were buried underground, while buildings, electrical instrumentation, and gas discharge towers were installed above-ground within the valve station yards.

5.3.1.2 Earthquake Damage at the Sendai New Port Delivery Facility

Figure 5.3.3(a) shows the damage to the delivery facility (VS29 in Figure 5.3.1), which was inundated by the tsunami on March 11, 2011. This section of the pipeline was exposed to liquefaction and tsunami erosion for approximately 1 km. While there was no damage to the pipeline itself, the pipeline sheathing was slightly damaged in several areas due to collisions with nearby tank fires and tsunami runoff. Various inspections, including pigging surveys, were conducted to assess the functional performance of the pipeline, confirming that there were no leaks after the earthquake. Additionally, despite the delivery facilities being inundated by the tsunami, they avoided fatal damage, allowing for the conditions to resume operation within a short period.

5.3.1.3 Earthquake Damage in Coastal Areas

Valve station VS26, located in the coastal area, was undamaged. However, valve stations VS27 to VS32 experienced damage to the valve station buildings and electrical instrumentation equipment due to the impact of floating debris and spills caused by the tsunami. As an example, Figure 5.3.3(b) illustrates the damage to VS27 after the tsunami. The fence and buildings on the ground surface sustained minor damage, and the foundation of the dissipation tower was displaced. However, the underground trunk pipeline itself remained intact. Additionally, while the coastal area was at high risk of liquefaction, the presence or absence of ground liquefaction could not be confirmed due to the tsunami.

5.3.1.4 Earthquake Damage in Mountainous Areas

The area between VS18 and VS21 is located in a mountainous region, but significant damage occurred along the road where the pipeline is buried. The section between VS19 and VS20 experienced severe ground damage, as shown in Figure 5.3.3(c). In this section, the pipeline was exposed in three locations due to distinct arc slips in the road embankment, while in other areas, it moved with the ground displacement. Despite this significant ground damage, no damage to the pipeline itself was observed. The ratio of landslide section length to the total length of the mountain section in the landslide-affected area between the valve stations was approximately 0.0058 per km.

A detailed site survey was conducted after the earthquake to assess the effects of embankment collapse, road surface subsidence, and landslides. Specifically, the

FIGURE 5.3.3 Facility damages by tsunami and road collapse and pipeline exposure in mountainous areas: (a) delivery facilities (VS2) flooded by the tsunami; (b) situation in the valve building (VS27) after the tsunami; (c) road collapse and pipeline exposure.

center position of the pipeline was determined using a pipe locator in the affected area, and both vertical movement and horizontal displacement from the road surface were measured. The results, shown in Figure 5.3.4, include two cases selected for regression analysis. A regression equation for pipeline deformation was calculated for two models: one using measurements from a relatively wide area (Model 1) and another using measurements from a narrower section (Model 2).

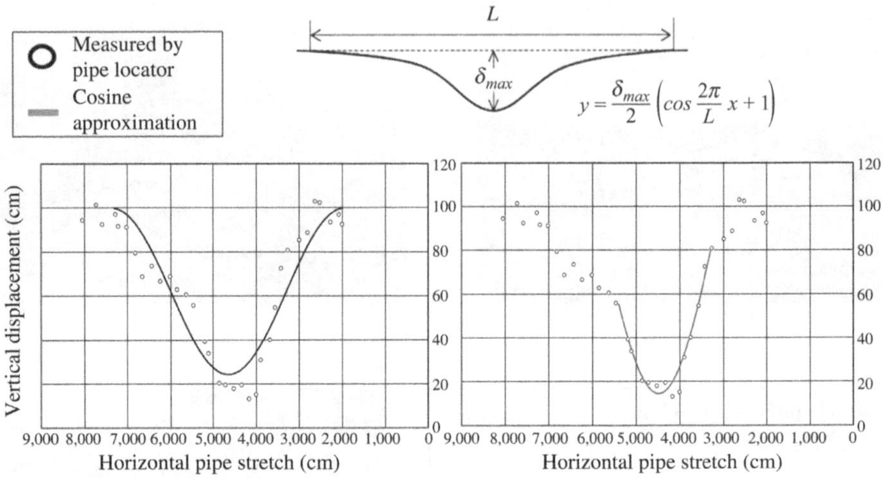

FIGURE 5.3.4 Strain distribution obtained by measuring the position of the pipe center using the pipeline locator.

$$y=\frac{\delta_{max}}{2}\left(cos\frac{2\pi}{L}x+1\right). \tag{5.3.1}$$

The regression equation for the red line in the figure was obtained by applying the results from both models. Table 5.3.1 shows that the maximum bending strain is roughly estimated to be 0.0026. It is therefore presumed that the pipes in this range have reached an elastoplastic state, slightly exceeding the yield strain.

5.3.2 Comparison of New and Old Seismic Design Methods

5.3.2.1 Basic Design Approach for Pipelines

The basic design of high-pressure gas pipelines is to determine a pipe thickness that can resist internal pressure, and specifically, it is requested to check that the circumferential stress intensity generated by the internal pressure does not exceed the allowable stress intensity, using the following formula:

$$\sigma_{hoop} \le \sigma_a, \tag{5.3.2}$$

TABLE 5.3.1
Measured Pipe Strains

Model	Length (m)	Displacement (cm)	Maximum Bending Strain
Model 1	53	76	0.0014
Model 2	20	21	0.0026

TABLE 5.3.2
Pipeline Specifications and Characteristic Values

Item	Symbol	Unit	Value
Diameter	D	Mm	508
Thickness	t	Mm	11.91
Min. stress	σ_y	MPa	422
Design factor	C		0.4
Pressure	P_r	MPa	7

where

$$\sigma_a = C \cdot \sigma_y = 0.4 \times 422MPa = 168.8MPa \tag{5.3.3}$$

$$\sigma_{hoop} = \frac{P_r(D-t)}{2t} = 1458kgf/cm^2 = 146MPa, \tag{5.3.4}$$

in which Table 5.3.2 shows the characteristics and values of the design coefficient C, yield stress σ_y (MPa), internal pressure P_r (MPa), pipe diameter D (mm) and pipe thickness t (mm).

5.3.2.2 Seismic Design Methods Before the 1995 Hyogo-ken Nanbu Earthquake

5.3.2.2.1 Seismic Design Methods Before the Earthquake

The *Technical Standard for Oil Pipelines*,[19] established in 1974, was the first seismic design method for oil pipelines in Japan, introducing the so-called response displacement method for the first time. This design method was systematized using the allowable stress design approach and proved effective within the elastic design range for small and medium earthquakes. However, it was not necessarily appropriate for large earthquakes that induce nonlinear responses.

Following the 1995 Hyogo-ken Nanbu Earthquake, the Japan Society of Civil Engineers (JSCE) proposed a new seismic design method using Level 1 ground motion (EQ1) and Level 2 ground motion (EQ2). Corresponding seismic design methods for major lifelines (such as water, sewerage, and gas pipelines)[20,21] were revised in Japan. However, the previously established technical standards for oil pipelines remained unchanged and were regarded as appropriate for Level 1 ground motion but not for Level 2 ground motion.

The pipeline in question – the Niigata-Sendai high-pressure gas pipeline – was constructed between 1990 and 1994, just before the Hyogo-ken Nanbu Earthquake. Its seismic design was primarily based on the technical standards for oil pipelines. Nevertheless, to ensure greater seismic performance, a *high-pressure gas pipeline technical guideline*[22] was specifically developed for this project, setting the seismic motion to 1.5 times the value used in the oil pipeline technical standard, and the implementation design was carried out accordingly.

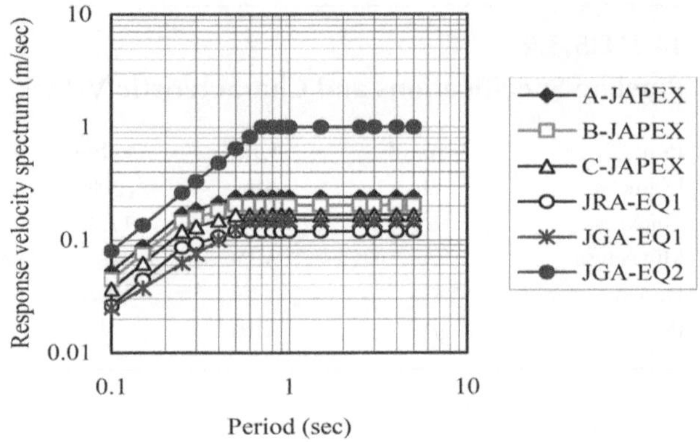

FIGURE 5.3.5 Design velocity response spectrum used in the response displacement method.

Figure 5.3.5 compares the seismic loads from the oil pipeline technical standard (JRA-EQ1), the high-pressure gas pipeline seismic design guidelines of the Japan Gas Association (Level 1 ground motion JGA-EQ1, Level 2 ground motion JGA-EQ2), and the high-pressure gas pipeline technical guidelines (soft ground A - JAPEX, medium ground B - JAPEX, and hard ground C - JAPEX) in the form of design velocity response spectra.

The seismic design equations used for the seismic design of the pipelines in question for Level 1 ground motions can be summarized as follows:

$$\text{Design basis} \quad \sigma_S \le \sigma_{cr} \tag{5.3.5}$$

$$\text{Combined stress} \quad \sigma_S = \sqrt{3.12\sigma_L^2 + \sigma_B^2}, \tag{5.3.6}$$

in which

$$\text{Axial stress} \quad \sigma_L = \alpha_1 E \varepsilon_G, \tag{5.3.7a}$$

$$\text{Bending stress} \quad \sigma_B = \frac{2\pi D}{L} E \alpha_2 \varepsilon_G, \tag{5.3.7b}$$

$$\text{Ground strain} \quad \varepsilon_G = \frac{\pi}{L} U_h, \tag{5.3.7c}$$

$$\text{Ground displacement} \quad U_h = \frac{2}{\pi^2} S_V(T) \cdot T \cdot \cos\left(\frac{\pi}{2H} z\right), \tag{5.3.7d}$$

where

$$\alpha_1 = \frac{1}{1 + \left(\dfrac{\sqrt{2}\pi}{\lambda_1 L}\right)^2}, \quad \alpha_2 = \frac{1}{1 + \left(\dfrac{2\pi}{\lambda_2 L}\right)^4} \tag{5.3.8a}$$

$$\lambda_1 = \sqrt{\frac{K_1}{EA}}, \qquad \lambda_2 = \sqrt[4]{\frac{K_2}{EI}}, \tag{5.3.8b}$$

where σ_{cr}, L, E, T, H, S_V, K_1, K_2, A, I, and z are, respectively, yield stress (MPa), seismic wavelength (m), Young's modulus of the steel pipe (MPa), ground period (sec), ground thickness (m), design velocity response spectrum (m/sec), axial ground constant (MPa), transverse axial ground constant (MPa), pipe cross-sectional area (m^2), pipe cross-sectional secondary moment (m^4), and pipe top depth (m).

5.3.2.2.2 Seismic Design Methods after the Earthquake

After the Hyogo-ken Nanbu Earthquake, the Japan Gas Association revised the seismic design guidelines for high-pressure gas pipelines and introduced the following safety assessment method using allowable strain for Level 2 ground motion:

$$\text{Design criteria} \qquad \varepsilon_S \le \varepsilon_{cr}, \tag{5.3.9}$$

where

$$\text{Pipe strain} \qquad \varepsilon_S = q\alpha_1 E\varepsilon_G \tag{5.3.10}$$

$$\text{Ground strain} \qquad \varepsilon_G = \frac{2\pi}{L}U_h \tag{5.3.11}$$

In this context, ε_{cr} represents the critical axial strain for Level 2 ground motion, specified 1–3% for ground shaking and 3–5% for ground deformation.[23] Additionally, q denotes the slippage factor used to evaluate the slip effect between the pipe circumference and the surrounding ground under Level 2 ground motion conditions.

Figure 5.3.6 compares the stresses and strains experienced by the pipeline under internal pressure and seismic loads for Level 1 and Level 2 ground motions. Specifically, Figure 5.3.6(1) illustrates the pipe stresses for the Level 1 ground motion case, displaying both the internal pressure scenario, calculated using Equation (5.3.4), and the ground motion scenario, determined via Equation (5.3.6), within the same diagram. Figure 5.3.6(2) presents the ground strain (labeled "ground") as per Equation (5.3.11), the pipe axial strain (labeled "pipe") according to Equation (5.3.10), and the yield strain ε_{cr} (labeled "yield"), which corresponds to the yield stress (the stress limit due to internal pressure) in the pipe when subjected to Level 2 ground motion. These elements are depicted together within the same graph. Initially, the pipeline was in the stress state as depicted in Figure 5.3.6(1), and the pipe thickness design, based on internal pressure considerations, was expected to ensure adequate seismic performance. However, as shown in Figure 5.3.6(2), when accounting for Level 2 ground motion, it became evident that the pipeline could experience significant strains exceeding the yield strain. At that time, due to the lack of experience with Level 2 ground motions, the seismic performance concerning plastic strain could not be verified.

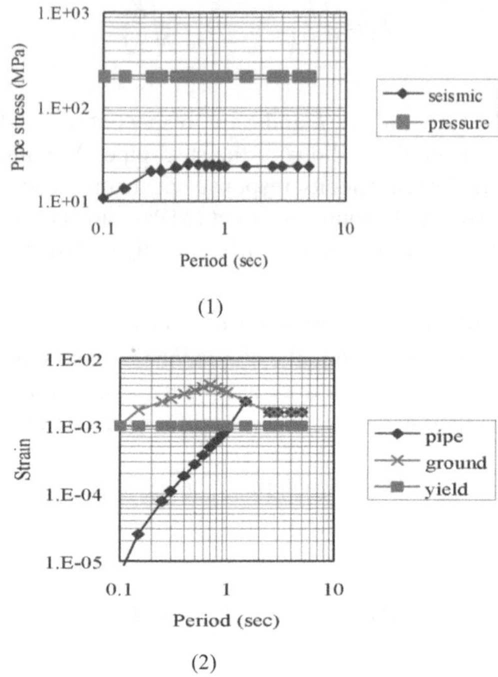

FIGURE 5.3.6 Comparison of seismic pipe stress and circumferential stress due to internal pressure for Level 1 ground motion, and comparison of seismic pipe strain, ground strain, and pipe yield strain for Level 2 ground motion: (1) stress comparison for Level 1 ground motion; (2) axial strains comparison for Level 2 ground motion.

5.3.3 Case Study of Seismic Safety Assessment of a Pipeline Operating After an Earthquake

This section quantitatively assesses the safety indices targeted at each stage of the design.

5.3.3.1 Seismic Safety Assessment in the Initial Design

In the initial design, the safety index was based on the allowable stress design method. Therefore, the safety factor was applied to the characteristic values of the design variables, such as the nominal values of the load L and strength R. If the coefficient of variation for each design variable is V_S, V_R, and the one-sided confidence interval width from the median (μ_S, μ_R) to the nominal value (S_k, R_k) and $k_S V_S$ – S, $k_R V_R$, then the central safety factor λ_0, the nominal safety factor λ_k of the design variable, and the reliability index are related[24] by the following equation.

$$\lambda_0 = \frac{\mu_R}{\mu_S} = \frac{1+\beta\sqrt{V_R^2+V_S^2-\beta^2 V_R^2 V_S^2}}{1-\beta^2 V_R^2}, \qquad \lambda_k = \frac{R_k}{S_k} = \frac{1-k_R V_R}{1+k_S V_S}\lambda_0, \tag{5.3.12a}$$

where

$$R_k = (1 - k_R V_R)\mu_R$$
$$S_k = (1 - k_S V_S)\mu_S. \quad (5.3.12b)$$

Once the reliability indices for the specified central safety factor and coefficient of variation values have been obtained using Equation (5.3.12a), it follows that the probability of failure can be calculated using Equation (5.3.13):

$$p_f = \Phi(-\beta). \quad (5.3.13)$$

When certain values are assumed, such as the safety factor λ_0, the reliability index β, and the probability of failure p_f, their relationship can be illustrated in Figure 5.3.7. Therefore, if the central safety factor is 2.0, the reliability index is 3.75, and the probability of failure is 8.84×10^{-5}.

Since pipelines function as series systems, failures occur due to latent defects that are randomly distributed along the pipeline. The pipeline traverses mountainous, plain, and coastal areas, and failures can result from various causes, including external damage by third parties, corrosion, construction-related damage, and seismic ground deformation. The results of an international survey on the frequency of these damage events have been compiled by EGIG in Table 5.3.3 as statistical data.[25] According to the same table, the damage incidence rate ranges from 0.01 to 0.17 per 1,000 km per year.

These statistical values may increase if the deterioration conditions near the pipeline are considered. Conversely, they may decrease if the effects of frequent site

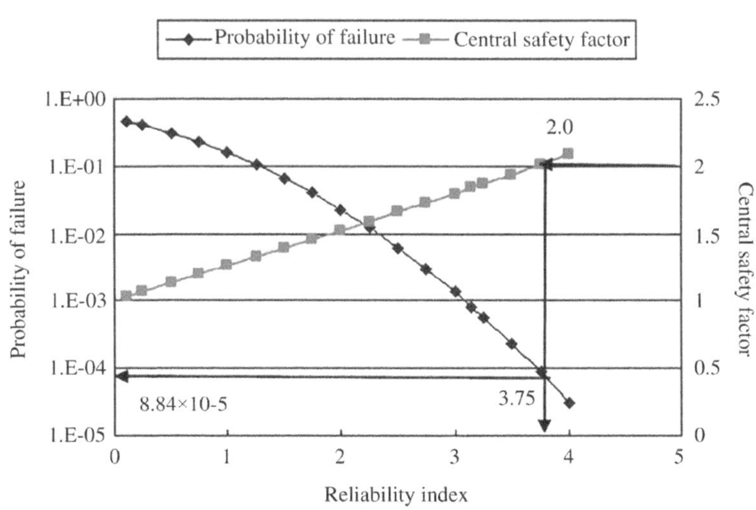

FIGURE 5.3.7 Interrelationships for several safety indicators.

TABLE 5.3.3
Cause-Specific Damage Rates[25]

Cause	Failure Frequency from 1970 to 2011 per 1000 km.yr
External interference	0.17
Corrosion	0.057
Construction defect/material failure	0.059
Hot tap made by error	0.017
Ground movement	0.026

patrols, regular inspections, and repair activities are considered. Such effects can be evaluated using Equation (5.3.14), which models degradation characteristic curves[26] concerning time variables.

$$g(t) = \left(1 + b \cdot \frac{t}{T_d}\right)^a. \tag{5.3.14}$$

Here T_d represents the service period, while a and b are control variables. In this study, a simplified deterioration model is adopted based on previous research on the deterioration of pipe bodies.[23] Assuming that the preconstruction pipeline was initially designed with a central safety factor of 2.0, the probability of pipeline failure due to internal pressure, considering maintenance conditions and the degraded environment after completion, can be calculated using Equation (5.3.15):

$$\begin{aligned} P\left[D_{pr}\right] &= 1 - exp\left[-L \cdot T_v \cdot \nu_0 g(t) P\left(C \cdot \sigma_y \le \sigma_{hoop}\right)\right] \\ &= 1 - exp\left(-112.7/1000 \times 1 \times 0.1 \times 8.84 \times 10^{-5}\right) \\ &= 9.95 \times 10^{-7}. \end{aligned} \tag{5.3.15}$$

Here, the pipe length (km), unit time step, and annual occurrence rate of potential damages per 1,000 km, provided in Table 5.3.3, are considered. The initial value of the degradation function was assumed to be 0.1 based on the results presented in Table 5.3.3.

5.3.3.2 Seismic Safety Assessment Under Revised Design Conditions after the Hyogo-ken Nanbu Earthquake

The *Guidelines for Seismic Engineering Methods for Water Supply Facilities*[21] present two curves for the design velocity response spectrum of Level 2 earthquake motion. These curves are based on the envelope of the velocity response spectrum from strong-motion records in Kobe City during the Hyogo-ken Nanbu Earthquake, corresponding to 90% and 70% non-exceedance probabilities. Based on this information, the mean value and coefficient of variation for the natural logarithm of the design velocity response spectrum are roughly estimated as follows:

$$E\left[\ln S_V^{EQ2}\right] = \ln S_V^{90} \Big/ \left\{1 + \mathrm{cov}\left(\ln S_V^{EQ2}\right) \cdot \Phi^{-1}(0.9)\right\} \tag{5.3.16a}$$

$$E\left[\ln S_V^{EQ2}\right] = \ln S_V^{70} \Big/ \left\{1 + \mathrm{cov}\left(\ln S_V^{EQ2}\right) \cdot \Phi^{-1}(0.7)\right\}. \tag{5.3.16b}$$

Here $cov(S)$ and Φ^{-1} represent the coefficient of variation of the random variable S and the inverse function of the standard normal distribution function, respectively. Additionally, S_V^{EQ2}, S_V^{90}, and S_V^{70} denote the design velocity response spectrum for Level 2 ground motion, the corresponding spectrum for a 90% non-exceedance probability, and the corresponding spectrum for a 70% non-exceedance probability, respectively.

Given the probability statistics S_V^{EQ2} from Equations (5.3.16a) and (5.3.16b), the probability of pipeline failure due to a latent defect caused by ground shaking can be calculated using the following equation:

$$P\left[D_{shake}\right] = P\left[\varepsilon_{cr}^{shake} \le \varepsilon_p^{EQ2}\right] = \Phi\left[-\frac{\mu_{lnZ_S}}{\sigma_{lnZ_S}}\right], \tag{5.3.17}$$

where ε_{cr}^{shake} is the critical pipe strain and ε_p^{EQ2} is the pipe strain response to Level 2 ground motion, which can be calculated as follows:

$$\varepsilon_p^{EQ2} = q\alpha_1 \cdot \frac{2\pi}{L} \cdot \frac{2}{\pi^2} S_V^{EQ2}(T) \cdot T. \tag{5.3.18}$$

Furthermore, the mean and standard deviation of the natural logarithm of the random variable Z_S are given by

$$\mu_{lnZ_S} = E\left[lnZ_S\right] = E\left[ln\varepsilon_{cr}^{shake} - ln\varepsilon_G^{EQ2}\right], \tag{5.3.19a}$$

$$\sigma_{lnZ_S} = \sqrt{\sigma^2_{ln\varepsilon_{cr}^{shake}} + \sigma^2_{ln\varepsilon_G^{EQ2}}}. \tag{5.3.19b}$$

Similarly, the probability of failure for ground deformation (PGD) can be calculated as

$$\left[D_{PGD}\right] = P\left[\varepsilon_{cr}^{PGD} \le \varepsilon_{PGD}\right] = \Phi\left[-\frac{\mu_{lnZ_d}}{\sigma_{lnZ_d}}\right]. \tag{5.3.20}$$

Here, ε_{cr}^{PGD} represents the critical strain for ground deformation, while ε_{PGD} denotes the pipe strain at that time. Similarly, the mean and standard deviation of the natural logarithm of the random variable for ground deformation are given by

$$\mu_{lnZ_d} = E\left[lnZ_d\right] = E\left[ln\varepsilon_{cr}^{PGD} - ln\varepsilon_{PGD}\right], \tag{5.3.21a}$$

$$\sigma_{lnZ_d} = \sqrt{\sigma^2_{ln\varepsilon^{PGD}_{cr}} + \sigma^2_{ln\varepsilon_{PGD}}}. \quad (5.3.21b)$$

Since latent defects in a pipeline can be assumed to occur independently and given that the defect distribution follows a Poisson distribution, the probability of failure at a defect can be evaluated using Equations (5.3.15) and (5.3.17). Based on this, the probability of failure due to ground shaking and ground deformation along the pipeline at any time can be calculated using Equations (5.3.22a) and (5.3.22b):

$$P\left[D_{EQ}^{shake}\right] = 1 - exp\, exp\left[-LT_v \nu_0^{shake} g(t) P\left(D_{shake}\right)\right], \quad (5.3.22a)$$

$$P\left[D_{EQ}^{PGD}\right] = 1 - exp\, exp\left[-L\nu_0^{PGD} g(t) P\left(D_{PGD}\right)\right]. \quad (5.3.22b)$$

Here, ν_0^{shake} and ν_0^{PGD} represent the potential defect incidence rates per unit distance for ground shaking and ground deformation, respectively. Specifically, the potential defect incidence rate for ground shaking is considered as the incidence rate per unit distance per year.

5.3.3 Seismic Safety Assessment Immediately After the Tohoku Earthquake

5.3.3.3.1 Review of Design Variables Under Post-Earthquake Nondestructive Conditions

The fact that the pipeline was able to supply gas to Sendai City approximately two weeks after the Tohoku–Pacific Ocean Earthquake without suffering catastrophic damage is considered a seismic nondestructive event, serving as a validation of the pipeline's post-earthquake seismic performance.

In other words, for the pipeline in question, the Tohoku Earthquake functioned as a form of assurance test, which fortunately confirmed its seismic performance. Prior to the earthquake, there may have been uncertainty regarding whether the pipeline's seismic performance was lower than the seismic load (X^*) of the Tohoku Earthquake. However, the absence of destructive events, as confirmed by the assurance test, indicates that the seismic performance exceeded the seismic load of the earthquake. This situation is schematically represented in Figure 5.3.8.

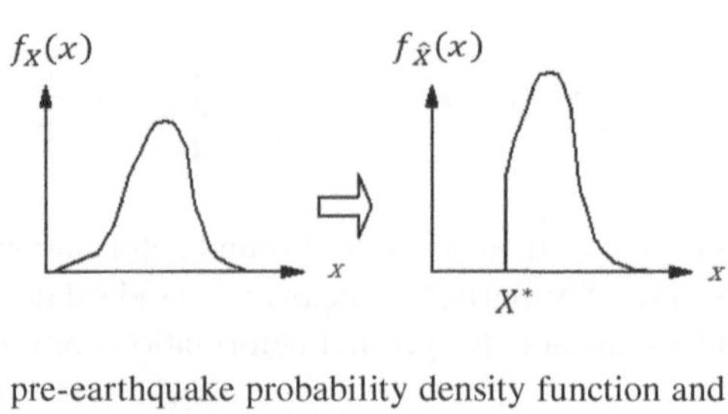

FIGURE 5.3.8 The pre-earthquake probability density function and the conditional probability density function account for the fact that the pipeline withstood the earthquake load X*: (left) before the earthquake; (right) after the earthquake.

As a result, the probability of failure can be expressed as the conditional failure probability of the seismic load EQ_2^* in the following equation:

$$P\left[D_{shake}\middle|EQ_2^*\right]=P\left[\varepsilon_{cr}^{shake}\leq\varepsilon_G^{EQ2}\middle|\varepsilon_G^{EQ2*}<\varepsilon_{cr}^{shake}\right], \quad (5.3.23a)$$

$$P\left[D_{PGD}\middle|EQ_2^*\right]=P\left[\varepsilon_{cr}^{PGD}\leq\varepsilon_{PGD}\middle|\varepsilon_{PGD*}<\varepsilon_{cr}^{PGD}\right]. \quad (5.3.23b)$$

Here, ε_G^{EQ2*} and ε_{PGD*} represent the measured values of the pipeline response during the Tohoku–Pacific Ocean Earthquake. Due to the tsunami, it was not possible to confirm liquefaction damage in the coastal area of the pipeline, preventing the determination of ground deformation incidence. Therefore, assuming a potentially damaged section exists at a distance of 1 m per km along the pipeline, the basic incidence of damage is estimated at 10–3 per km. Assuming ground liquefaction occurs over a width of 2–5 m per section, the liquefaction rate is estimated at 2–5 per 1,000 km; for landslides, the incidence rate is estimated at 6 per 1,000 km, based on the ratio of landslide section lengths in Table 5.3.4.

Table 5.3.4 summarizes the potential damage occurrence rates per 1,000 km per year for historical high-pressure gas pipelines by year. Here, ν_0^{crack}, ν_0^{shake}, and ν_0^{PGD} denote the damage incidence due to defects, ground shaking, and ground deformation per 1,000 kilometers per year, respectively.

Figure 5.3.9 compares the probability of failure for each valve section of the pipeline before and after the Tohoku–Pacific Ocean Earthquake (with sections defined as follows: VS19 to VS20 corresponds to Section 1, VS20 to VS21 to Section 2, and VS31 to VS32 to Section 13). Figure 5.3.9(1) illustrates the probability of failure due to ground shaking, while Figure 5.3.9(2) depicts the probability of failure due to ground deformation. In both panels, the lighter line (representing post-Tohoku Earthquake data) is generally lower than the darker line (representing pre-Tohoku Earthquake data), indicating that the uncertainty regarding seismic performance has been reduced by observing the pipeline's survival during the earthquake. Notably, the significant deviation between the light and dark lines in the mountainous area in Figure 5.3.9(2) reflects the pipeline's resilience to ground deformation.

TABLE 5.3.4
Potential Damage Incidence per 1,000 km per Year

	ν_0^{crack} (per 1,000 km.yr)	ν_0^{shake} (per 1,000 km.yr)	ν_0^{PGD} (per 1,000 km)		
	Site Classification				
Occurrence Rate	**General**	**General**	**Liquefaction**	**Fault**	**Landslide**
Before 1995	0.1	0.1	2–5	0	6
1995–2011	0.1	0.1	2–5	0	6
After 2011	0.1	0.1	The critical value is revised with the observed data		

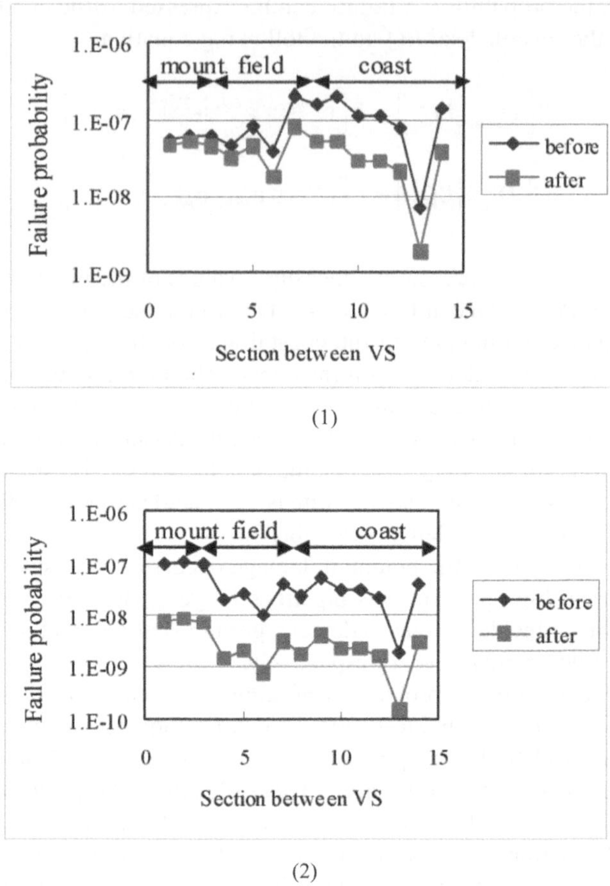

FIGURE 5.3.9 Probability of failure of each section of pipeline before and after the earthquake: (1) ground shaking; (2) ground deformation.

5.3.3.3.2 Seismic Safety Assessment of Pipelines Under Nondestructive Conditions after an Earthquake

Figure 5.3.10 illustrates the probability of pipeline failure before and after the earthquake. The 15th year in the diagram represents the number of years the pipeline had been in operation at the time of the Tohoku Earthquake, indicating that the earthquake occurred in its 15th year. Figure 5.3.10(1) demonstrates how the seismic performance of the pipeline against ground shaking and ground deformation changed before and after the earthquake, without accounting for the effects of pipeline aging. The reduction in the probability of failure due to ground deformation after the earthquake reflects the assessment that the pipeline remained intact despite landslides in mountainous areas.

Figure 5.3.10(2) compares the probability of failure due to ground shaking and internal pressure while considering the effects of aging before and after the

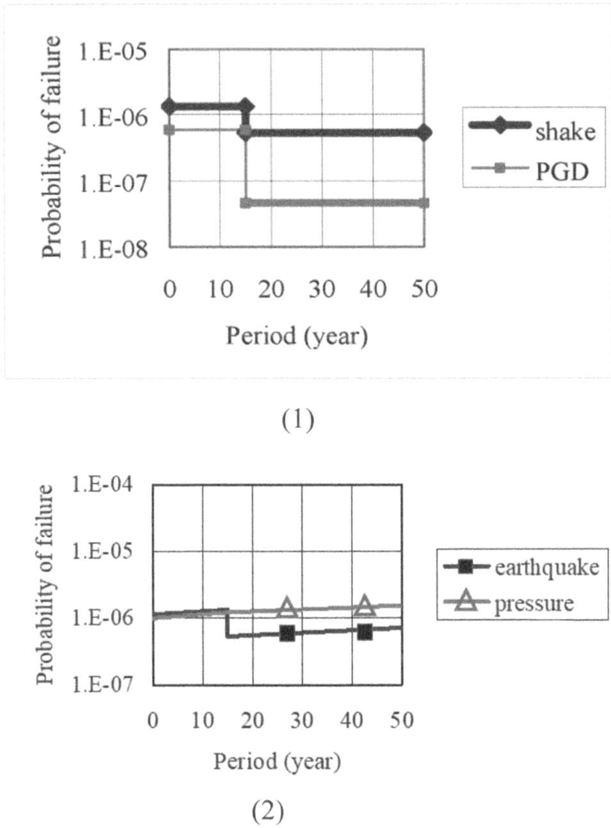

FIGURE 5.3.10 Probability of pipeline rupture before and after the Tohoku earthquake: (1) comparison of failure probabilities for ground shaking and ground deformation under nondegraded conditions; (2) comparison of failure probabilities for seismic loads and internal pressure under degraded conditions.

earthquake. The slight increase in both probabilities over time reflects the impact of aging. The figure indicates that before the earthquake, the probabilities of failure due to ground shaking and internal pressure were nearly identical. However, after the earthquake, the probability of failure due to ground shaking was lower than that due to internal pressure, demonstrating that the pipeline's seismic performance had been confirmed. The pipeline, originally designed to be earthquake-resistant before the Hyogo-ken Nanbu Earthquake, was able to maintain its required supply function despite earthquake damage in mountainous and coastal areas during the Tohoku Earthquake. This resilience is attributed to the high elongation capacity, or ductility, of the steel pipes used in its basic design. These pipes were capable of absorbing axial strain caused by Level 2 ground motion.

Additionally, the pipeline's structural integrity was preserved throughout its operational period through rigorous maintenance activities. Routine patrols helped eliminate potential hazards along long-distance sections, ensuring spatial safety, while periodic maintenance efforts focused on corrosion prevention and repairing deteriorated or damaged sections. The fact that the pipeline resumed operation within a short period after the earthquake highlights the effectiveness of both its structural design and maintenance management in ensuring its long-term integrity.

5.3.3.4 Lessons Learned

In this section, the study examined the seismic performance of the pipeline, which was originally designed to be seismically resistant before the Hyogo-ken Nanbu Earthquake and demonstrated the required functional performance despite the earthquake damage in mountainous and coastal areas during the Tohoku–Pacific Ocean Earthquake.

The results obtained are as follows:

1. A method to rationally evaluate the seismic performance of long-distance pipelines before and after the Tohoku earthquake was formulated, and the seismic safety of the pipelines after the earthquake was quantitatively evaluated in terms of fracture probability.
2. Despite the pipeline not being originally designed to withstand Level 2 ground motion, it avoided damage from the Tohoku Earthquake due to the sufficient earthquake resistance provided by the ductility of the steel pipes selected for the basic design. Additionally, maintenance and management activities, such as appropriate patrols and anticorrosion management, were carried out. The prescribed installation conditions and the integrity of the pipeline were maintained through these appropriate patrols and maintenance activities.
3. As a future seismic measure, regular monitoring activities should be conducted along mountainous and coastal areas where there is a potential risk of landslides and tsunamis. Furthermore, plant facilities that must be installed near the coast should receive special consideration for tsunami countermeasures and liquefaction measures.

5.4 RISK MITIGATION MEASURES UNDER DAILY OPERATIONS TO ENSURE SEISMIC SAFETY OF LIFELINES

To ensure the safety of pipelines during earthquakes, implementing appropriate daily maintenance and management measures is essential. This section outlines maintenance and management strategies in pipeline operations prior to an earthquake, serving as prerequisites for ensuring seismic safety.

High-pressure gas pipelines, suitable for long-distance transport, function as trunk lines originating from energy terminals and connecting cities. Their routes traverse various roads and occupied lands, including multiple river, railway, and road crossings. Ideally, based on preliminary investigations, pipeline routes should avoid areas with potential hazards. However, in Japan, where many cities are situated on alluvial plains and industrial activities and urban life thrive on reclaimed land, it is often unavoidable for pipeline routes to pass through soft ground areas.

Consequently, pipelines may need to be constructed in regions susceptible to liquefaction. Additionally, laying pipelines under roads always carries the risk of damage due to future excavation work by third parties.

Therefore, the structural safety assessment[27] of long-distance pipelines must encompass not only verification against general external loads – such as internal pressure, earth pressure, traffic loads, and temperature loads – but also considerations for crack propagation from corrosion and welding defects, damage caused by third-party actions, unequal settlement due to varying ground characteristics, earthquake-induced damages, and the seismic response of ancillary facilities.

However, existing assessment methods have not been uniformly developed based on reliability theory for all the loads and failure modes. As a result, the safety or failure probability of pipeline systems during their service period, as ensured by such assessments, remains uncertain. Determining the necessary design parameters to maintain the in-service system's failure probability below a specified threshold or controlling it to meet a target level during the maintenance policy-setting process is imperative.

The concept of performance-based design, increasingly practiced in recent years, involves first establishing the required performance and then implementing a design that meets the target failure probability, thereby ensuring its realization. This approach quantitatively clarifies the design goals of a pipeline project and its feasibility, enabling operators to estimate the magnitude of risk they must assume. It also allows project contractors to demonstrate the advantages of adopting new technologies or proposing cost reductions using a risk-based framework.

In the performance-based design method, the responsibility for ensuring the desired performance lies with the project proponent, who decides whether to adopt or reject specific methods. Thus, while the facility's designer or builder may guarantee its performance during the completion period, the operator retains responsibility for managing the facility and mitigating future accident risks throughout its operational lifespan.

To reduce or eliminate the risks that operators must bear and to achieve low-risk operations, proactive measures[28] should be adopted to address potential risks at the time of facility completion or during its operation. Even with a strategy of periodic inspection and repair, residual risk remains with the operator, and the possibility exists that, in the event of an accident, substantial costs may be incurred for remediation and compensation. Alternatively, employing non-life insurance schemes can transfer this risk to other parties. Operators must evaluate which risk management measures are appropriate as part of a maintenance strategy aimed at optimizing life cycle costs from a managerial perspective. Decision-making regarding risk reduction and transfer clarifies the acceptable failure probability for the operator, thereby informing the target failure probability in performance-based design based on the manufacturer's acceptable risk levels.

This section addresses the following topics: (1) analysis of pipeline damage modes and risks, and development of maintenance strategies for risk reduction; (2) formulation of pipeline failure probabilities before and after periodic inspection and repair work; (3) determination of inspection and repair intervals throughout the life cycle and corresponding pipeline failure probabilities; and (4) elucidation of the relationship between maintenance costs and risk transfer costs through non-life insurance, detailing the decision-making process for maintenance and management strategies using risk management methods.

5.4.1 Damage Risk Analysis of Pipelines

5.4.1.1 Accident Analysis

Analyzing pipeline leakage accidents is crucial, especially when hazardous events such as explosions or fires may occur due to the release of petroleum products or high-pressure gases into the atmosphere.

Cracks can develop in pipelines due to internal or external corrosion or mechanical damage. Over time, operational loads may exceed the pipe's leakage strength, leading to rapid crack propagation and potential leakage. Additionally, during seismic events, sudden increases in axial loads can surpass the pipeline's strength limits, resulting in leakage incidents.

It is important to note that, depending on the surrounding conditions, leakage accidents can escalate into fires or explosions. Such incidents may not be confined to the pipeline's right of way (ROW) and could potentially cause harm to nearby third parties.

To systematically analyze these accident scenarios, tools like event tree analysis (ETA) and fault tree analysis (FTA) are employed. ETA models possible pathways from an initiating event to various outcomes, helping to assess the effectiveness of safety barriers and the probability of different consequences. FTA, on the other hand, starts with a potential undesirable event and works backward to identify all possible causes and their interrelationships. These methodologies provide a structured approach to understanding and mitigating risks associated with pipeline operations.

5.4.1.2 Estimation of Losses Associated with Accidents

When a pipeline leakage accident occurs, restoration work at the leakage point, fire and explosion accident treatment, and, in some cases, countermeasures against damage to third parties must be carried out. If the costs are C_{3j}, C_{4j}, and C_{5j} as shown in Figure 5.4.1, expenses will arise according to the scale of the accident. The amount of loss H corresponding to each damage mode is shown on the right-hand side of the same figure.

When a pipeline leakage accident occurs, the risk of the accident developing into an explosion, fire, or damage to third parties in the vicinity depends on the scale of the leakage. Therefore, a major leakage accident, in which crack propagation leads to pipeline collapse, is defined here as a major damage mode. A moderate leakage accident that does not lead to a major damage mode is defined as a moderate damage mode, and the smallest leakage or lower damage level is classified as a minor damage mode.

To simplify the discussion, it is assumed that the modes of pipeline leakage damage follow the accident occurrence tree shown in Figure 5.4.1 and that the relationships shown in Table 5.4.1 exist for the probabilities of explosion, fire, and third-party damage for each leakage size. That is, the probability of explosion and fire under a leakage incident is P_K, and the probability of damage to third

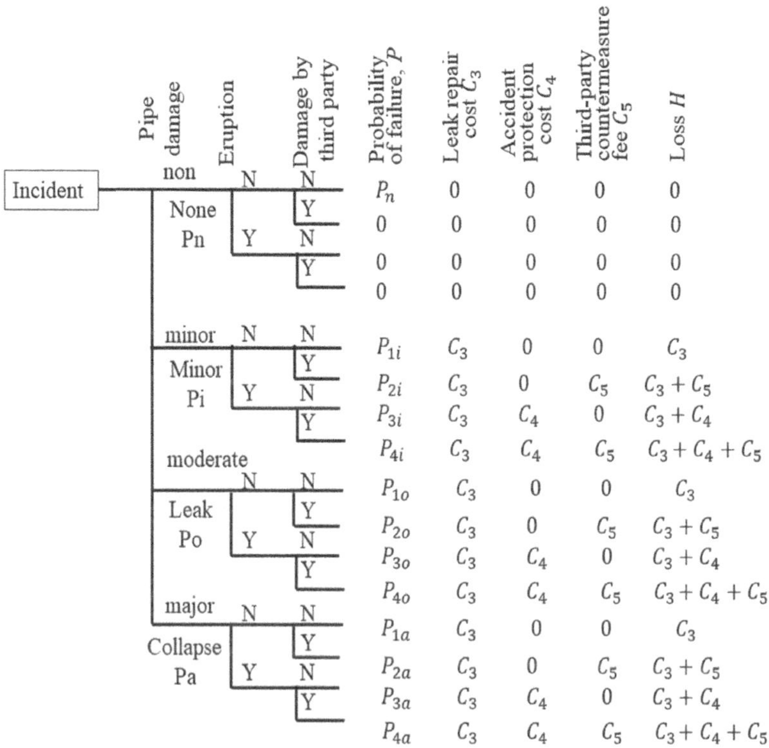

FIGURE 5.4.1 Pipeline leakage incident analysis (Y: Yes; N: No).

parties is P_T. The probabilities of occurrence for each damage mode are listed in Table 5.4.1. Additionally, there is likely to be a significant difference in the amounts C_{3j}, C_{4j}, and C_{5j} required for accident restoration and subsequent countermeasures when the damage to the pipeline is minor compared to when there is a leak or collapse. Taking this into account, it is decided here to set the cost of the former at one tenth of the cost of the latter.

Using the probabilities of occurrence in Table 5.4.1, the conditional expected value H_A and its standard deviation σ_{H_A} of the total cost of leakage restoration, accident

TABLE 5.4.1
Incident Probability Against Leak Size

Damage Mode	Probability of Eruption and Fire Occurrence	Probability of Third-Party Accident Occurrence
Minor (1)	0.01	0.0001
Moderate (2)	0.1	0.01
Major (3)	1	1

remediation, and third-party remediation under the conditions of a leakage accident can be calculated from the following equation:

$$\begin{aligned} H_A(j) &= \overline{P_K}(j)\overline{P_T}(j)C_{3j} + \overline{P_K}(j)P_T(j)\left(C_{3j}+C_{5j}\right) \\ &\quad + P_K(j)\overline{P_T}(j)\left(C_{3j}+C_{4j}\right) + P_K(j)P_T(j)\left(C_{3j}+C_{4j}+C_{5j}\right), \end{aligned} \tag{5.4.1}$$

$$\sigma_{H_A}(j) = \sqrt{A_1 + A_2 + A_3 + A_4}, \tag{5.4.2}$$

where

$$\begin{aligned} A_1 &= \overline{P_K}(j)\overline{P_T}(j)\left\{C_{3j} - H_A(j)\right\}^2, \\ A_2 &= \overline{P_K}(j)P_T(j)\left\{C_{3j} + C_{5j} - H_A(j)\right\}^2, \\ A_3 &= P_K(j)\overline{P_T}(j)\left\{C_{3j} + C_{4j} - H_A(j)\right\}^2, \\ A_4 &= P_K(j)P_T(j)\left\{C_{3j} + C_{4j} + C_{5j} - H_A(j)\right\}^2, \end{aligned} \tag{5.4.3}$$

where $j = 1,2,3$ is a symbol to distinguish between pipe damage modes
$j = 1$, *minor* (no damage or minor damage)
$j = 2$, *moderate* (leakage damage)
$j = 3$, *major* (collapse damage)

If we adopt the idea that the damage modes corresponding to no damage and minor damage are covered by the deductible in non-life insurance, the amount of damage *H_DL* given by the following equation can provide an actuarial basis for the determination of the deductible limit:

$$H_{DL} = H_A(1) + \xi\sigma_{H_A}(1), \tag{5.4.4}$$

where ξ is a parameter that specifies the limit. Although this is normally a final decision made by the management of the insurance company, $\xi = 0.1$ is assumed here because there is no clear numerical basis for this parameter.

If the insurance coverage is limited to leakage incidents and the collapse of the pipeline system is excluded from the insurance coverage, the insured loss H_B and the corresponding probability of occurrence P_B of a disruptive event are given by the following equation using the probability of occurrence P_{mo} for leakage damage:

$$\begin{aligned} H_B &= H_A(2) - H_{DL} \\ P_B &= P_{mo}. \end{aligned} \tag{5.4.5}$$

On the other hand, when the insured loss is a leakage or collapse accident, the insured loss and the corresponding probability of failure are given by the following equations using the probabilities of occurrence P_{mo} and P_{ma} for the respective damage:

$$H_B = H_A(3) - H_{DL}$$
$$P_B = P_{ma} + P_{mo} - P_{ma} \cdot P_{mo}. \tag{5.4.6}$$

5.4.2 Maintenance Management Strategies for Risk Mitigation

5.4.2.1 Maintenance Management Strategy

The maintenance and management strategy adopted by an operator is influenced on the decision to retain or transfer the risk of damage in the event of an accident. This decision should consider factors such as the service period length (T), the facility's deterioration characteristics, and associated maintenance and management costs.

If the service life is short, the system hasn't significantly deteriorated, and setting an optimal periodic inspection and repair interval is challenging, one viable option is to utilize non-life insurance to transfer the risk of leakage incidents. Conversely, neglecting maintenance management activities may lead to an excessive probability of failure in the event of a leakage accident. Therefore, a combined approach – reducing the probability of failure through proactive maintenance and transferring residual risk via non-life insurance – is effective in minimizing overall risk. Considering these factors, four fundamental maintenance management strategies emerge, combining initial investment, non-life insurance, and maintenance expenditure. In the subsequent discussion, the probability of failure in Equation (5.4.6) corresponding to strategies 1 and 3 will be denoted as P_{B1} and P_{B2}, respectively.

First,

1. Strategy 1
   A strategy in which the probability of guaranteed performance during the period of service is set to $(1 - P_{B1})$ and an initial investment is made that can guarantee that performance.
2. Strategy 2
   A strategy to adopt a property insurance scheme to transfer the risk of the failure probability P_{B1} held by the operator to others.
3. Strategy 3
   A strategy in which maintenance activities are carried out to reduce the risk due to the destruction probability P_{B1} held by the operator and to reduce the destruction probability to P_{B2}. However, in this case the risk due to the failure probability P_{B2} is owned by the operator.
4. Strategy 4
   A strategy in which the operator uses a non-life insurance scheme to transfer the risk held by the operator in Strategy 3 to another party.

Figure 5.4.2 conceptually illustrates the changes in the probability of non-failure during the period of service for maintenance strategies 1 and 3.

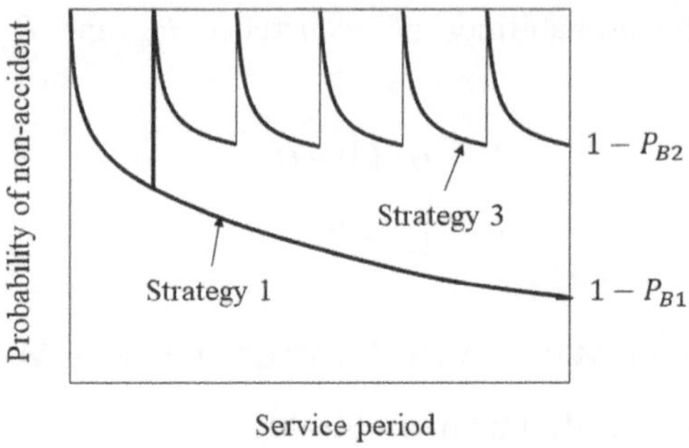

FIGURE 5.4.2 Various maintenance strategies.

5.4.2.2 Concept of the Level of Indemnity Insurance Coverage and Premium Rate for Loss Due to an Accident

In the pipeline damage scenarios depicted in Figure 5.4.1, losses from minor incidents are typically indemnified. However, for medium to severe leakage incidents or significant pipeline collapses, compensation is provided. Let P_B represent the probability of such damage, H_B the associated loss amount, Y the insurer's coverage limit,[29] m the insurance premium rate, and additional costs like insurer expenses are α. Assuming the coverage limit does not exceed the loss amount, these variables satisfy the following inequality:

$$\begin{gathered} mY + \alpha \le Y \le H_B \\ P_B \le m < 1. \end{gathered} \tag{5.4.7}$$

Based on the relationship in Equation (5.4.7), the limit of indemnity insurance coverage and the premium rate when the pipeline operator is the insured are calculated as follows.[7] In other words, the costs incurred by the insured in the event of an accident are given by

$$x_a = H_B - Y + mY + \alpha. \tag{5.4.8}$$

On the other hand, the burden in the absence of accidents is

$$x_b = mY + \alpha. \tag{5.4.9}$$

Therefore, assume that the insured person's expected contribution indicator can be expressed using an evaluation function as follows:

$$E[U_1] = P_B U_1(x_a) + (1 - P_B) U_1(x_b). \tag{5.4.10}$$

The evaluation function for the insured assumes Equation (5.4.11) as shown in Figure 5.4.3. The equation was developed considering the condition that the

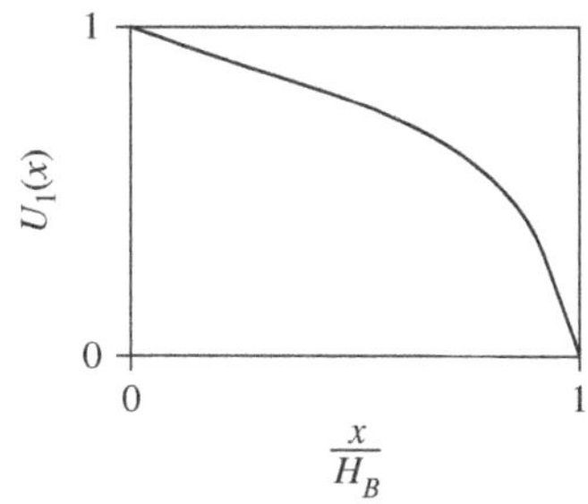

FIGURE 5.4.3 Evaluation functions for insured persons.

evaluation index declines as the burden increases, but the upper limit of the compensation amount does not exceed the loss amount.

$$U_1 = \left(1 - \frac{x}{H_B}\right)^{n_1}. \tag{5.4.11}$$

In order to minimize the insured's burden in the event of an accident, it is necessary to find a combination of coverage amount and premium rate that minimizes the expected loss amount. The minimum value of the expected loss $E[U_1]$ is obtained from the condition that the derivative coefficient of the expected loss differentiated by the guaranteed amount Y is zero, and the following relation is obtained:

$$\frac{P_B}{1-P_B} \cdot \frac{U_1(x_a)}{U_1(x_b)} = \frac{m}{1-m}. \tag{5.4.12}$$

On the other hand, from the insurer's income and expenditure x_a at accident and x_b at nonaccident, the expected income $E[x]$ when the probability of an accident is P_B is obtained by the following equation:

$$\begin{aligned} x_a &= mY - Y, \\ x_b &= mY, \\ E[x] &= P_B \cdot [x_a] + (1-P_B) \cdot [x_b] = mY - P_B Y. \end{aligned} \tag{5.4.13}$$

If the rate of return of the insurance company is now determined as a percentage β of the coverage amount, then $E[x]$ in Equation (5.4.13) is equal to βY, and the required premium rate of the insurance company is obtained by the following equation:

$$m_{opt} = P_B + \beta. \tag{5.4.14}$$

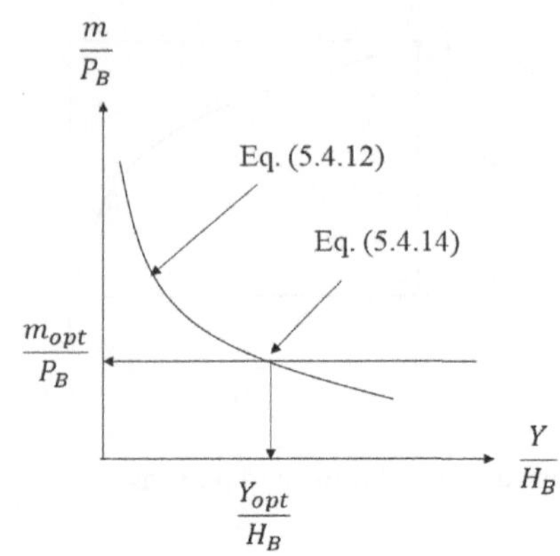

FIGURE 5.4.4 Conceptual diagram regarding optimum compensation costs and premium rates.

From the above, the optimal compensation cost Y_{opt} and the optimal premium rate m_{opt}, which satisfy the insurance conditions of both insured and insurance company, can be obtained as the intersection of the two curves as shown in Figure 5.4.4.

5.4.3 Pipeline Failure Modes

5.4.3.1 Corrosion Damage Modes

Structural damage includes damage to the surface of the pipe due to corrosion or dents. Pipes with these structural damages are not immune to strength deterioration when compared to sound pipes. This section outlines methods for estimating the residual strength of structurally damaged pipelines.[30,31]

Figure 5.4.5 shows that corrosion damage has occurred on the pipe surface, with the pipe thickness corroding by d^c at the width W of the pipe surface. The corrosion depth progresses with the elapsed time τ. It is assumed that the progression of the corrosion depth, d^c, can be expressed by the following equation proposed by Ahammed and Melchers:[32]

$$d^c = \varsigma \tau^a, \tag{5.4.15}$$

where the coefficients a and ς are the corrosion deterioration parameters, respectively, and Ahammed et al. adopted $a = 0.53$ and $\varsigma = 0.1$.

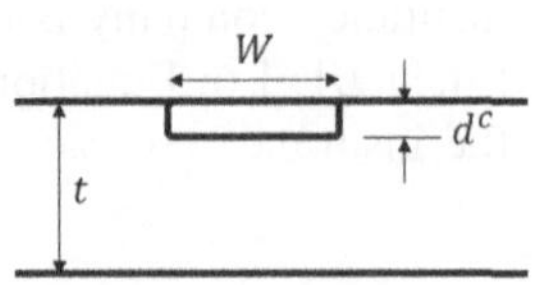

FIGURE 5.4.5 Corrosion cracks in the pipe body.

In the presence of corrosion or welding defects on the pipe surface, the circumferential stress at the onset of failure is lower than in the defect-free case due to stress concentration around the defect. According to Mok et al.,[33] the fracture initiation circumferential stress σ_{cr}^{c} of a pipe with long corrosion defects is given by

$$\sigma_{cr}^{c}(\tau)=\sigma_{flow}\frac{1-\dfrac{A(\tau)}{A_0}}{1-\dfrac{A(\tau)}{MA_0}}, \tag{5.4.16}$$

where

σ_{flow}: critical strength of pipe material
A: area of defect
W: length of defect on pipe surface
t: pipe thickness
M: Folias factor[33]

and is given by

$$A_0=t\cdot W,\qquad A(\tau)=d^{c}(\tau)\cdot W,$$
$$M=\sqrt{1+\frac{1.255}{2}\left(\frac{W}{\sqrt{Dt}}\right)^2-\frac{0.0135}{4}\left(\frac{W}{\sqrt{Dt}}\right)^4}, \tag{5.4.17}$$

where D: pipe outer diameter.

5.4.3.2 Dent Damage Mode

A typical example of pipeline damage caused by a third party is a dent in the pipe body. Consider a deep dent with gouge, as shown in Figure 5.4.6. Assume a case where a scratch of depth L_d occurs and a further scratch of depth d occurs on its

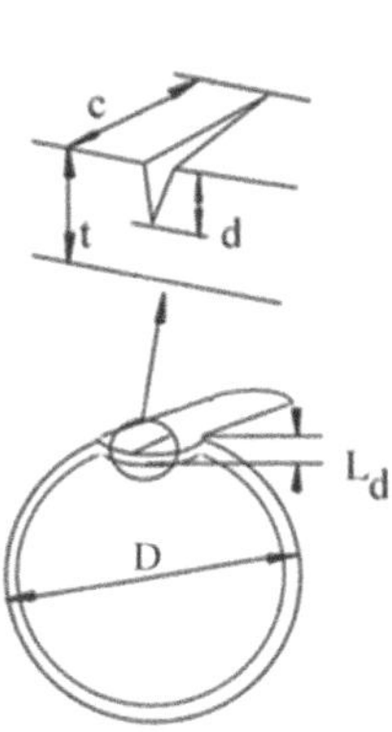

FIGURE 5.4.6 Dents in the gouged pipe body.

surface. For the circumferential strength σ_{cr}^{d} of such a dent with gouge, British Gas[34] proposed the following estimation equation based on experimental results for a typical high-pressure gas pipeline:

$$\sigma_{cr}^{d} = \sigma_{flow} \frac{2}{\pi} \cos^{-1}(s), \tag{5.4.18}$$

where

$$s = exp\left[-\frac{1.5\pi E}{\sigma_{flow}^{2} Ad} \cdot \frac{exp\left\{ \dfrac{ln(C_v) - K_1}{K_2} \right\}}{\left\{ Y_1\left(1 - \dfrac{1.8 l_d}{2R}\right) + Y_2\left(10.2 \dfrac{L_d}{2t}\right)^2 \right\}} \right],$$

$$Y_1 = 1.12 - 0.23\left(\frac{d}{t}\right) + 10.6\left(\frac{d}{t}\right)^2 - 21.7\left(\frac{d}{t}\right)^3 + 30.4\left(\frac{d}{t}\right)^4, \tag{5.4.19}$$

$$Y_2 = 1.12 - 1.39\left(\frac{d}{t}\right) + 7.32\left(\frac{d}{t}\right)^2 - 13.1\left(\frac{d}{t}\right)^3 + 14.4\left(\frac{d}{t}\right)^4,$$

in which

σ_{flow}: limit strength of pipe material (lbf/in^2)
E: Young's modulus $(30Mlbf/in^2)$
A: area of failure zone of 2/3 Charpy specimen (= $0.083in^2$)
d: depth of failure zone (inch)
Ld: depth of dent zone measured at zero pressure (inch)
R: radius of tube
Cv: toughness value (2/3 Charpy toughness value, *ftlbf*)
K_1, K_2: nonlinear parameters with $K_1 = 1.9$, $K_2 = 0.57$

Although the applicability of Equation (5.4.18) is limited, it is adopted here as the critical strength estimation equation because there is currently no simpler equation.

5.4.3.3 Damage Mode Due to Crack Propagation

In high-pressure gas pipelines, crack propagation occurs when the material's toughness is insufficient. Once a crack forms, the compressive energy stored under high pressure continuously drives the crack's expansion. Fracture mechanics evaluation methods, such as the Charpy Test and the Drop Weight Tear Test (DWTT), can determine whether crack propagation can be prevented. Table 5.4.2 presents the estimated equation for the Charpy Toughness required to prevent ductile fracture crack propagation in natural gas pipelines.

The limit strength, σ_{cr}^{P}, necessary to prevent crack propagation should be estimated from various parameters, including pipe material, pipe geometry, crack

TABLE 5.4.2
Charpy Toughness Value Estimation Equations for Various Standards[36]

Specification	Charpy Toughness
	$C_v = f(R,t)\cdot(\sigma_h)^\eta$
Battele	$C_v = 1.5\times 2.382\times 10^{-5}\sigma_h^2(Rt)^{1/3}$
AISI	$C_v = 1.5\times 2.377\times 10^{-4}\sigma_h^{3/2}(Rt)^{1/2}$
BG	$C_v = 1.5\times\left(2.08\frac{R}{t^{0.5}} - 0.001v_0\frac{R^{1.25}}{t^{0.75}}\right)\sigma_h\times 10^{-3}$
BS	$C_v = 1.5\times 1.34\times 10^{-6}\sigma_h^2 R$
Mannesmann	$C_v = 1.5\times 19.99\times 10^{-8}e^{0.287}\sigma_h^{1.76}D^{1.09}t^{0.585}$
Japanese	$C_v = 1.5\times 2.49810^{-4}e^{2.33}\sigma_h^{1.76}D^{0.3}t^{0.47}$

geometry, and operational load characteristics. However, as a simplified approach, Equations (5.4.16) and (5.4.17), along with the estimation equations[35] in Table 5.4.2, provide the following formula for σ_{cr}^P:

$$\sigma_{cr}^P = \left\{\frac{C_v}{f(R,t)}\right\}^{1/\eta}\cdot\frac{1-\dfrac{A_p(\tau)}{A_0}}{1-\dfrac{A_p(\tau)}{MA_0}}, \tag{5.4.20}$$

where the function $f(R,\tau)$ is the coefficient part of the circumferential stress-related term $(\sigma_h)^\eta$ from Table 5.4.2, and $A_p(\tau)$ is the area of the defect zone according to crack length. Furthermore, v_0 is the sound velocity of the gas (396 m/s) and σ_h is the circumferential stress occurring in a pipe subjected to an internal pressure P_r, given by

$$\sigma_h = \frac{P_r R}{t}. \tag{5.4.21}$$

5.4.4 System Failure Probability and Maintenance Cost Evaluation under Inspection and Repair Activities[26]

5.4.4.1 Inspection and Repair Intervals and System Failure Probability

Upon commencing pipeline operations, various latent defects may develop due to environmental influences. Periodic inspections and subsequent repairs can reduce the number of these defects but cannot eliminate them entirely due to limitations in detection accuracy and repair efficacy, as conceptually illustrated in Figure 5.4.7. If the detection and repair performance characteristic curve for a latent defect *D* is represented by *P(I|D)* in Figure 5.4.7, and the prior probability distribution of

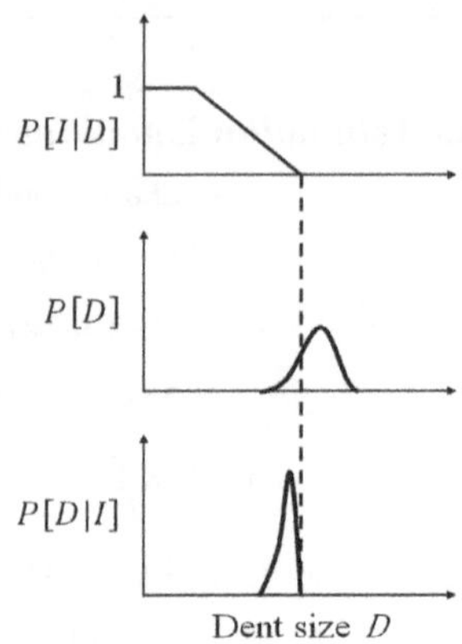

FIGURE 5.4.7 Probability density function of defect size to be corrected before and after inspection and repair.

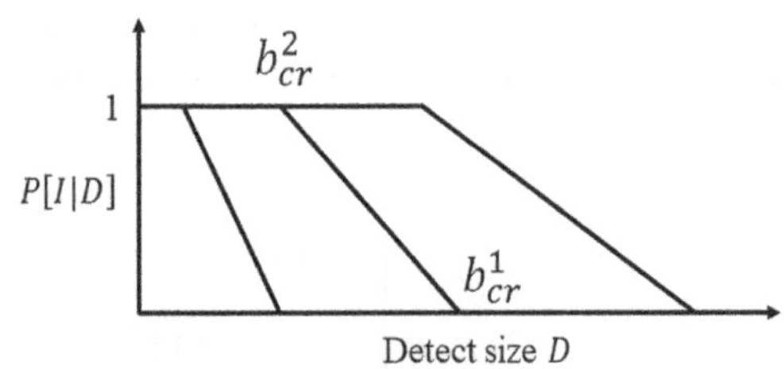

FIGURE 5.4.8 Conceptual diagram of different defect detection accuracy.

latent defects before inspection is *P(D)*, then the probability distribution of latent defects after inspection and repair, *P(D|I)*, can be determined using Bayes' theorem as follows:

$$P[D|I] = \frac{P[I|D]P[D]}{\sum P[I|D]P[D]}. \tag{5.4.22}$$

The accuracy of defect detection and the certainty of repairs during pipeline inspections can be categorized into three levels, high, medium, and low, as illustrated in Figure 5.4.8. These levels depend on the chosen maintenance management strategies.

Now, the fracture limit state $Z(\tau)$ of the pipe at time τ is given by

$$Z(\tau) = \sigma_{cr}(\tau) - \sigma_h. \tag{5.4.23}$$

When the random variable Z follows a normal distribution, the probability of fracture – corresponding to instances where Z is negative – can be determined using the standard normal cumulative distribution function:

$$P[Z(\tau) \le 0] = \Phi\left[-\frac{\mu_Z(\tau)}{\sigma_Z(\tau)}\right], \tag{5.4.24}$$

where

$$\mu_Z = E[Z], \qquad \sigma_Z = \sqrt{Var[Z]}. \tag{5.4.25}$$

Pipelines in service are subjected to various external loads, including corrosion on the outer surface, dents caused by third-party accidents, seismic effects, and unequal settlement. For simplicity, this discussion focuses on corrosion damage, dent damage, and crack propagation modes. Although periodic inspections and subsequent repairs enhance pipeline safety, latent defects that remain undetected can develop during the interval until the next inspection, thereby increasing the pipeline's probability of failure. Therefore, it is necessary to assess the probability of system failure due to residual defects at a specific time after periodic inspection and repair. As a pipeline operates, various hazards emerge and become spatially distributed along its length over time. Let $\lambda_k(\tau, x)$ denote the hazard function, representing the probability of occurrence of pipeline damage per unit time and unit length. According to the definitions in Section 5.4.1 (2), k = 1 corresponds to the corrosion damage mode in Section 5.4.3 (1), k = 2 to the dent damage mode in Section 5.4.3 (2), and k = 3 to the crack propagation damage mode in Section 5.4.3 (3). In this context, $\lambda_k(\tau, x)$ represents the hazard function at elapsed time τ and position x due to latent cracks generated by damage mode k. Utilizing the defect detection probability from Equation (5.4.22), the probability distribution of latent defects after inspection and repair can be updated accordingly:

$$\lambda_k(\tau, x; t) = \nu_k(x; t) \int_0^{D_{max}} P\left[Z_k(\tau), \le 0\right] f_{D_k|I}(\delta)\, d\delta. \tag{5.4.26}$$

where τ means the time elapsed since the pipeline was completed, and $F_{D_k|I}(\delta)$ *and* $f_{D_k|I}(\delta)$ defined by the following equations are the probability distribution function and its density function of the latent defect length by damage mode after inspection and repair:

$$F_{D_k|I}(\delta) = P\left[D_k < \delta | I\right], \qquad f_{D_k|I}(\delta) = \frac{dF_{D_k|I}(\delta)}{d\delta}, \tag{5.4.27}$$

ν_k is the probability of the existence of latent cracks due to damage factor k at unit distance and unit time in the area including location x. It is assumed to be given as the product of the function $Q_k(t)$, which expresses the tendency of latent cracks to increase with time since the initial construction as shown in Figure 5.4.9, and ν_{0k}, which expresses the spatial occurrence rate of latent cracks:

$$\nu_k(x; t) = Q_k(t)\, \nu_{0k}(x). \tag{5.4.28}$$

The hazard rate, denoted as τ, L; t, represents the probability of failure per unit time over the entire length of a pipeline at a given time t after periodic inspection and repair. If the damage modes – such as corrosion, dents, and crack propagation – operate independently, the overall hazard rate can be expressed as the sum of the

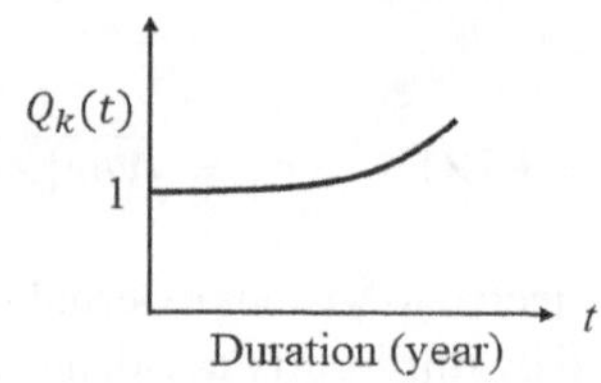

FIGURE 5.4.9 Conceptual diagram of defect progressive process for the kth damage mode.

individual hazard rates for each damage mode. Specifically, by distinguishing between moderate damage modes (e.g., leakage) and major damage modes (e.g., collapse), the total hazard rate can be formulated as follows:

$$\Lambda(\tau, L; t) = \begin{cases} \Lambda_{mo} = \sum_{k=1}^{2} \int_0^L \lambda_k(\tau, x; t)dx \\ \Lambda_{ma} = \int_0^L \lambda_3(\tau, x; t)dx \end{cases} \tag{5.4.29}$$

The reliability of a pipeline segment of length L over a time interval from 0 to t, denoted as $R(t, L)$, can be formulated by considering the reliability immediately after the ith periodic inspection/repair, $(0 \le \tau \le \Delta\tau)$, the periodic inspection interval T, and the hazard function $\tau(L; t)$ at time t after inspection:

$$R_0[T_i, L] = exp\left\{-\int_0^{\Delta\tau} \Lambda(\tau, L;\ T_i)d\tau\right\}. \tag{5.4.30}$$

The reliability at τ hours after the ith periodic inspection/repair, i.e., t hours after completion of construction, is given by

$$R[t,L] = R_0[T_i, L]exp\{-\tau \cdot \Lambda(\tau, L; T_i + \tau)\}, \tag{5.4.31}$$

where

$$T_i = \sum_{j=1}^{i} \tau_j, \qquad 0 \le \tau \le \Delta T, \qquad t = T_i + \tau. \tag{5.4.32}$$

From the above, the probability of failure of the pipeline system in section L, at the end of the ith periodic inspection, $\tau = \Delta T$, is obtained by the following equation:

$$p_f[\Delta T; T_i, L] = 1 - R[T_i + \Delta T, L]. \tag{5.4.33}$$

5.4.4.2 Maintenance Cost Assessment

The total number of potential defects of the kth damage mode to be repaired detected in section L at inspection time τ is given by

$$M_k(\tau) = \int_0^L \nu_k(x; T_i + \tau)\,dx \cdot \int_0^{D_{max}} \{1 - P|I|d_k(\tau) = \delta\} \cdot f_{D_k(\tau)} d\delta. \qquad (5.4.34)$$

where D_{max} is the magnitude of the maximum damage.

To simplify the analysis, consider the scenario where at most one medium-sized or larger leakage incident occurs during the pipeline's service period. Using Equation (5.4.34), the cumulative maintenance cost C – comprising both periodic inspection and repair expenses and the expected restoration cost in the event of an accident within section L from construction time to time T – can be calculated using the following equation:

$$C(T,L,n) = \sum_{i=1}^{n}\left(C_i + \sum_{k=1}^{2} M_k(\Delta T)C_{2k}\right) + \bigcup_{i=1}^{n} p_f\left[\Delta T; T_i, L_i\right](H_B + H_{DL}), \qquad (5.4.35)$$

where n is the number of periodic inspections from the start of service to point T and is given by

$$n = \frac{T}{\Delta T}. \qquad (5.4.36)$$

C_p, C_{2k}, and $(H_B + H_{DL})$ are the inspection cost per inspection, the repair cost of the kth damage mode, and the expected loss in the event of a leakage incident, respectively. In Equation (5.4.35), C represents the cumulative maintenance cost, which depends on the probability of system failure calculated using Equation (5.4.33) for a chosen inspection and repair level, assuming periodic inspection and repair intervals of n years over a service period of T years. The pipeline's deterioration progresses with the elapsed time T since system completion. Consequently, the probability of failure in Equation (5.4.33), corresponding to Equation (5.4.29), can be determined for moderate and major damage modes as follows:

$$\begin{aligned} P_{mo}(T) &= p_f[\Delta T;\ T_i,\ L]; \qquad \Lambda = \Lambda_{mo} \\ P_{ma}(T) &= p_f[\Delta T;\ T_i,\ L]; \qquad \Lambda = \Lambda_{ma} \end{aligned} \qquad (5.4.37)$$

Typically, future costs are discounted to present value to account for the time value of money. However, since maintenance expenses are labor-intensive and likely to rise over time, discounting may underestimate actual future costs. Therefore, in this discussion, maintenance costs calculated using Equation (5.4.35) are considered at their nominal value without discounting.

5.4.5 Decision-making on Maintenance Strategies Using Risk Management Methods

An operator's fundamental strategy in pipeline transport is to minimize risks associated with both revenue and expenditure to maximize profit during the service period. The value of the pipeline facility at time T can be expressed using revenue $I(T)$, operating costs $K(T)$, and maintenance costs $S(T)$.

$$Q(T)=I(T)-K(T)-S(T), \tag{5.4.38}$$

where the maintenance costs can be expressed as the sum of the initial construction costs and the accumulated maintenance costs from Equation (5.4.35) by the following equation:

$$S(T)=C_0+C(T,\ L,\ n). \tag{5.4.39}$$

For simplicity, the income and operating costs are assumed to be constant. Risk minimization strategies are considered, regarding various uncertainties related to maintenance and management activities.

5.4.5.1 Strategy 1 (Initial Investment Method)

The initial investment method refers to a scenario where a pipeline, constructed in accordance with appropriate design guidelines, operates without any maintenance activities after commissioning. This approach may be considered reasonable for facilities that have been in operation for a relatively short period and where there is no concern about progressive strength deterioration or corrosion development. However, if periodic inspections are not conducted, the probability of an accident occurring over the service period T increases due to the lack of detection and mitigation of potential defects.

$$P_{B1}=p_f(T;T,\ L). \tag{5.4.40}$$

The expected project cost $S_1(T)$ according to Strategy 1 is equal to the sum of the initial construction cost C_0 and the potential risk, using the following equation:

$$S_1(T)=C_0+P_{B1}\cdot(H_B+H_{DL}). \tag{5.4.41}$$

This means that the operator has, as a potential risk, the expected losses due to corrosion during the period of service and leakage accidents caused by third-party causes.

5.4.5.2 Strategy 2 (Non-Life Insurance Method)

When the service period T is short, the system has not significantly deteriorated, and determining an optimal interval for periodic inspection and repair is challenging, operators may consider transferring the risk of leakage accidents through non-life insurance. This risk management decision involves comparing the uncertainty of retaining damage risk in the event of an accident with the cost of risk transfer via non-life insurance

Under the conditions outlined in Section 5.4.2.2 for establishing non-life insurance, the expected operating cost $S_2(T)$ and the annual premium $h_B(P_{B1}, H_B)$ for Strategy 2 can be expressed as follows, respectively, based on the failure probability P_{B1}, the optimal insurance coverage corresponding to the amount of damage Y_{opt} and the insurance premium rate m_{opt}:

$$S_2(T) = C_0 + \sum_{j=1}^{T} \frac{h_B(P_{B1}, H_B)}{(1+r)^j}$$
$$h_B(P_{B1}, H_B) = \frac{m_{opt}(P_{B1}, H_B) \cdot Y_{opt} + \alpha}{T}, \tag{5.4.42}$$

where r is the discount rate to present value.

5.4.5.3 Strategy 3 (Maintenance and Management Activities Method)

Implementing maintenance and management activities is an effective strategy to reduce the risk of damage from potential accidents in existing pipeline facilities. Determining the appropriate investment of management resources requires accurately assessing the reduction in failure probability achieved through these activities and the consequent decrease in damage risk.

Consider a scenario where the probability of a leakage failure during the service period decreases from P_{B1} to P_{B2} due to the introduction of periodic inspection and repair activities. In this case, the expected project cost for Strategy 3, denoted as 3, $S_3(T)$, can be expressed by Equation (5.4.43).

$$S_3(T) = C_0 + \sum_{i=1}^{n_D} \left\{ C_1 + \sum_{k=1}^{2} M_k(\tau_i) C_{2k} \right\} + \sum_{i=1}^{n_D} P_{B2}(T_i) \cdot (H_B + H_{DL}), \tag{5.4.43}$$

where

$$P_{B2}(T_i) = p_f(\Delta T;\ T_i, L). \tag{5.4.44}$$

5.4.5.4 Strategy 4 (Combined Method of Non-Life Insurance and Maintenance Activities)

When combining maintenance and management activities with non-life insurance to mitigate and transfer pipeline failure risks, the expected project cost, denoted as $S_4(T)$, can be calculated using the following equation:

$$S_4(T) = C_0 + \sum_{i=1}^{n_D} \left\{ C_1 + \sum_{k=1}^{2} M_k(\tau_i) C_{2k} \right\}$$
$$+ \sum_{i=1}^{n_D} P_{B2}(T_i) \cdot (H_B + H_{DL} - Y_{opt}) + \sum_{j=1}^{T} \frac{h_B(P_{B2}, H_B)}{(1+r)^j}, \tag{5.4.45}$$

where

$$h_B(P_{B2}, H_B) = \frac{m_{opt}(P_{B2}, H_B) \cdot Y_{opt} + \alpha}{T}. \tag{5.4.46}$$

5.4.5.5 Examples of Numerical Calculations

1. Numerical specifications

In this numerical example, we consider a long-distance gas pipeline with a diameter of 600 mm, a length of 100 km, and an internal pressure of 7 MPa. The pipeline specifications are detailed in Table 5.4.3. Table 5.4.4 lists the limit values for defect detection accuracy, with associated costs varying depending on the inspection accuracy and repair level employed.

Table 5.4.5 lists the statistics assumed for the defect dimensions with gouges on the dent surface. The probability distribution of the latent crack length for each

TABLE 5.4.3
Pipeline Specifications

Item	Symbol	Unit	Value
Pipe diameter	D	Mm	610
Thickness	t	Mm	8.7
Internal pressure		MPa	7
Failure strength		MPa	520
Charpy toughness value		J	500

TABLE 5.4.4
Detection Limits of Latent Defects

		Bound of Inspection		
Quality of Inspection	Unit	Upper	Lower	Cost Ratio
High	Cm	2.0	0.1	5
Medium	Cm	5.0	0.5	1
Low	Cm	10.0	1.5	1/5

TABLE 5.4.5
Mean and Standard Deviation of Latent Cracks for Each Damage Mode

Item	Symbol	Unit	Mean	Standard Deviation
Depth of corrosion	d^c	cm	1.0	0.5
Depth of dent	d	cm	1.2	0.6
Depth of crack	d_p	cm	1.5	0.5

TABLE 5.4.6
Latent Crack Occurrence Rates for Each Damage Mode

Item	Symbol	Unit	Rate of Occurrence
Fatigue crack	v_{01}	1/km	0.0001
Dent crack	v_{02}	1/km	0.002
Crack propagation	v_{03}	1/km	0.000003

damage mode is assumed to be the Rayleigh distribution, and the shape parameter of the Rayleigh distribution is determined from the standard deviation in Table 5.4.5.

$Q_k(t)$, introduced in Equation (5.4.28), is a function that controls the increasing trend over time of the spatial incidence of latent cracks corresponding to the damage modes, and is assumed here to be expressed by the following equation:

$$Q_k(t) = \left\{1 + \left(\frac{t}{T}\right)^{\gamma_2}\right\}^{\gamma_1}, \tag{5.4.47}$$

where T and γ_1, and γ_2 are the service period and parameters, respectively, and $\gamma_1 = 1.0$, $\gamma_2 = 1.5$. The potential crack occurrence rate per unit distance for each damage mode immediately after pipeline completion is assumed to be the values shown in Table 5.4.6.

In Section 5.4.2.2, the coverage level and premium rate of non-life insurance were discussed. The variable n_1, insurance company expense α, and rate of return β of the valuation function U_1 introduced there are assumed to be as shown in Table 5.4.7. In addition, the discount rate r to present value used in 5.4.5 is assumed to be 4%.

Table 5.4.8 presents various cost components as proportions of the initial construction cost C_0, including the inspection cost per inspection C_1, repair cost per damage mode C_{2k}, restoration cost for a major leakage accident C_{3j}, compensation cost to users C_{4j}, and compensation cost to third parties C_{5j}. These values are hypothetical and intended for numerical calculations; they should be adjusted to reflect the specifics of the target system in practical applications. In this context, Case 1 serves as the standard scenario. Case 2 assumes that engineering costs in the event of an accident are twice the initial construction cost, while maintenance costs are half those of the standard case. Conversely, Case 3 posits that accident compensation costs are half those of the standard case, and maintenance costs are twice the initial construction

TABLE 5.4.7
Insurance Design Conditions

Item	Symbol	Value
Parameter	n_j	0.25
Expense ratio	α	1% of insurance fee
Profit ratio	β	2% of insurance fee

TABLE 5.4.8
Maintenance Costs Relative to Initial Investment

	Standard Cost	Expensive Repair Cost	Expensive Maintenance Cost
Cost Ratio	**Case 1**	**Case 2**	**Case 3**
C_1/C_0	2.5/100	2.5/200	2.5/50
C_{21}/C_0 (corrosion)	2/100	2.5/200	2/50
C_{22}/C_0 (dent)	5/100	5/200	5/50
C_{23}/C_0 (crack)	3/100	3/200	3/50
$(C_{3j}+C_{4j}+C_{5j})/C_0$	1	2	1/2
$C_{3j}/C_0,\ j=2,3$	0.4	0.8	0.2
$C_{4j}/C_0,\ j=2,3$	0.4	0.8	0.2
$C_{5j}/C_0,\ j=2,3$	0.2	0.4	0.1

Note: $\frac{C_{3j}}{C_0}, \frac{C_{4j}}{C_0}, \frac{C_{5j}}{C_0}$ for $j=1$ are 10% of those for $j=2,3$.

cost. The insured loss coverage corresponding to each damage mode, as indicated in Equations (5.4.5) and (5.4.6), is normalized by the initial construction cost, yielding the following values:

$$\frac{H_B(moderate)}{C_0}=0.396, \qquad \frac{H_B(major)}{C_0}=0.952.$$

5.4.5.5.1 Characteristics of Quantities Necessary for Decision-Making on Maintenance Levels

In Strategy 3, referred to as the risk retention method, the operator aims to reduce the probability of failure through inspection and repair while retaining the associated risk. The operator's burden ratio, R, under this method is defined as a dimensionless quantity calculated by dividing the sum of the operator's periodic inspection and repair costs up to time T during the service period and the expected accident burden – determined by the accident probability P_B – by the loss covered by property insurance, H_B:

$$R=\frac{S_3(T)-C_0}{H_B}. \tag{5.4.48}$$

Therefore, the operator's burden ratio, R, serves as an indicator for Strategy 3. Similarly, Strategy 4, discussed in the previous Section 5.4.5.4 is termed the risk-shifting method, which involves transferring risk through non-life insurance. Under this approach, the operator's burden ratio is defined as a dimensionless quantity calculated by dividing the sum of the expected periodic inspection and repair costs up to time T during the service period, the non-life insurance premium, and the expected operator's own cost in the event of an accident (excluding insurance compensation from the total loss) by the loss amount H_B covered by non-life insurance, as follows:

$$F = \frac{S_4(T) - C_0}{H_B}. \tag{5.4.49}$$

Therefore, the operator's burden ratio, F, serves as an indicator for Strategy 4. Conversely, the out-of-pocket ratio, H, is a dimensionless quantity calculated by dividing the operator's out-of-pocket loss at the time of an accident – determined by subtracting the compensation provided by non-life insurance from the total loss – by the loss amount H_B covered by non-life insurance, as defined by the following equation:

$$H = \frac{H_B - Y_{opt}}{H_B}. \tag{5.4.50}$$

a. Relationship between inspection/repair intervals and failure probability, risk retention, and risk transfer methods
   In the standard scenario outlined in Table 5.4.8, inspection intervals of 3, 4, 5, 6, 7, 8, 9, 10, 15, 20, 25, and 50 years are evaluated. Figure 5.4.10 illustrates the relationship between inspection/repair intervals and three key metrics: probability of failure, risk retention (R), and risk transfer (F). The probability of failure decreases with shorter inspection intervals and increases with longer ones. Both the risk retention and risk transfer curves exhibit higher values at very short and very long inspection intervals, reaching their lowest points around a 20-year interval. As the inspection interval approaches 50 years, the probability of failure and risk retention curves

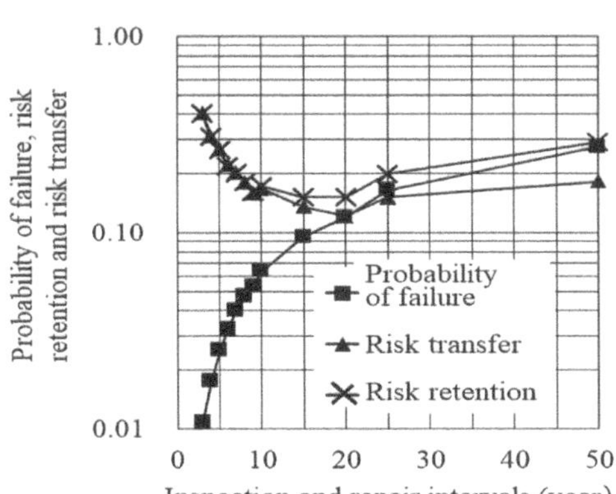

FIGURE 5.4.10 Characteristics of failure probability, risk retention, and risk transfer methods for inspection and repair intervals.

converge. Notably, the risk transfer curve lies below the other two, indicating the effect of discounting the premium burden to present value (at a discount rate of 4%). This suggests that, when evaluated in present value terms, Strategy 4 (risk transfer) can reduce the operator's risk more cost-effectively than Strategy 3 (risk retention), highlighting the efficacy of the non-life insurance approach in mitigating risk.

b. Relationship between inspection intervals and in-service period
Figure 5.4.11 analyzes the relationship between inspection and repair intervals during the service period and the associated burden of the risk-shifting method. Overall, the burden tends to increase with longer inspection intervals. This trend is attributed to the accumulation of inspection and repair costs, the extended periods between inspections, and the corresponding rise in insurance premium rates at the time of policy renewal.

It is important to note that the apparent zero cost immediately following an inspection and repair is a result of graphical processing, reflecting the discontinuity in the curve for that specific year. The curve representing a five-year inspection interval is higher than those for other intervals, indicating that frequent inspections and repairs elevate maintenance costs, which can offset the benefits of risk reduction achieved through repairs. Conversely, the curves for 10-, 15-, and 20-year intervals show a reduction in burden immediately after inspection and repair, suggesting that the risk reduction benefits are more pronounced with these intervals.

c. Relationship between maintenance and accident costs and inspection/repair intervals
Figure 5.4.12 compares the operator's burden (F) across three scenarios with varying maintenance and management costs, as detailed in Table 5.4.8. The analysis reveals that the relative magnitudes of the three curves invert around a 25-year inspection interval. For shorter intervals, the high maintenance cost scenario (Case 3) exhibits a relatively higher burden. Conversely, with longer intervals, the high compensation cost scenario (Case 2) shows the greatest burden. Specifically, the medium-cost scenario (Case 1) reaches

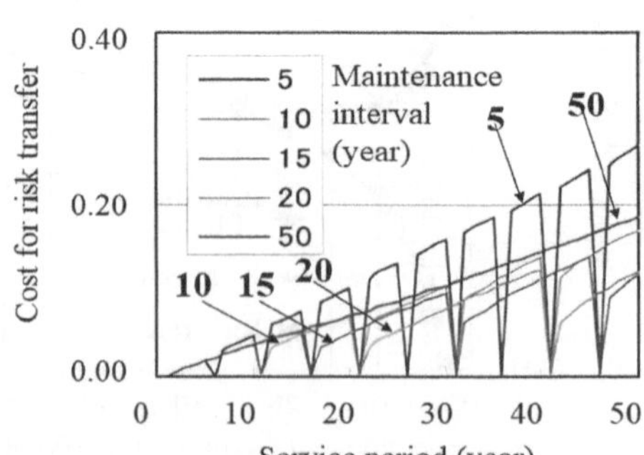

FIGURE 5.4.11 Relationship between service period and inspection interval for risk-shifting methods.

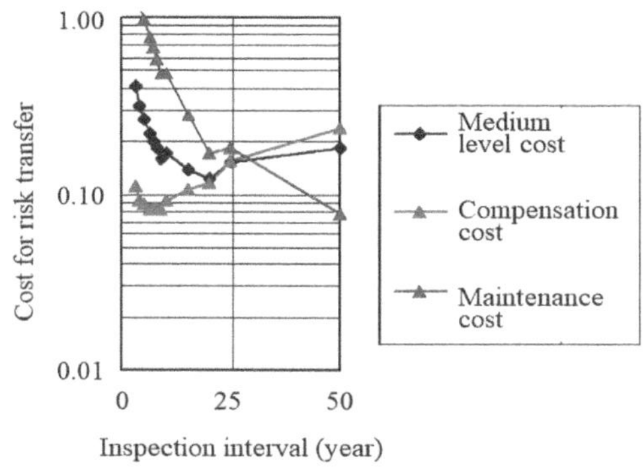

FIGURE 5.4.12 Impact of increased costs of non-life insurance compensation and maintenance.

its minimum burden around a 20-year inspection interval, while the high compensation cost scenario (Case 2) does so around seven years. Notably, the high maintenance cost scenario (Case 3) lacks a clear minimum burden point.

These findings suggest that when maintenance expenses are substantial, an optimal inspection interval may not be evident. However, in cases where maintenance costs are moderate, and considering the risk mitigation benefits of property insurance, an optimal interval between 7 and 20 years can minimize life cycle costs. Conducting case studies on actual systems can provide more precise determinations of these optimal intervals.

d. Coverage rates for operators for different damage criteria
In Section 5.4.1, pipeline damage is categorized into moderate (leakage incidents) and major (collapse incidents), with insurance options: (M_o) (covering only leakage) and (M_a) (covering both). Figure 5.4.13 illustrates that under (M_a) coverage, the operator's burden ratios (R,F) are minimized with a

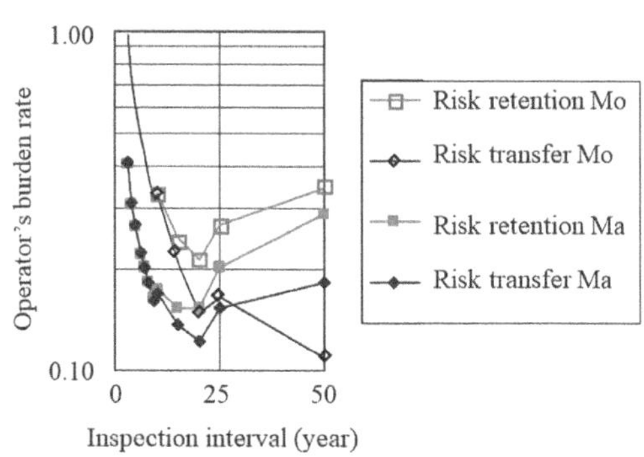

FIGURE 5.4.13 Risk retention and risk transfer schemes for leakage and collapse damage levels impact on operator's burden rates.

20-year inspection interval. Conversely, with (M_o) coverage, Strategy 2 – which relies solely on non-life insurance without inspections – results in the lowest operator burden. This suggests that comprehensive coverage benefits from regular inspections, while limited coverage may favor risk transfer through insurance alone.

e. Relationship between the operator's burden ratio (R, F) and the probability of system failure
Figure 5.4.14 shows the relationship between the operator's burden ratio and the probability of system failure for the risk retention and risk transfer methods, with the inspection/repair interval as a parameter.
While the inspection and repair intervals are short, the burden rates of the two curves are close to each other, but as the inspection and repair intervals increase, there is a divergence between the two. This difference indicates the amount of insurance required to transfer the risk through non-life insurance. The results also show that the operator's burden ratio (R,F) is relatively lowest at an inspection interval of around 20 years.

f. Comparison of insured modes
Figure 5.4.15 shows the relationship between the self-insurance rate (H) and the burden at risk transfer (F) for the insured M_o and M_a. A smaller own-load ratio indicates higher non-life insurance coverage during accidents. Longer inspection intervals, associated with increased system failure probability, lead to a higher insurance coverage ratio and a lower co-payment ratio. Conversely, shorter intervals result in reduced insurance coverage.

g. Effect of inspection and repair accuracy
Figure 5.4.16 illustrates that enhancing inspection accuracy for latent cracks yields minimal improvement in reducing system failure probability. This outcome may stem from the insensitivity of the crack effect evaluation function used in the limit strength estimations of Equations (5.4.16), (5.4.18), and (5.4.20) to crack propagation.

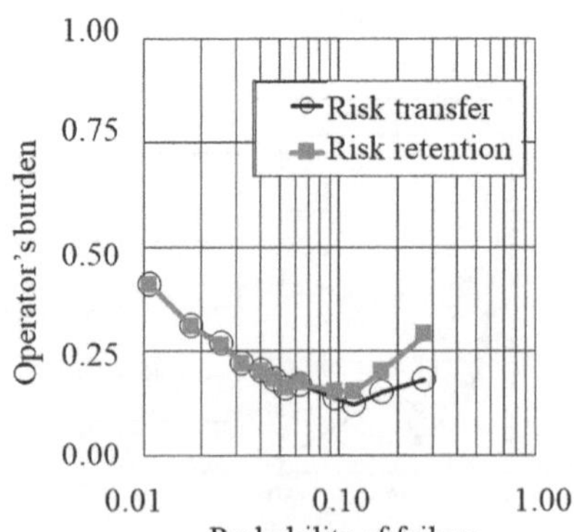

FIGURE 5.4.14 Relationship between the operator's burden rate and probability of failure for the risk retention and risk transfer method (in major mode).

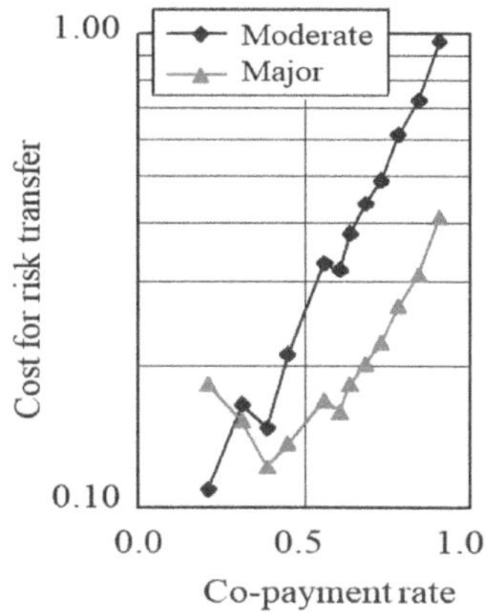

FIGURE 5.4.15 Relationship of co-payment rates to covered modes.

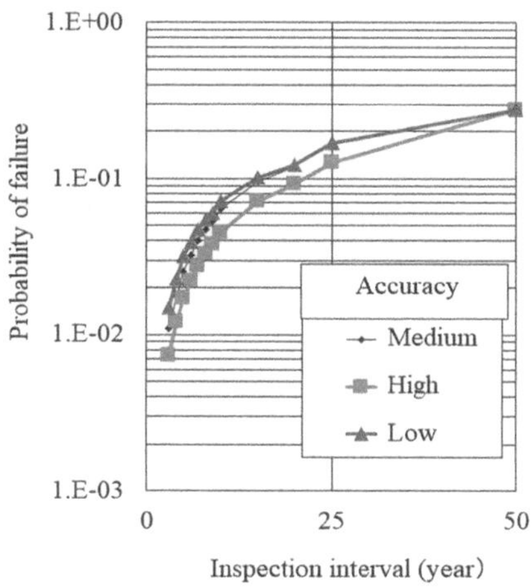

FIGURE 5.4.16 Effect of inspection accuracy.

5.4.5.6 Lessons Learned

Section 5.4 introduces a reliability analysis method with numerical examples to help operators develop maintenance strategies for long-distance pipelines. However, the pipeline parameters, maintenance costs, and insurance variables used are hypothetical, constraining the results to these predefined conditions. For practical application, operators should adapt the analysis to reflect their specific pipeline systems, enhancing the accuracy and relevance of the reliability assessments.

The results obtained in this section can be summarized as follows.

1. A system safety assessment method could be developed for pipeline damage caused by natural and social environmental conditions surrounding long-distance pipelines, when periodic inspections and repairs are carried out.

2. By introducing a simple non-life insurance model, a method could be developed to quantitatively evaluate the insurance conditions and their effectiveness in transferring the risks owned by the operator to the insurance company.
3. Using the present value discount method, it was found that the risk transfer method (Strategy 4 in the main text), which combines inspection and repair with non-life insurance, can reduce the risk of ownership at relatively lower costs than the risk retention method (Strategy 3), which does not use non-life insurance.
4. It was found that the optimum inspection interval may not exist when maintenance costs are high, but when maintenance costs are not necessarily high and when the risk-shifting effect of non-life insurance is taken into account, there is an optimum inspection and repair interval that minimizes life cycle costs, and this interval is between seven and 20 years.
5. The relationship between the risk retention/transfer method and the inspection interval for insurable medium damage type M_o and medium/major damage type M_a, respectively, shows that for insurable M_a, the operator's burden rate is at the minimum at the 20-year inspection interval, while for insurable M_o, Strategy 2 using non-inspection/repair and non-life insurance with regard to the risk transfer method minimizes operator burden.
6. The small effect of changing the inspection accuracy level on the probability of failure is considered to be due to the fact that the formula for evaluating the deterioration of critical strength in relation to potential crack propagation reflects the results showing that the formula is insensitive to crack propagation.
7. Based on the quantitative evaluation of maintenance strategies 1–4, it was possible to develop a method that enables operators to make decisions (management decisions) on the acceptable level of insured burden, i.e., the corresponding failure probability (target failure probability).

REFERENCES

1. Japan Highway Association: *Road Bridge Specifications and Commentary I Common Edition*, Japan Highway Association, 2017.
2. Japan Gas Association: *High-Pressure Gas Pipeline Seismic Design Guideline*, JGA Finger-206-03, Committee for Investigation of Technical Standards for Gas Facilities, Japan Gas Association, 2004.
3. Japan Water Works Association: *Guidelines and Commentary on Earthquake Resistant Construction Methods for Water Supply Facilities*, 2009 Edition, Japan Water Works Association, 2009.
4. Japan Ports and Harbors Association: *Technical Standards for Port and Harbor Facilities and Commentary (Upper and Lower)*, Japan Ports and Harbors Association, 1999.
5. Cornell, C. A.: A probability based structural code, *Journal of American Concrete Institute*, Vol. 66, No. 12, pp. 974–985, 1969.
6. CEB: Common Unified Rules for Different Types of Construction and Material (3rd draft), Bulletin d'Information No. 116-E, Comite Europeen du Beton, Paris, 1976.

7. ISO: *ISO 2394, General Principles on Reliability for Structure*, International Standard Organization, 1998.
8. Mabuchi, T., T. Tamakoshi, and K. Ikuta: Method for setting load factors based on the partial factor design method, *Proceedings of the Technical Research Report Meeting of the Ministry of Land, Infrastructure and Transport*, 2009, pp. 159–164, 2009.
9. Ravindra, M. K. and T. V. Galambos: Load and resistance factor design for steel, *Journal of Structural Division, ASCE*, Vol. 104, No. ST9, 1337–1353, 1978.
10. Japan Gas Association: *High-Pressure Gas Pipeline Liquefaction Resistance Seismic Design Guidelines*, JGA Finger-207-01, Committee for Investigation of Technical Standards for Gas Facilities, Japan Gas Association, 2001.
11. Koike, T., M. Hamano, T. Watanabe, and N. Hasegawa: Performance design of earthquake-resistant construction methods for water supply facilities, *JSCE, Journal of JSCE A1 (Structural and Earthquake Engineering)*, Vol. 76, No. 4, L_196–I_204, 2020.
12. Ang, A. H. S. and W. H. Tang: *Probability Concepts in Engineering Planning and Design, Volume II: Decision, Risk, and Reliability*, John Wiley & Sons, 1984.
13. Charney, F. A.: *Seismic Loads, Guide to the Seismic Load Provision of ASCE 7-05*, ASCE, 2010.
14. Japan Association of Polyethylene Pipe Systems for Water Distribution: *Guide to Seismic Design of Polyethylene Pipes for Water Distribution*, Seismic Resistance Evaluation Committee for Polyethylene Pipes for Water Distribution, 2018.
15. Japan Ductile Pipe Association: *Guidance of Ductile Iron Pipe Conduits*, JDPA T 26, 2015.
16. Mori, S., K. Chiba, and T. Koike: Seismic performance of long-distance gas pipelines during the 2011 off the Pacific coast of Tohoku Earthquake, *Journal of JSCE A1 (Structural and Earthquake Engineering)*, Vol. 69, No. 4, I_111–I_120, 2013.
17. Japan Meteorological Agency: Estimated Seismic Intensity Distribution Map 2011 off the Pacific Coast of Tohoku Earthquake (Detailed Map of Miyagi Prefecture), 2011.
18. JMA: Strong-Motion Seismic Waveform 2011 Tohoku Surface Pacific Ocean Earthquake (Miyagi Prefecture, Zao Town, Enda), 2011.
19. Japan Road Association: *Oil Pipeline Technical Standard*, 1974.
20. Japan Gas Association: *High Pressure Gas Pipeline Seismic Design Guidelines*, 2000.
21. Japan Water Works Association: *Guidelines for Earthquake-Resistant Construction Methods for Water Supply Facilities and Commentary*, 1996.
22. National Land Development Technology Research Centre: *Report of the Subcommittee for Technical Review of Gas Pipelines (Technical Guidelines for High Pressure Gas Pipelines)*, 1991.
23. Suzuki, N., S. Igi, and K. Masamura: Seismic integrity of high-strength pipelines, *JFE Technical Report*, No. 12, pp. 14–19, 2008.
24. Melchers, R. E.: *Structural Reliability Analysis and Prediction*, 2nd edition, Wiley, 1999.
25. EGIG: *Gas Pipeline Incidents, 8th Report of the European Gas Pipeline Incident Data Group*: Document Number EGIG 11.R.0402 (version 2), 2011.
26. T. Koike: Pipeline maintenance and management strategies based on risk management methods, *Journal of the Japan Society of Civil Engineers*, No. 794/I-72, pp. 189–202, 2005.
27. T. Koike: On lifetime risk assessment of long-distance pipelines, *Materials*, Vol. 48, No. 1, pp. 83–88, 1999.
28. Hoshitani, M.: Development of probabilistic DCF method of asset management considering risk, *JSCE 59th Annual Conference*, pp.175–176, 2004.
29. Nawata, K.: *Economics and Finance Theory*, Toyo Keizai Shinposha, 2003.
30. Det Norske Veritas: Submarine Pipeline Systems, Offshore Standard OS-F101, DNV, 2000.

31. Stephens, D. R., R. J. Olsen, and M. J. Rosenfeld: Topical Report of Pipeline Monitoring – Limit State Criteria, NG-18 Battelle Report, No.188, 1991.
32. Ahammed, M. and R. E. Melchers: Probabilistic analysis of underground pipe lines subjected to combined stress and corrosion, *Engineering Structures*, Vol.19, No.12, pp. 988–994, 1997.
33. Mok, D. R. B., R. J. Pick, and A. G. Glover: Behavior of line pipe with long external corrosion, general interest, *MP*, pp. 75–79, 1990.
34. Hopkins, P., D. G. Jones, and A. C. Clyne: The significance of dents and defects in transmission pipelines, conference on pipework engineering and operation, IMechE, London, Paper C376/049, 1989.
35. Kanninen, M. F., T. S. Grant, G. Demonfonti, and S. Venzi: The development and validation of a theoretical ductile fracture model, Paper 12, 8th Symposium on Line Pipe Research, Pipeline Research Committee, AGA, Catalog No. L51680, Houston, 1993.
36. Vogt, G. H., M. Bramante, D. G. Jones, F. O. Koch, J. Kugel, H. Pero, and G. Re: *EPRG Report on the Toughness for Crack Arrest in Gas Transmission Pipelines*, 3R International, Essen, Germany, 1983.
37. Shinozuka, M. and Koike, T.: Seismic risk of underground lifeline system resulting from fault movement, Proceedings of the 2nd U.S. National Conference on Earthquake Engineering, Stanford University, California, USA, pp. 663–672, 1979.
38. Shinozuka, M. and Tan, R.: Estimation and optimization of the serviceability of underground water transmission network systems under seismic risk, Proceedings of review meeting of U.S.-Japan Cooperative Research on Seismic risk analysis and its application to reliability-based design of lifeline systems, Honolulu, Hawaii, USA, pp. 207–228,1981.
39. Streeter, V. L.: Fluid mechanics, McGraw-Hill, New York, 7th edition, 1979.

Appendix A1: Fundamentals of Structural Design Methods

A1.1 TYPES OF STRUCTURAL DESIGN METHODS

A1.1.1 Allowable Stress Design

Allowable stress design (ASD) is a design method for verifying safety and serviceability by confirming that the stress acting on each section of a member is less than or equal to the allowable stress of the materials constituting the member.

A1.1.2 Load and Resistance Factor Design (LRFD)

LRFD is a design method adopted in the 1998 revision of the American Specification for Highway Bridge Design (AASHTO).

LRFD is a design method based on probability and reliability and requires direct consideration of statistical mean resisting forces and loads, nominal or arbitrary resisting forces and loads, and their standard deviations and coefficients of variation.

The LRFD specification requires that some or all of the four following limit states be verified in the design of each section of the bridge:

a. Serviceability limit state
b. Fatigue failure limit state
c. Capacity limit state
d. Extreme limit state

A1.1.3 Limit State Design Method

The limit state design method is a design method in which the limit state to be checked is clarified. In other words, it is a design method that sets each limit state focusing on various required performances such as usability, safety, recoverability, and durability of the structure and verifies that the limit values set for each limit state are satisfied.

The following limit states are commonly used:

a. Serviceability limit state
   A limit state beyond which a structure or structural element cannot satisfy the required performance in terms of serviceability.
b. Restorability limit state
   The limit state at which the structure can be brought back into continuous service by repairing the damage expected to result from the anticipated action, using applicable technology and within a reasonable cost and time frame. It can also be understood as one of the limit states for serviceability.

c. Ultimate limit state
The limit state that leads to collapse or similar destruction of the structure.

A1.1.4 Partial Safety Factor Design Method

A design method that uses partial factors to take into account the margin required for a structure to satisfy a specified limit state with a reasonable probability against uncertainties in various actions acting on the structure, geotechnical parameters, structure dimensions, accuracy of design calculation models, and reference values for verifying the limit state in design calculations.

There are six types of partial safety factors: member coefficient, material coefficient, structure coefficient, structural analysis coefficient, load coefficient, and load combination coefficient.

A1.1.5 Reliability-Based Design

A design method in which the possibility of a structure reaching its limit state is probabilistically verified.

Level I reliability design method: design method using partial safety coefficients.
Level II reliability design method: design method using safety indices.
Level III reliability design method: design method using failure probability.

A1.1.6 Performance-Based Design

There are two types of performance-based design methods: performance-based design methods and performance-based design codes.

A1.1.6.1 Performance-Based Design Method

A design method in which any structural form, structural material, design method, or construction method may be used as long as the designed structure satisfies the required performance. More specifically, it is a design method that specifies the purpose of the structure and the functions to be met, specifies the performance required to provide the functions, and ensures that the functions are satisfied by ensuring the specified performance throughout the service life of the structure. Similar terms include performance-dictated design, performance-explicit design, and performance-oriented design.

Verification is the act of determining whether a structure satisfies the performance specifications. In the case of the limit state design method, it is the act of determining $S \leq R$ between the response value S and the corresponding limit value R.

A1.1.6.2 Performance-Based Design Codes

In design codes for structures that specify structures not according to their specifications but according to their socially required performance, a performance-based design code, or a performance-explicit design code is "a design method that clarifies the relationship between the level of performance required to ensure the function of a structure and the level of action used to verify it." The design method is defined as a performance-specific design method or a performance-explicit design method.

A1.2 RELIABILITY DESIGN METHODS[1]

A1.2.1 Terminology

A1.2.1.1 Safety

A property or degree of property that does not cause danger to human health or damage or destruction of the structure.

A1.2.1.2 Reliability

Reliability is a concept consisting of safety, serviceability, and durability. Safety is a part of reliability, and reliability is a broader concept than safety.

Reliability is expressed using a reliability function L.

A1.2.1.3 Probability of Failure

The probability of failure p_f is the probability of the occurrence of a damage or failure event in a structure.

A1.2.1.4 Safety Index

As it is generally difficult to assess safety directly using the probability of failure, a safety index β using only the mean and variance, which is easy to calculate, has been proposed as an alternative.

A1.2.1.5 Safety Factor

The safety factor $S.F.$ is the reference strength of the structural material or the resisting strength of the structure divided by the load action. Since the strength and load are random variables, dividing by the mean values of strength and load is called the central safety factor, while dividing by the nominal values of strength and load is called the nominal safety factor.

A1.2.1.6 Failure Criterion Function

A failure phenomenon is defined as the value of a function expressed in terms of a number of random variables, such as strength and load, falling below a certain value. This function is called the failure criterion function. For example, if R and S are strength and load, respectively, the failure criterion function Z is defined as

$$Z = R - S. \tag{A1.2.1}$$

In this case, the failure condition equation is expressed as

$$Z < 0. \tag{A1.2.2}$$

The probability of failure p_f is given by

$$p_f = P[Z < 0]. \tag{A1.2.3}$$

A1.2.1.7 Nominal Values

The strength and load values specified in the conventional allowable stress design method are referred to as nominal values. They are also referred to as nominal strengths and nominal loads. From a probabilistic level, nominal strengths are considerably lower than the average values, and nominal loads are higher than the average values. It should be noted, however, that nominal strengths and nominal loads are not probabilistically distanced from the mean on the basis of their respective means and standard deviations.

A1.2.1.8 Characteristic Values

In the limit state design method, characteristic values for strengths and loads are defined as follows: *Characteristic strengths* are strength values for which the level of non-exceedance probability is sufficiently small to reach a certain value (e.g., 5%). *Characteristic loads* are load values for which the level of non-exceedance probability is sufficiently high for a given value (e.g., 5%).

The values used in the design of strength and loads should be sufficiently safe, taking into account their variability. The characteristic values are defined in such a way that the different variability is probabilistically balanced to the same level.

It is sometimes said that the conventional nominal values (design basis strength or standard value) are sufficiently safe as characteristic values, but such a view is not appropriate. It should be noted that the characteristic value not only has the role of making the value sufficiently safe, but also has the role of adjusting for differences in variability, e.g., between loads, by aligning values such as exceedance probability.

A1.2.1.9 Uncertainty, Variability, and Fuzziness

Uncertainty refers to the property or degree of being indefinite. Uncertainty can also include the meaning of being uncertain and possibly wrong. Variability is the property or degree to which changing movements are not constant. Fuzziness is the quality or degree of being blurred or unclear.

Uncertainty in structural relationships is broadly classified into physical uncertainty, statistical uncertainty, model uncertainty, and human uncertainty. Physical uncertainty can be broadly classified into three categories: load, statistical uncertainty, model uncertainty, and human uncertainty. Physical uncertainty is due to the variability of physical quantities such as loads, material strength, and dimensions. Statistical uncertainty is due to evaluation errors in the variability of physical quantities (generally due to a limited number of samples). Model uncertainties are due to the various assumptions and approximations made in modeling in structural analysis. Human uncertainty is due to human error or gross error.

A1.2.2 Secondary Moment Method

The calculation of failure probability requires the preparation of probability distribution functions for strength and load and the integration of both, and the development of design methods using failure probability has been extremely difficult to introduce into practical design. Against this background, there has been a search for a method to easily calculate the probability of failure at an early stage and for a safety evaluation index that can be easily related to the probability of failure using probability statistics quantities such as mean value and standard deviation. In particular, the quadratic moment method as a design method using only limited probability statistics such as mean and standard deviation was proposed by several researchers until the 1960s.

The design method based on the method of quadratic moments proposed by Cornell in 1969 was a practical design method that satisfied the aforementioned requirements. His proposed safety index β is defined by the following equation:

$$\beta = \mu_Z / \sigma_Z, \tag{A1.2.4}$$

where Z is the failure criterion function, μ_Z is its mean value, and σ_Z is its standard deviation.

Depending on how the failure criterion function Z is expressed, the safety index is calculated as follows:

$$Z = R - S \rightarrow \quad \beta = \frac{\mu_R - \mu_S}{\sqrt{\sigma_R^2 + \sigma_S^2}} \tag{A1.2.5}$$

$$Z = lnR - lnS \rightarrow \beta = \frac{ln\lambda_R - ln\lambda_S}{\sqrt{\varsigma_R^2 + \varsigma_S^2}} \tag{A1.2.6}$$

where the mean and variance with respect to the logarithm of the random variable X are given by

$$\begin{aligned} \lambda_X &= E[lnX] = \frac{\mu_X}{\sqrt{1+\delta_X^2}} \\ \zeta_X &= Var[lnX] = ln\left(1+\delta_X^2\right). \end{aligned} \tag{A1.2.7}$$

If the coefficients of variation of R and S are relatively small (e.g., below 0.05), the safety index of Z in logarithmic representation can be approximated by the following equation:

$$Z = lnR - lnS \rightarrow \beta = \frac{ln\mu_R - ln\mu_S}{\sqrt{\delta_R^2 + \delta_S^2}} \tag{A1.2.8}$$

The relationship between safety indicators and the probability of failure is explained below.

First, Figure A1.2.1 shows the relationship between the probability density function of the failure criterion function Z and the safety index. In this figure, the probability of failure corresponds, by definition, to the area of the region $z < 0$ of the probability density function of Z. The safety index $\beta\sigma_Z$, which is composed of the safety index β and the standard deviation σ_Z of Z, is the distance from the mean value of Z to the origin. Therefore, a larger safety index means a larger distance from the mean value to the origin and a smaller probability of failure.

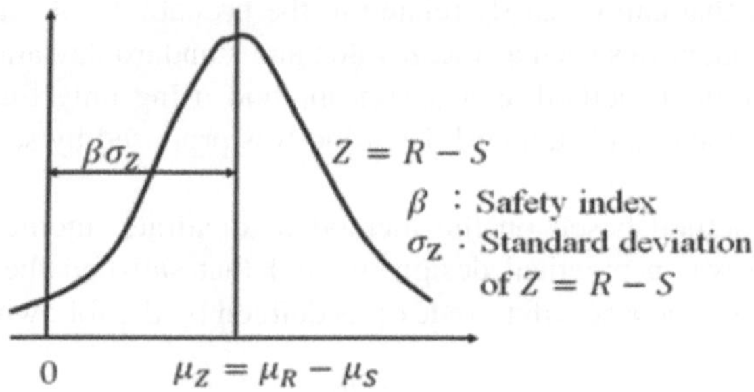

FIGURE A1.2.1 Relationship between the safety index and probability of failure.

In fact, the fracture probability and the safety index are linked by the following relationship when the fracture criterion function Z can be assumed to follow a normal distribution:

$$p_f = \Phi(-\beta), \tag{A1.2.9}$$

where Φ is the standard normal distribution function.

This means that there is a one-to-one correspondence between the safety index and the probability of failure. In this sense, the safety index can be said to play an equivalent role to the failure probability.

Figure A1.2.2 shows the geometric relationship of the safety indices in the design space. The vertical axis is the normalized load axis, and the horizontal axis is the normalized strength axis. The figure shows the boundary line dividing the failure and nonfailure regions, which is called the failure curve, and the relational equation representing it. This equation represents the failure criterion function $Z = 0$ using the normalized variables R and S. The foot of the perpendicular line (shortest distance) from the origin to the failure curve in this diagram corresponds to the safety index β.

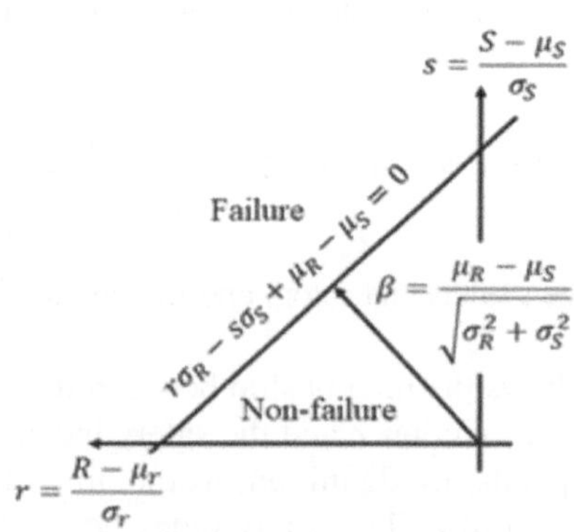

FIGURE A1.2.2 Geometric interpretation of the safety index.

When the failure criterion function is given by a nonlinear relationship between R and S, the shortest distance from the origin to the failure curve is different from that indicated below because the failure curve is a curve instead of a straight line, but the fact that the safety index is the shortest distance from the origin to the failure phase (Hasofer–Lind's transformation) is constant.

The case where the design variables are not mutually independent (correlated varieties) corresponds to the case where the horizontal and vertical axes below are not necessarily orthogonal. In such cases, it is necessary to ensure the orthogonality of the coordinate system by transforming the covariance matrix, which represents the relationship between the design variables, into independent design variables through an orthogonalization transformation. Furthermore, if the design variables are not normal distribution functions, one method is to approximate the probability distribution function of the design variables to an equivalent normal distribution function (Rosenblatt transformation) to ensure the orthogonality of the coordinate system.

A1.2.3 Level I, II, and III Reliability Design Methods

Level I reliability uses probabilistically defined characteristic values for the strength and load used in the design and tries to take various uncertainties into account using a partial safety factor, also known as the partial safety factor method. In this case, safety is determined by the following equation:

$$S < R. \tag{A1.2.10}$$

The Level II reliability design method attempts to design using safety indices instead of probabilities to assess safety and reliability. In this case, safety is determined by the following equation:

$$\beta_{cr} < \beta, \tag{A1.2.11}$$

where β_{cr} is the critical safety index.

The Level III reliability design method attempts to design directly using probabilities to assess safety and reliability. In this case, safety is determined by using the reliability function L and the probability of failure p_f as follows:

$$L = 1 - p_f = P[Z > 0]. \tag{A1.2.12}$$

A1.2.4 Classification of Limit States

The first step is to clearly define the limit states in which the structure will not fulfil the functions it is expected to perform during construction and use. The limit state design method is a method of designing a structure with the help of reliability theory so that the possibility of this happening is sufficiently small.

The limit states are the ultimate limit state, serviceability limit state, and fatigue limit state. Instead of the fatigue limit, a restoration limit state is sometimes set.

A1.2.5 Partial Safety Factor Design Methods in Various Countries Around the World[2,3]

In the Level I reliability design method, among the various uncertainties, the variable factors of load and material strength are taken into account using their characteristic values. Other uncertainties, such as the social impact when the limit state is reached, are taken into account by multiplying or dividing by several coefficients in the verification equation. Such coefficients are called partial safety factors.

Research and practice of design code formats in Europe and the United States started in the 1970s, and the following partial safety evaluation formulae were proposed in the design guidelines of Eurocode, Canada, and the United States in the late 1970s.

CEB (Comite European de Beton)

$$g_R\left(\frac{f_k}{r_{m1}r_{m2}r_m}\right) \geq g_S\left(\gamma_{f1}\gamma_{f2}\gamma_{f3}Q_k\right), \tag{A1.2.13}$$

where g_R and g_S are functions that convert strength and load into their effects. f_k and Q_k are characteristic values of strength and load. γ_X is the partial safety factor for variable X.

γ_{mi} is the resistance coefficient.
γ_{f1} is a coefficient that evaluates the deviation from the characteristic value (load coefficient).
γ_{f2} is a coefficient to assess the combined load effect (combination coefficient).
γ_{f3} is a coefficient to evaluate load imprecision (structural analysis coefficient).
NRCC (National Research Council of Canada)
CSA (Canadian Standard Association)

$$\phi R \geq g_S\left[\gamma_D D_n + \psi\left(\gamma_L L_n + \gamma_W W_n + \gamma_T T_n + \cdots\right)\right], \tag{A1.2.14}$$

where R is the characteristic value of the resisting force, ϕ is the partial safety factor, and D_n, L_n, W_n, and T_n are the nominal values of the dead, live, wind, and thermal loads. The variables D_n, L_n, W_n, and T_n on the right-hand side are the loads, γ_D, γ_L, γ_W, and γ_T are the safety factors for the respective loads and γ_S is a function that converts the loads into load effects. ψ is a coefficient evaluating the probability of all the component loads reaching their nominal values simultaneously, which is 1.0 for one load and increases with two or three loads. The value decreases to 0.7 and 0.6 as the number of loads increases by two or three.

Load and Resistance Factor Design (LRFD)
ACI (American Concrete Institute)

$$\phi R_n \geq \sum_{k=1}^{i} \gamma_k S_{km} \tag{A1.2.15}$$

where ϕ and R_n are the partial safety factor of the resisting force and the nominal value of the resisting force. γ_k and S_{km} on the right-hand side are the partial safety factor of the load and the nominal value of the load.

The following equation is proposed as a more specific expression:

$$\begin{aligned}
&\phi R_n \geq \gamma_D \bar{D} + \gamma_L \bar{L} \\
&\phi R_n \geq \gamma_D \bar{D} + \gamma_{opt} \bar{L}_{opt} + \gamma_W \bar{W} \\
&\phi R_n \geq \gamma_D \bar{D} + \gamma_{opt} \bar{L}_{opt} + \gamma_S \bar{S} \\
&\phi R_n \geq \gamma_W \bar{W} - \gamma_D \bar{D},
\end{aligned} \tag{A1.2.16}$$

where $\overline{D}$, $\bar{L}$, $\overline{W}$, and $\bar{S}$ are the load effects of the average dead load and the average values of the maximum live load, wind load, and snow load during service, respectively. γ_D, γ_L, γ_W, and γ_S are the partial safety factors for the dead load, live load, wind load, and snow load. L_{opt} is the live load at an arbitrary point in time.

In particular, the *California Structural Design Guidelines* present the following combinations of loads as specific expressions on the load side:

$$\begin{aligned}
&1.4(D+F) \\
&1.2(D+F)+1.6(L+H)+0.5(L_r \text{or } S \text{ or } R) \\
&1.2(D+F)+1.0E+f_1L+1.6H+f_2S \\
&1.2(D+F)+1.0W+f_1L+1.6H+0.5(L_r \text{or } S \text{ or } R) \\
&1.2(D+F)+1.6(L_r \ or\ S\ or\ R)+1.6H+0.5(f_1L \text{ or } 0.5W) \\
&0.9D+1.0W+1.6H \\
&0.9(D+F)+1.0E+1.6H,
\end{aligned} \tag{A1.2.17}$$

where D, L, L_r, W, F, H, S, R, and E are the dead load, live load, roof load, wind load, water pressure, earth or ground pressure, snow load, rain load, and earthquake load, respectively.

Japanese Highway Bridges Specifications

$$\xi_1 \cdot \xi_2 \cdot \Phi_R \cdot R(f_c, \Delta_c) \geq \sum S_i (\gamma_{pi} \cdot \gamma_{qi} \cdot P_i), \tag{A1.2.18}$$

where P_i is the characteristic value of the action, S_i is the action effect, R is the resistance characteristic value of the member etc., f_c is the characteristic value of the material, Δ_i is the characteristic value of the dimension, γ_{pi} is the load combination factor, γ_{qi} is the load factor, ζ_1 is the investigation and analysis coefficients, ζ_2 are material and structural coefficients, and Φ_R are resistance coefficients.

According to Appendix 4 of the Ministry of Land, Infrastructure, Transport, and Tourism, the following observations[4] are made on the two coefficients in the *Specification for Highway Bridges*.

The investigation and analysis coefficients ζ_1 (or structural analysis coefficients) are coefficients to ensure an appropriate safety margin for the bridge under consideration, depending on the variations associated with the calculation of the response values of the structure or members due to the effects of the action.

Structural coefficients ζ_2 are coefficients to ensure an appropriate margin of safety for the subject bridge, depending on the difference in the effect on the condition of the bridge when the structure or members reach their limit state, the cost and social impact of restoring the function if the prescribed performance is not met, etc.

However, as it is expected to be difficult to set any of these coefficients rationally by statistical treatment, it is envisaged that they will be set in consideration of existing performance and the maintenance targets to be achieved by the standards.

Guidelines for Seismic Design of High-Pressure Gas Pipelines against Liquefaction (Japan Gas Association).

These guidelines provide safety assessment methods for gas pipelines buried in liquefied ground.

The partial safety factor is the same as the partial safety factor method in Eurocode (CEB) and the *Japanese Road and Bridge Specifications*. There are six types of partial safety factors: member factor γ_b, material factor γ_m, structure factor γ_i, structural analysis factor γ_a, load factor γ_l, and load combination factor γ_c.

In the *Japanese Highway Bridges Specification*, only seismic loads are taken out here, although combinations of multiple loads are covered in the *Japanese Highway Bridges Specification*.

The safety of a structure is determined by the following equation:

$$S_d \le R_d, \tag{A1.2.19}$$

where S_d and R_d are the load and the strength for verification, respectively. and

$$\begin{aligned} S_d &= \gamma_a \cdot S\left(\gamma_f \cdot F_k\right) \\ R_d &= R\left(\frac{f_k}{\gamma_m}\right)/\gamma_b, \end{aligned} \tag{A1.2.20}$$

where

$S(*)$: load calculation function
F_k: characteristic value of the load
γ_f: load factor (variation of the characteristic value of the load)
γ_a: structural analysis factor (variation of the load calculation formula)
$R(*)$: resistance calculation function
f_k: characteristic value of resistance
γ_m: material factor (variation of the characteristic value of the material)
γ_b: member factor (variation of the limit value calculation formula)

The safety assessment formulae in the *Seismic Design Guidelines for Liquefaction Resistance of High-Pressure Gas Pipes* are designed to verify the seismic safety of high-pressure gas pipes in the event of liquefaction (ground failure) of the ground around buried gas pipes due to Level 2 ground motion. Overlap with other loads is not considered.

Guidelines and Commentary on Seismic Design Methods for Water Supply Facilities

In the current design guidelines (*Guidelines for Seismic Engineering Methods for Water Supply Facilities and Commentary*, 2009), the same safety assessment methods are basically used as in the *Guidelines for Seismic Design of High-Pressure Gas Pipes against Liquefaction.*

$$S_d \leq \frac{R_d}{\gamma_i} \tag{A1.2.21}$$

where S_d is the load to be verified, R_d is the strength to be verified, and γ_i is the structure factor, which "is determined by taking into account the importance of the structure and the social impact when the limit state is reached, and should generally be 1.0~1.2." It is explained as follows.

Table A1.2.1 summarizes the relationship between the load combinations for buried pipes and Levels 1 and 2 ground motions.

TABLE A1.2.1
Comparison of Seismic Design Items for Level 1 and Level 2 Ground Motions

	Seismic Loads		
Item	**Level 1 Ground Motion**	**Level 2 Ground Motion**	
US classification	MOE	DBE or MCE	MCE
Limit state	Serviceable	Repairable	Ultimate
Performance level	Seismic performance 1	Seismic performance 2	Seismic performance 3
Ground condition	Elastic	Elastoplastic	Rupture
Design strength	Elastic buckling limit strength	Repairable limit displacement	Ultimate limit displacement
Design load	Ordinary and seismic loads	Response displacement	Ground collapse displacement
Load combination	Combination of ordinary and seismic loads	Seismic load	Ground displacement
Partial safety factor	Partial safety factor for ordinary and seismic loads	Partial safety factors of member, material, structure, structural analysis, load and load combination for seismic load	Partial safety factors of member, material, structure, structural analysis, load and load combination for seismic load

A1.2.6 Probability Distributions and Statistics of Random Variables

The following representative values are generally used to express the probability distribution characteristics of the random variable X.

Probability density function:

$$f_X(x) = \frac{P[x < X \leq x + dx]}{dx} \tag{A1.2.22}$$

Probability distribution function:

$$F_X(x) = \int_{-\infty}^{x} f_X(x)\,dx \tag{A1.2.23}$$

Mean value, expected value:

$$E[X] \equiv \mu_X = \int_{-\infty}^{\infty} x f_X(x)\,dx \tag{A1.2.24}$$

Variance:

$$E\left[(X - \mu_X)^2\right] \equiv \sigma_X^2 \tag{A1.2.25}$$

Standard deviation:

$$\sqrt{E\left[(X - \mu_X)^2\right]} = \sigma_X \tag{A1.2.26}$$

Coefficient of variation:

$$\delta_X \equiv \sigma_X / \mu_X \tag{A1.2.27}$$

A1.3 PERFORMANCE-BASED DESIGN METHOD

A1.3.1 Performance Prescriptive Design Standards

The specification rules that have been in operation for some time have the advantages of being concrete, easy for anyone to understand, requiring no special skills in design, and facilitating a true assessment of conformity with the standards, because they specify the materials and dimensions to be used. On the other hand, it is difficult to respond to the emergence of new materials and structures; it is rigid, and the goals to be achieved are not clear. A further disadvantage is the lack of international compatibility when different specifications exist internationally, which results in international tariff barriers.

On the other hand, performance specifications are characterized by a large degree of freedom and flexibility in design, and have the advantage that the target performance to be achieved is clear, and it is easy to respond to technological advances and to contribute to international harmonization of standards. However, there are demerits in that advanced technology is required to examine the performance design and to judge conformity, and that a high level of competence is required in design.

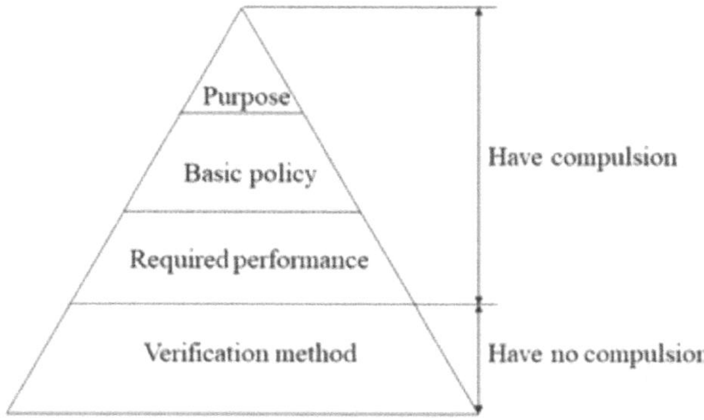

FIGURE A1.3.1 Hierarchy model of performance prescriptive technical standards.

From a comprehensive perspective as Japan, our country ratified the World Trade Organization (WTO) Convention in 1966 and introduced the performance design method as a world-standard structural design method, and since then, domestic design standards have been developed toward performance design.

The framework of the performance-specific performance design method that has become common in Japan is shown in Figure A1.3.1. Here, the requirements in the top three levels are mandatory, and only the fourth level, the verification method, is not mandatory and can be reviewed and enhanced as technology advances.

A1.3.2 Structural Safety Achieved by the Partial Safety Factor Design Method

For example, in the revised edition of the *Specification for Highway Bridges*, structural safety is judged to have been achieved if the resistance characteristic value multiplied by various partial factors exceeds the action effect, according to the following design verification formula:

$$\xi_1 \cdot \xi_2 \cdot \Phi_R \cdot R(f_c, \Delta_c) \geq \sum S_i(\gamma_{pi} \cdot \gamma_{qi} \cdot P_i), \qquad \text{(A1.2.28)}$$

where P_i is the characteristic value of the action, S_i is the action effect, R is the resistance characteristic value of the member, etc., f_c is the characteristic value of the material, Δ_i is the characteristic value of the dimension, γ_{pi} is the load combination factor, γ_{qi} is the load factor, ζ_1 are the investigation and analysis coefficients, ζ_2 are material and structural coefficients, and Φ_R are resistance coefficients.

However, when structural safety verification is carried out using only inequality-based judgments, designers are not able to quantitatively grasp the degree of safety achieved by the performance design. In addition, the values of the various partial safety factors were originally set in advance by the code writers who established the design criteria, and practical designers cannot set them to arbitrary values. The existing partial safety factor design method is at a level where safety verification (judgment) is possible, but quantitative evaluation of safety is not possible. In this sense, it is exactly why it is called the Level I reliability design method.

On the other hand, if data on the probability statistics of the characteristic values of action and resistance are available, these characteristic values can be redefined using the mean and standard deviation, and the safety index can be calculated according to the method of quadratic moments. Furthermore, if the normal distribution of the design variables is assumed, the probability of failure can also be calculated.

Therefore, in order to advance the reliability design level from Level I to Level II and Level III, it is essential to obtain data on the probability statistics of the characteristic values of the action and the resistance characteristic values.

REFERENCES

1. JSCE Subcommittee on Structural Safety Research: *Assessment of Lifetime Risk of Structures*, JSCE Structural Engineering Committee, JSCE, 1988.
2. Melchers, R. E.: *Structural Reliability Analysis and Prediction*, John Wiley & Sons Ltd, 1999.
3. Architectural Institute of Japan: *Limit State Design Guidelines for Buildings, Architectural Institute of Japan*, Maruzen Corporation, 2002.
4. Mabuchi, T., T. Tamakoshi, and K. Ikuta: Method for setting load coefficients based on partial coefficient design method, *Proceedings of the Technical Research Report Meeting of the Ministry of Land, Infrastructure, Transport and Tourism*, 2009, pp. 159–164, 2009.

Appendix A2: Seismic Performance and Design Ground Motions

A2.1 DEFINITION OF SEISMIC PERFORMANCE

A2.1.1 LIMIT STATES (SERVICE LIMIT, RESTORATION LIMIT, AND ULTIMATE LIMIT) AND DESIGN GROUND MOTION

The limit state design method for structures establishes three limit states for structures: (a) the use limit, (b) the restoration limit, and (c) the ultimate limit.

These limit states are defined as follows:

a. Serviceability limit state
   The limit state at which the response of the structure is in an elastic state and, if this limit is not exceeded, the structure can be used immediately without causing damage.
b. Restorability limit state
   The limit state in which the response of the structure is in an elastoplastic state and, if this limit is not exceeded, the structure can continue to be used by repairing minor damage.
c. Ultimate limit state
   The limit state in which the response of the structure is in an elastoplastic state and, if this limit is not exceeded, the structure can be spared from collapse even though considerable damage may occur to the structure.

On the other hand, in the *Guidelines for Seismic Engineering Methods for Water Supply Facilities*, seismic performance 1, 2, and 3 are defined as corresponding to the aforementioned limit states as follows:

Seismic performance 1:	Performance in which an earthquake does not cause damage to the existing sound condition.
Seismic performance 2:	Performance in which the damage caused by the earthquake is minor, and the required repair after the earthquake is only minor and does not seriously affect the functionality of the building.
Seismic performance 3:	Performance where damage caused by the earthquake is minor and requires repair after the earthquake but does not significantly affect the function.

Figure A2.1.1 illustrates the relationship between the limit states (a), (b), and (c) using external force-displacement curves for structures. For comparison, Figure A2.1.2, which illustrates the limit states in the United States, is also included.

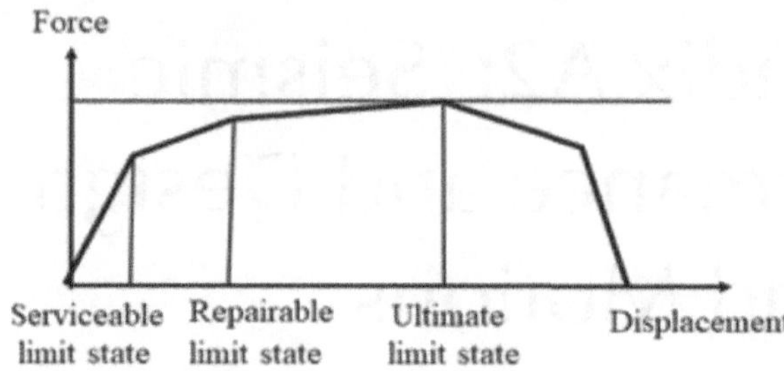

FIGURE A2.1.1 Conceptual diagram of structural limit state classification in Japan.

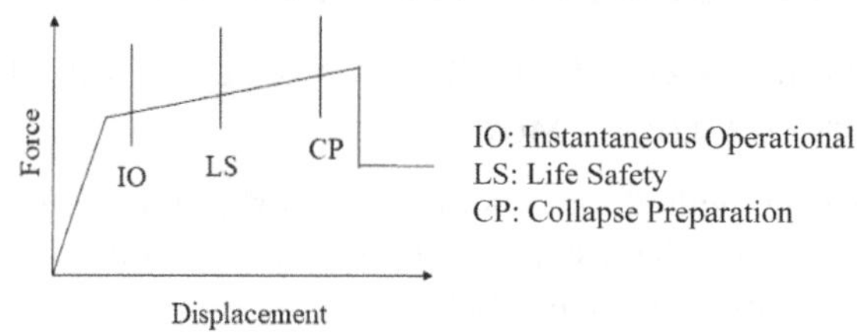

FIGURE A2.1.2 Conceptual diagram of structural limit state classification in the United States.

Figure A2.1.3 illustrates the relationship between the three limit states and the following three seismic performance levels.

Discussion of seismic performance provisions:

Problem 1: "Seismic performance 3, with only minor damage caused by earthquakes" does not mean that the damage is minor if the repair limit is exceeded by Level 2 ground motion, but rather that considerable damage is expected. Wouldn't this mean that a considerable amount of time and cost would be required to repair the damage?

Problem 2: Seismic performance 2 and 3 both acknowledge the occurrence of damage after the earthquake. However, the presence or absence of leakage should be clearly stated.

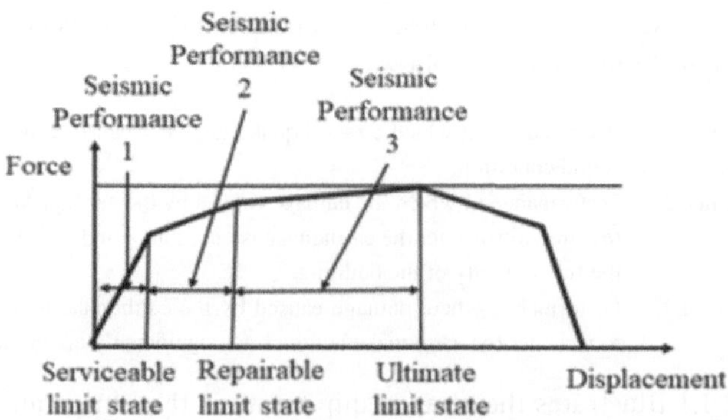

FIGURE A2.1.3 Relationship between limit states and seismic performance.

In order to solve problems 1 and 2, we propose the following modifications to the seismic performance rules. In other words:

Modified seismic performance 2:	Performance in which the damage caused by the earthquake is minor, no leakage occurs, and the repair required after the earthquake is only minor and does not seriously affect the function.
Modified seismic performance 3:	Significant earthquake-induced damage but no leakage requires repair after the earthquake but does not seriously affect the function.

The current *Guidelines for Seismic Engineering Methods for Water Supply Facilities* state that "For pipelines and water bridges, the performance to be maintained is limited to seismic performance 2, since the seismic performance is defined by the presence or absence of leakage. This judgement is not necessarily correct. Seismic safety verification for seismic performance 3 can be added by adding the words 'no leakage' to the provisions for seismic performance 3, as in the revised provisions for seismic performance 2 and 3."

A2.2 DESIGN GROUND MOTIONS FOR SEISMIC PERFORMANCE 1, 2, AND 3

Seismic motions for verification should be prepared for each seismic performance 1, 2, and 3. In the United States, design earthquakes maximum operational earthquake (MOE), base design basis earthquake (DBE), and maximum considered earthquake (MCE) are prepared for IO (service limit), LS (repair limit), and CP (terminal limit). On the other hand, in Japan, for example, for critical water supply facilities, Level 1 ground motion is provided for seismic performance 1 (service limit) and Level 2 ground motion for seismic performance 2 (repair limit), but there is no mention of verification for seismic performance 3 (ultimate limit).

This problem is caused by the fact that only two verification ground motions, Level 1 and Level 2, are provided for the three seismic performance categories. As a solution to this problem, it is proposed that three types of ground motions for verification should be prepared for the time being: Level 1, Level 2L ground motion equivalent to a 70% non-exceedance probability, and Level 2U ground motion equivalent to a 90% non-exceedance probability.

Table A2.2.1 is prepared based on this policy. Table A2.2.2 shows the relationship between limit states and ground motions in Japan and the United States.

TABLE A2.2.1
Comparison of US and Japanese Classifications for Limit States

Country	Limit State	Serviceable Limit State	Repairable Limit State	Ultimate Limit State
Japan	Seismic Performance	1	2	3
	Ground Motion	Level 1 ground motion	Level 2L ground motion	Level 2U ground motion
USA	Limit State	IO	LS	CP
	Earthquake	MOE	DBE	MCE

TABLE A2.2.2

Relationship Between Limit States and Ground Motions in Japan and the United States

		Limit State				Design Loads	
Exceedance Probability	**Return Period**	**S-1 IO**	**S-2**	**S-3 LS**	**S-4 CP**	**Japan**	**USA**
50%/50yrs	72yrs	●				Level 1	MOE
20%/50yrs	225yrs	■	●				
10%/50yrs	475yrs		■	●		Level 2L	DBE
5%/50yrs	970yrs			□	○		MCE
2%/50yrs	2475yrs			■	●	Level 2U	

Definition of Limit State

S-1	Serviceability limit
S-2	Repairable limit
S-3	Life safety limit
S-4	Ultimate limit
●	Basic structure
■	Important structure

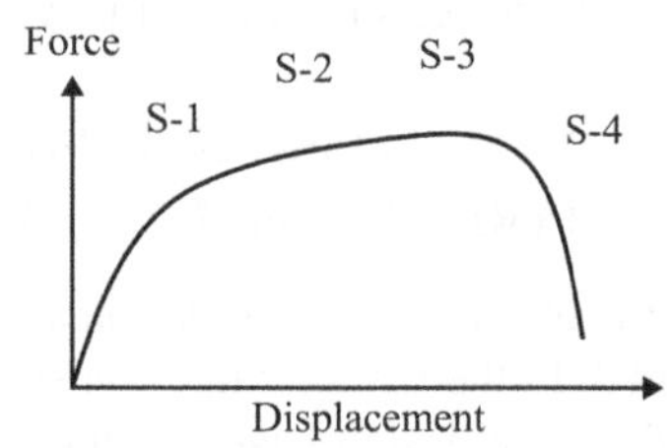

Appendix A3: Surface Wave Transfer Matrix

This appendix details the contents of the matrix Equation (2.3.22), where the surface wave transfer matrix appears in the case of Rayleigh waves in Section 2.3.2, "Vibration Analysis and Dispersion Characteristic Equations for Surface Waves."

$$
D_m(z_m)=\begin{bmatrix}
-\left(\frac{v_{pm}}{c}\right)^2 cos\,P_m & i\left(\frac{v_{pm}}{c}\right)^2 sin\,P_m & -\Gamma_m r_{\beta m} cos\,Q_m & i\Gamma_m r_{\beta m} sin\,Q_m \\
i\left(\frac{v_{pm}}{c}\right)^2 r_{am} sin\,P_m & -\left(\frac{v_{pm}}{c}\right)^2 r_{am} cos\,P_m & -i\Gamma_m sin\,Q_m & \Gamma_m cos\,Q_m \\
-\rho_m v_{pm}^2(\Gamma_m-1)cos\,P_m & i\rho_m v_{pm}^2(\Gamma_m-1)sin\,P_m & -\rho_m c^2\Gamma_m r_{\beta m} cos\,Q_m & i\rho_m c^2\Gamma_m^2 sin\,Q_m \\
i\rho_m v_{pm}^2\Gamma_m r_{\alpha m} sin\,P_m & \rho_m v_{pm}^2\Gamma_m r_{\alpha m} cos\,P_m & i\rho_m c^2\Gamma_m(\Gamma_m-1)sin\,Q_m & -i\rho_m c^2\Gamma_m(\Gamma_m-1)cos\,Q_m
\end{bmatrix}
$$

$$
E_m=\begin{bmatrix}
-\left(\frac{v_{pm}}{c}\right)^2 & 0 & -\Gamma_m r_{\beta m} & 0 \\
0 & -\left(\frac{v_{pm}}{c}\right)^2 r_{\alpha m} & 0 & \Gamma_m \\
-\rho_m v_{pm}^2(\Gamma_m-1) & 0 & -\rho_m c^2\Gamma_m r_{\beta m} & 0 \\
0 & \rho_m v_{pm}^2\Gamma_m r_{\alpha m} & 0 & -\rho_m c^2\Gamma_m(\Gamma_m-1)
\end{bmatrix}
$$

where ρ_m: density, H_m:thickness of layers, $v_{pm}=\sqrt{(\lambda_m+2\mu_m)/\rho_m}$: phase velocity of dilatational wave, $v_{sm}=\sqrt{\mu_m/\rho_m}$:phase velocity of shear wave, λ_m. μ_m:*Lamé* constants, $\Gamma_m=2(v_{pm}/c)^2$; $P_m=kr_{\alpha m}z_m$; $Q_m=kr_{\beta m}z_m$; $k=\omega/c$: wave number, and

$$
r_{\alpha m}=\sqrt{\left(\frac{c}{v_{pm}}\right)^2-1}\ \text{ for } c\geq v_{pm}\qquad r_{\alpha m}=-i\sqrt{1-\left(\frac{c}{v_{pm}}\right)^2}\ \text{ for } c<v_{pm}
$$

$$
r_{\beta m}=\sqrt{\left(\frac{c}{v_{sm}}\right)^2-1}\ \text{ for } c\geq v_{sm}\qquad r_{\beta m}=-i\sqrt{1-\left(\frac{c}{v_{sm}}\right)^2}\ \text{ for } c<v_{sm}
$$

Appendix A4: Guidance on Design Formulae for Bend Pipes and T-Junctions

This section describes the detailed process of formulation for calculating the pipe strain of bend pipes and T-junctions.

A4.1 PIPE STRAIN OF BEND PIPES

Consider the case of a 90°bend pipe with traveling waves incident along the pipe axis (element 1), as shown in Figure A4.1.1.

For ease of analysis, it is assumed here that the curved pipe section of element 3 itself is not in contact with the surrounding ground and can be treated in the same way as an above-ground pipe. When bending moments M_1 and M_2 act on the end of element 3 in the diagram, the deflection angle θ_B of the curved pipe and the bending deformation Δ_F at the end of the pipe are obtained using Casteliano's theorem as follows:

$$\theta_B = \frac{\pi}{2}\frac{nR}{EI^*}\left\{M_x + RS_1\left(\frac{2}{\pi} - 1\right) - \frac{2}{\pi}RS_2\right\} \tag{A4.1}$$

$$\Delta_F = \frac{nR^2}{2EI^*}\left(-\frac{2M_2}{R} + S_1 + \frac{\pi}{2}S_2\right) + \frac{R}{4EA^*}\left(-2S_1 + \pi S_2\right), \tag{A4.2}$$

where A^* is the cross-sectional area of the bend pipe, I^* is the second moment of section of the bend pipe, n is the flexibility factor, R is the radius of curvature, and E is Young's modulus of the pipe material.

On the other hand, the boundary conditions at bend pipe member ends I and II are given by

$$\left.\begin{aligned} v_1 &= 0 \\ v_2 &= \Delta_{B1} \\ \theta_1 - \theta_2 &= \theta_B \\ M_2 - M_1 &= R(S_1 + S_2) \end{aligned}\right\} \tag{A4.3}$$

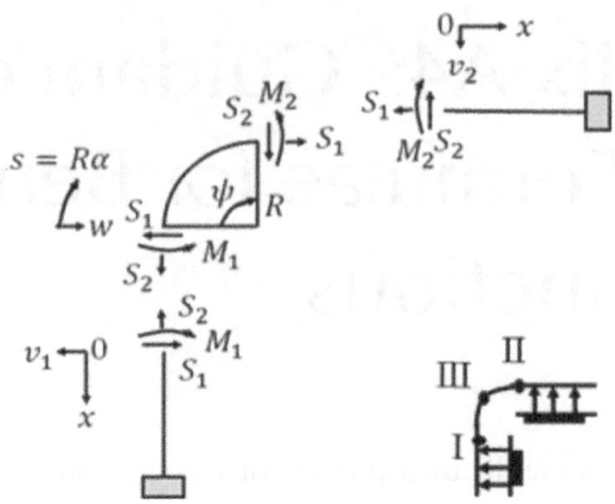

FIGURE A4.1.1 Bending moment and axial and shear forces acting on a 90° bend pipe with the sleeve pipes supported on an elastic bed.

Substituting Equation (A4.3) into Equations (A4.1) and (A4.2), the bending deformation Δ_{B1} is obtained as follows:

$$\Delta_{B1} = \frac{\Delta_{R1}}{1+\dfrac{L_{a1}I\lambda^2}{2A}\left|1+b_2\right|+b_3} \tag{A4.4}$$

where

$$b_1 = -\frac{1+2R\lambda+(\pi-2)nR^2\lambda^2}{(1+R\lambda)\left|2+\pi nR\lambda+(4-\pi)nR^2\lambda^2\right|} \tag{A4.5}$$

$$b_2 = \frac{1-2nR^2\lambda^2-(4-\pi)nR^3\lambda^3}{(1+R\lambda)\left|2+\pi nR\lambda+(4-\pi)nR^2\lambda^2\right|} \tag{A4.6}$$

$$b_3 = nR^3\lambda^3\left\{\frac{\pi}{2}+\frac{\pi I_r}{2nA_rR^2}+\left(1-\frac{I_r}{nA_rR^2}\right)b_1+\left(\frac{2}{R\lambda}+\frac{\pi}{2}+\frac{\pi I_r}{2nA_rR^2}\right)b_2\right\} \tag{A4.7}$$

On the other hand, the shear forces and bending moments at the ends of bend pipe members are given in the following way:

$$\left.\begin{aligned} S_1 &= -2\lambda^3EIb_1\Delta_{B1} \\ S_2 &= -2\lambda^3EI\left(1+b_2\right)\Delta_{B1} \\ M_1 &= 2\lambda^2EIb_1\Delta_{B1} \\ M_2 &= 2\lambda^2EIb_2\Delta_{B1} \end{aligned}\right\} \tag{A4.8}$$

The shear forces and bending moments at an angle α in element 3 are calculated as follows:

$$S(\alpha) = S_1 sin(\alpha) - S_2 cos(\alpha) \tag{A4.9}$$

$$M(\alpha) = M_2 - S_1 R(1 - sin(\alpha)) - S_2 R cos(\alpha). \tag{A4.10}$$

Therefore, the pipe strains at points I($\alpha = 0$), II ($\alpha = \pi/2$), and III ($0 < \alpha < \pi/2$) shown in Figure A4.1.1 can be expressed respectively as

$$\left.\begin{aligned} \varepsilon_B^I &= \frac{D}{2EI}|M_1| + \frac{1}{AE}|S_2| \\ \varepsilon_B^{II} &= \frac{D}{2EI}|M_2| + \frac{1}{AE}|S_1| \\ \varepsilon_B^{III} &= \frac{iD}{2EI^*}|M(\alpha)| + \frac{1}{A^*E}|S(\alpha)| \end{aligned}\right\} \tag{A4.11}$$

Equation (A4.11) gives the bend pipe strain for incident waves along pipe axis element 1, but similar results can be obtained for incident waves along pipe axis element 2. However, in that case, instead of Δ_{B1} in Equation (A4.4), the following formula is used for Δ_{B2}:

$$\Delta_{B2} = \frac{\Delta_{R2}}{1 + \frac{L_{a2} I \lambda^3}{2A}|1 + b_2| + b_3}. \tag{A4.12}$$

Ultimately, the composite strain in each section of a 90° bend pipe due to traveling waves (waves propagating horizontally in the surface soil) incident at an angle φ is given by

$$\begin{aligned} \varepsilon_B^I(\varphi) &= \left\{ D\lambda^2|b_1| + \frac{2\lambda^3 I}{A}|1 + b_2| \right\}\Delta_{B1} + \left\{ D\lambda^2|b_2| + \frac{2\lambda^3 I}{A}|b_1| \right\}\Delta_{B2} \equiv \beta_B^I(\varphi)\varepsilon_{max} \\ \varepsilon_B^{II}(\varphi) &= \left\{ D\lambda^2|b_2| + \frac{2\lambda^3 I}{A}|b_1| \right\}\Delta_{B1} + \left\{ D\lambda^2|b_1| + \frac{2\lambda^3 I}{A}|1 + b_2| \right\}\Delta_{B2} \equiv \beta_B^{II}(\varphi)\varepsilon_{max} \\ \varepsilon_B^{III}(\varphi,\alpha) &= iD\lambda^2\frac{1}{I^*}|b_2 + \lambda R(1 - sin\,\alpha)b_1 + \lambda R\,cos\,\alpha(1 + b_2)|\Delta_{B1} \\ &\quad + 2\frac{\lambda^3 I}{A^*}|b_1\,sin\,\alpha - (1 + b_2)cos\,\alpha|\Delta_{B1} \\ &\quad + 2\frac{\lambda^3 I}{A^*}\left| b_2 + \lambda R(1 - cos\,\alpha)b_1 + \lambda R\,sin\,\alpha(1 + b_2) + 2\frac{\lambda^3 I}{A^*}|b_1\,cos\,\alpha - (1 + b_2)sin\,\alpha| \right|\Delta_{B2} \\ &\equiv \beta_B^{III}(\varphi,\alpha)\varepsilon_{max}, \end{aligned} \tag{A4.13}$$

where $\beta_B^I(\varphi), \beta_B^{II}(\varphi)$ and $\beta_B^{III}(\varphi,\alpha)$ are the conversion factors of a 90° bend pipe for a traveling wave of incident angle φ.

A4.2 T-JUNCTION PIPE STRAIN

Consider the case where a traveling wave is incident on a branch pipe at an angle φ.

The vibration components of the traveling wave can be decomposed into a component parallel to the branch pipe (element 1) and a component parallel to the main pipe (element 2), as shown in Figure A4.2.1. The branch pipe behavior due to each vibration component is denoted as Models A and B. That is, model A shows the deformation state when the relative deformation Δ_{R1} is forced along the branch pipe axis, while model B shows the deformation state when the relative deformation Δ_{R2} is forced along the main pipe axis. Therefore, the pipe strain at the end of the branch pipe material due to traveling waves generally incident at an angle φ can be obtained by superposition of the pipe strains of Models A and B respectively.

The relative deformation along the respective pipe axis is given by

$$\Delta_{R1} = (1-\beta_{D1})\frac{L_{a1}}{2\pi}\varepsilon_{G_1} \tag{A4.14}$$

$$\Delta_{R2} = (1-\beta_{D2})\frac{L_{a2}}{2\pi}\varepsilon_{G_2}, \tag{A4.15}$$

where $L_{a1} = L/\sin\varphi$, $L_{a2} = L/\cos\varphi$, and ε_{G1} is the apparent axial ground strain in the axial direction of the branch pipe under the deformed condition of model A in Figure A4.2.1 and ε_{G2} is the apparent axial ground strain in the axial direction of the main pipe under the deformed condition of model B. β_{D1} and β_{D2} are conversion factors for determining the relative deformation between the branch pipe and the main pipe and the surrounding ground, respectively, and are given by the following equations:

$$\beta_D = \begin{cases} \beta_S : \text{when no slip occurs in the straight section} \\ q^*\beta_S : \text{when slip occurs in the straight section} \end{cases} \tag{A4.16}$$

where β_{D1} and β_{D2} can be obtained from Equation (A4.16) using the apparent wavelengths L_{a1} *and* L_{a2}.

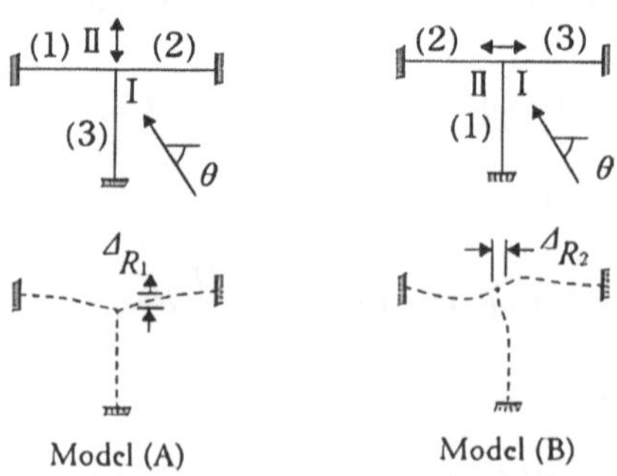

FIGURE A4.2.1 Typical deformation modes of T-junctions.

Branch pipe joints are assumed to be rigidly connected, as they are usually sufficiently reinforced or have a thicker pipe thickness than the straight pipe section. The shear force S_j and bending moment M_j $(j = 1,2,3)$ occurring at each end of the branch pipe are shown in Figures A4.2.2 and A4.2.3.

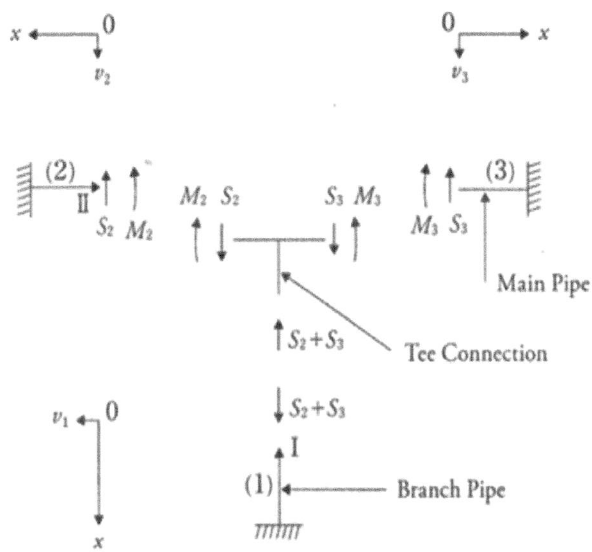

FIGURE A4.2.2 Bending moment and shear force of branch pipe model A.

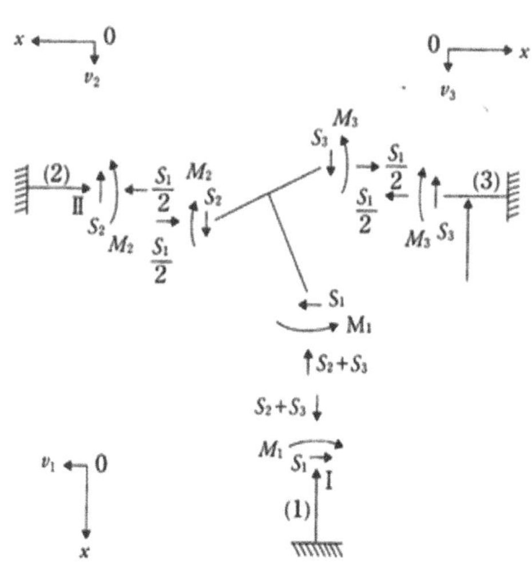

FIGURE A4.2.3 Bending moment and shear force of branch pipe model B.

The boundary conditions are therefore given by using the bending deformations Δ_{B1} and Δ_{B2} for the branch and main pipe axes, which are unknowns, as follows:

Model A:

$$\left.\begin{aligned} v_2 &= \Delta_{B1} \\ \theta_1 &= \theta_2 = 0 \\ M_1 &= 0 \\ M_2 &= M_3 \\ S_2 &= S_3 \end{aligned}\right\} \tag{A4.17}$$

Model B:

$$\left.\begin{aligned} v_1 &= \Delta_{B2} \\ v_2 &= v_3 = 0 \\ \theta_2 + \theta_3 &= 0 \\ \theta_2 &= \theta_3 \\ M_1 - M_2 + M_3 &= 0 \end{aligned}\right\} \tag{A4.18}$$

where v_j, θ_j, M_j, and S_j are the deflection, deflection angle, bending moment, and shear force for the deformed pipe element *j*.

$(S_2 + S_3)$ acts on the branch pipe in model A as an axial force, so that axial elongation Δ_{S1} occurs as given by

$$\Delta_{S1} = \frac{L_{a1}}{4}\frac{S_2 + S_3}{A_1 E}. \tag{A4.19}$$

From Equations (3.3.54), (A4.17), and (A4.19), the bending deformation Δ_{B1} of model A is given by

$$\Delta_{B1} = \frac{\Delta_{R1}}{1 + \dfrac{2L_{a1}I_2\lambda_2^3}{A_1}} \tag{A4.20}$$

where A_j, I_j, and λ_j are the cross-sectional area, second moment of section, and relative composite ground/pipe property values λ_j of the deformed pipe element *j* are given by

$$\lambda_j = \sqrt[4]{\frac{K}{4EI_j}}. \tag{A4.21}$$

Similarly, the bending deformation Δ_{B2} of model B is obtained by the following equation:

$$\Delta_{B2} = \frac{\Delta_{R2}}{1 + \frac{C_1 L_{a2} I_1 \lambda_1^3}{4A_2}} \tag{A4.22}$$

where

$$C_1 = \frac{1 + 4(\lambda_1 / \lambda_2)^3}{1 + 2(\lambda_1 / \lambda_2)^3} \tag{A4.23}$$

Solving the balance equation for beams on elastic foundations and considering the boundary conditions in Equations (A4.17) and (A4.18), the cross-sectional forces and bending moments at each end of the branch pipe are given by

Model A:

$$\left.\begin{aligned} S_1 &= 0 \\ S_2 = S_3 &= -4EI_2\lambda_2^3\Delta_{B1} \\ M_1 &= 0 \\ M_2 = M_3 &= 2EI_2\lambda_2^2\Delta_{B1} \end{aligned}\right\} \tag{A4.24}$$

Model B:

$$\left.\begin{aligned} S_1 &= -2EI_1\lambda_1^3\Delta_{B1}C_1 \\ S_2 = S_3 &= -2EI_2\lambda_2^3\Delta_{B2}C_2 \\ M_1 &= 4EI_1\lambda_1^2\Delta_{B2}C_2\left(\lambda_1/\lambda_2\right)^2 \\ M_2 = M_3 &= 2EI_2\lambda_2^2\Delta_{B2}C_2 \end{aligned}\right\} \tag{A4.25}$$

where

$$C_2 = \frac{\lambda_1 / \lambda_2}{1 + 2(\lambda_1 / \lambda_2)^3} \tag{A4.26}$$

Therefore, the composite strains at the branch pipe material end (I) and the main pipe material end (II) of both models are given by

Model A:

$$\begin{aligned} \varepsilon_T^{IA} &= -\frac{2S_2}{A_1 E} \\ \varepsilon_T^{IIA} &= \frac{M_2 D_2}{2EI_2} \end{aligned} \tag{A4.27}$$

Model B:

$$\begin{aligned} \varepsilon_T^{IB} &= \frac{M_1 D_1}{2EI_1} \\ \varepsilon_T^{IIB} &= \frac{M_2 D_2}{2EI_2} - \frac{S_1}{2A_2 E} \end{aligned} \tag{A4.28}$$

where D_j is the pipe diameter of the geometrical pipe element j.

Finally, the composite strain at each end of branch pipes I and II for traveling waves of arbitrary angle of incidence φ is given by summing the results of Equations (A4.27) and (A4.28) in the following way:

$$\begin{aligned} \varepsilon_T^{I}(\varphi) &= \varepsilon_T^{IA} + \varepsilon_T^{IB} = \frac{8I_2\lambda_2^3}{A_1}\Delta_{B1} + 2\left(\frac{\lambda_1^2}{\lambda_2}\right)^2 D_1 C_2 \Delta_{B2} \equiv \beta_T^{I}(\varphi)\varepsilon_{max} \\ \varepsilon_T^{II}(\varphi) &= \varepsilon_T^{IIA} + \varepsilon_T^{IIB} = \lambda_2^2 D_2 \Delta_{B1} + \left(\lambda_2^2 D_2 C_2 + \frac{I_1\lambda_1^3 C_1}{A_2}\right)\Delta_{B2} \equiv \beta_T^{II}(\varphi)\varepsilon_{max}, \end{aligned} \tag{A4.29}$$

where $\beta_T^{I}(\varphi)$ and $\beta_T^{II}(\varphi)$ are the conversion factors for obtaining the composite strain at the end of the material when a traveling wave with ground strain amplitude ε_{max} is incident on the branch pipe at angle φ.

Appendix A5: Flowchart for Connectivity Analysis

The various symbols shown in the flowchart in Figure A5.1 are defined as follows:

MCS: number of Monte Carlo simulations
NL: number of links in the target network
NN: number of nodes in the target network
NC: maximum number of transitions to find the water supply reachability
B_j: node arrival state vector at the j-th step
$(b_q)_j$: the q-th component of the node vector B_j, i.e., a variable indicating the arrival state from the water supply end (water treatment plant) to the q-th node at the jth step
C_i: Value of the link connection state of the ith link. The value $C_i = 1$ is given when the link is connected and $C_i = 0$ when the link is broken
A_j: Transition matrix for the transition from node vector B_j to node vector B_{j+1} in the jth step

When the link leading from the s-th node to the t-th node is the ith link, the first s-th component $(a_{st})_j$ of the matrix A_j is given by C_i.

Furthermore, the words and phrases in the flowchart are defined as follows:

Link breakage probability modification: The link breakage probability is modified by assuming that the connected link bypassing the broken node is increased by the node breakage probability.
Transition matrix: The transition matrix component is assumed to be 1 if the s-th node is connected to the t-th node, and 0 if it is not.
Water supply reachability: The water supply reachability is defined as *M/MCS*, where *M* is the number of times out of the number of trials *MCS* that the linkage path from the water supply end (the water treatment plant) to the critical point is secured.
Physical damage degree: Expected number of damaged pipes divided by the total number of pipes.
Functional damage degree: Expected number of non-reachable nodes divided by the total number of nodes.
Seismic resistance ratio: The value obtained by dividing the replaced pipe length by the total pipe length when ductile iron pipes with conventional joints are replaced by ductile iron pipes with earthquake-resistant joints (earthquake-resistant pipes) for each pipe section.

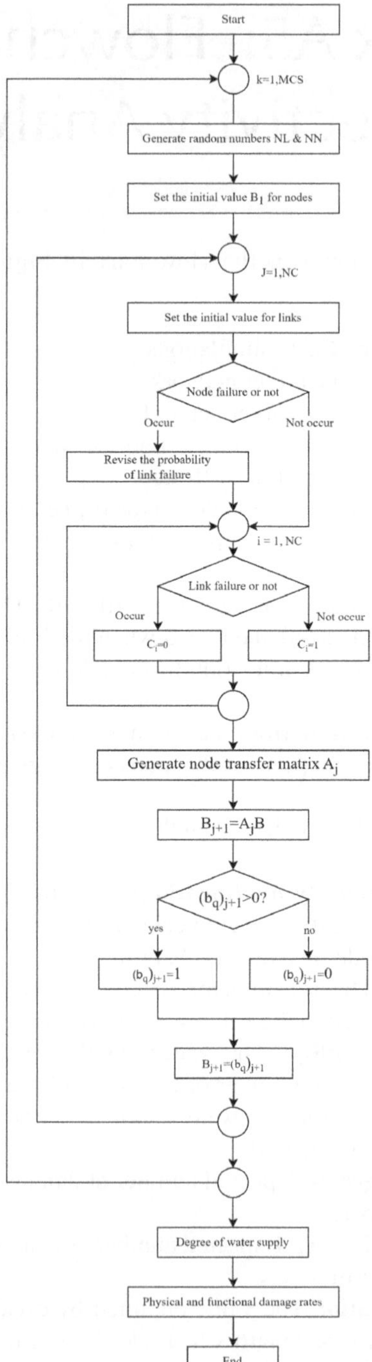

FIGURE A5.1 Flowchart for connectivity analysis.

Index

F

G

H

I

J

L

M

T

U

V

W

Y